Current Topics in
Developmental Biology

Volume 35

Current Topics in Developmental Biology

Volume 35

Edited by

Roger A. Pedersen

Reproductive Genetics Division
Department of Obstetrics, Gynecology,
and Reproductive Sciences
University of California, San Francisco
San Francisco, California

Gerald P. Schatten

Department of Obstetrics–Gynecology
and Cell and Developmental Biology
Oregon Regional Primate Research Center
Oregon Health Sciences University
Beaverton, Oregon

Academic Press

San Diego   London   Boston   New York   Sydney   Tokyo   Toronto

Cover photograph: The wild-type cuticular pattern of a first-instar larva is shown. In this darkfield micrograph, the areas covered with denticle belts and naked cuticle are clearly visible. The segmental expression fo *wingless* mRNA (in blue) and Engrailed protein (in rose) is shown.

Academic Press
a division of Harcourt Brace & Company
525 B Street, Suite 1900, San Diego, California 92101-4495, USA
http://www.apnet.com

Academic Press Limited
24-28 Oval Road, London NW1 7DX, UK
http://www.hbuk.co.uk/ap/

International Standard Book Number: 0-12-153135-X

PRINTED IN THE UNITED STATES OF AMERICA
97 98 99 00 01 02 BB 9 8 7 6 5 4 3 2 1

Contents

1

Life and Death Decisions Influenced by Retinoids
Melissa B. Rogers

2

Developmental Modulation of the Nuclear Envelope
Jun Liu, Jacqueline M. Lopez, and Mariana F. Wolfner

3

The EGFR Gene Family in Embryonic Cell Activities
Eileen D. Adamson and Lynn M. Wiley

v

4

The Development and Evolution of Polyembronic Insects

Michael R. Strand and Miodrag Grbić

5

β-Catenin Is a Traget for Extracellular Signals Controlling Cadherin Function: The Neurocan–GalNAcPTase Connection

Jack Lilien, Stanley Hoffman, Carol Eisenberg, and Janne Balsamo

6

Neural Induction in Amphibians
Horst Grunz

7

Paradigms to Study Signal Transduction Pathways in *Drosophila*
Lee Engstrom, Elizabeth Noll, and Norbert Perrimon

Contributors

Numbers in parentheses indicate the pages on which the authors' contributions begin.

Eileen D. Adamson (71), La Jolla Cancer Research Center, The Burnham Institute, La Jolla, California 92037

Janne Balsamo (161), Department of Biological Sciences, Wayne State University, Detroit, Michigan 48230

Carol Eisenberg (161), Department of Cell Biology, Medical University of South Carolina, Charleston, South Carolina 29425-2229

Lee Engstrom (229), Muncie Center for Medical Education, Indiana University School of Medicine, Ball State University, Muncie, Indiana 47306

Miodrag Grbić (121), Department of Entomology, College of Agriculture and Life Sciences, University of Wisconsin–Madison, Madison, Wisconsin 53706

Horst Grunz (191), Department of Zoophysiology, University GH Essen, 45117 Essen, Germany

Stanley Hoffman (161), Department of Medicine, Division of Rheumatology and Immunology, Medical University of South Carolina, Charleston, South Carolina 29425-2229

Jack Lilien (161), Department of Biological Sciences, Wayne State University, Detroit, Michigan 48230

Jun Liu (47), Section of Biochemistry, Molecular and Cell Biology, Cornell University, Ithaca, New York 14853-2703

Jacqueline M. Lopez (47), Section of Biochemistry, Molecular and Cell Biology, Cornell University, Ithaca, New York 14853-2703

Elizabeth Noll (229), Department of Genetics, Howard Hughes Medical Institute, Harvard Medical School, Boston, Massachusetts 02115

Norbert Perrimon (229), Department of Genetics, Howard Hughes Medical Institute, Harvard Medical School, Boston, Massachusetts 02115

Melissa B. Rogers (1), Departments of Biology, Pharmacology and Institute for Biomolecular Science, University of South Florida, Tampa, Florida 33620

Michael R. Strand (121), Department of Entomology, College of Agriculture and Life Sciences, University of Wisconsin–Madison, Madison, Wisconsin 53706

Lynn M. Wiley (71), Division of Reproductive Biology and Medicine, Department of Obstetrics and Gynecology, University of California, Davis, California 95616

Mariana F. Wolfner (47), Sections of Genetics and Development, Cornell University, Ithaca, New York 14853-2703

Preface

This volume continues the custom of addressing developmental mechanisms in a variety of experimental systems. The conceptual sequence of topics begins with "Life and Death Decisions Influenced by Retinoids," a contribution from Melissa Rogers at the University of South Florida. The article by Jun Liu, Jacqueline Lopez, and Mariana Wolfner from Cornell University, "Developmental Modulation of the Nuclear Envelope," highlights the newest discoveries that this structure is critical during development and differentiation. In "The EGFR Gene Family and Embryonic Cell Activities," Eileen Adamson and Lynn Wiley of The Burnham Institute and the University of California, Davis, respectively, describes the vital importance of this gene family in normal and abnormal development in a host of organisms. Mike Strand and Miodrag Grbić from the University of Wisconsin–Madison report on a fascinating topic in "The Development and Evolution of Polyembryonic Insects." Jack Lilien, Stanley Hoffman, Carol Eisenberg, and Janne Balsamo from Wayne State University and the Medical University of South Carolina show that "β-Catenin Is a Target for Extracellular Signals Controlling Cadherin Function: The Neurocan–GalNAcPTase Connection." The contribution from Horst Grunz from the University of Essen is titled "Neural Induction in Amphibians." This volume concludes with "Paradigms to Study Signal Transduction Pathways in *Drosophila*" by Lee Engstrom, Elizabeth Noll, and Norbert Perrimon from Indiana University School of Medicine and Harvard Medical School.

Together with other volumes in this series, this volume provides a comprehensive survey of major issues at the forefront of modern developmental biology. These chapters should be valuable to researchers in the fields of mammalian and nonmammalian development, as well as to students and other professionals who want an introduction to cellular, molecular, and genetic approaches to developmental biology, as well as neurobiology. This volume in particular will be essential reading for anyone interested in growth factors, apoptosis, transcription factors, signal transduction, intracellular structural modifications during development and differentiation, fertilization, cell cycle regulation, embryo formation, developmental mechanisms and evolution, morphogenetic movements, neural development, differentiation, cell adhesion molecules, and neurobiology.

This volume has benefited from the ongoing cooperation of a team of participants who are jointly responsible for the content and quality of its material. The authors deserve the full credit for their success in covering their subjects in depth, yet with clarity, and for challenging the reader to think about these topics in new ways. We thank the members of the Editorial Board for their suggestions of topics and authors. We thank Liana Hartanto, Heather Aronson, and Diana Myers for their exemplary administrative and editorial support. We are also grateful to the scientists who prepared articles for this volume and to their funding agencies for supporting their research.

Gerald Schatten
Portland, Oregon

Roger A. Pedersen
San Francisco, California

1

Life-and-Death Decisions Influenced by Retinoids

Melissa B. Rogers
Departments of Biology, Pharmacology and Institute for Biomolecular Science
University of South Florida
Tampa, Florida 33620-5150

I. Introduction

Cell proliferation and programmed cell death are highly regulated processes integral to animal development (Milligan and Schwartz, 1996). Since even subtle mistakes in cell division rates or cell survival can cause embryonic malformations or adult disease, the regulation is complex and probably uses redundant mechanisms. Great strides have been made regarding the basic biochemistry underlying these processes, but integrating the profusion of cell cycle and death regulatory genes with signal transduction pathways remains a major challenge.

   Vitamin A (retinol) and other retinoids modify the proliferation rate or cause programmed cell death by apoptosis in a host of different mammalian cell types

Current Topics in Developmental Biology, Vol. 35

1

in vivo and *in vitro*. Retinoid-induced decreases in cell proliferation rate are tightly linked to the well-known ability of retinoids to induce differentiation. In some cells, apoptosis follows this differentiation. Paradoxically, retinoids promote proliferation and survival in other cells. This review will focus on the known interactions between retinoids and their receptors and the regulators of cell proliferation and death. Retinoids exert many effects via transcriptional regulation, and so far over 200 known genes have been found to be retinoid-responsive (Gudas *et al.*, 1994). Some retinoid-responsive genes may act directly in the paths affecting cell proliferation and death.

In vertebrates, retinol is converted into a variety of metabolites. These have widely varied biologic activity, although retinoic acid (RA) is one of the most active compounds. Thousands of novel retinoids have also been synthesized to identify drugs that might have the beneficial effects of natural retinoids but lack the toxic effects. Many of these compounds bear little resemblance to retinol. Thus it was suggested that the definition of a retinoid include "substances that elicit specific biologic responses by binding to and activating specific receptors" (Sporn and Roberts, 1994). This review will address primarily the effects of naturally occurring retinoids, except where the synthetic retinoids provide specific information regarding retinoid signaling mechanisms.

II. Overview of the Cell Cycle and Apoptosis

A. Cell Cycle

Intertwined cellular signals confront cells with three primary choices: stop growing (desirable in most adult cells), proliferate, or die by apoptosis. Although the biochemical mechanisms underlying a cell's ultimate choice are inextricably linked to differentiation and many details are unique to a cell type, certain basic principles exist. This section will provide a brief overview of the cell cycle as understood from a few well-characterized cell types, for example, fibroblasts. The reader will be referred to some of the many excellent reviews on cell cycle and apoptosis.

The basic cell cycle mechanisms have been highly conserved and were identified from studies of mitogen-stimulated mammalian cells, yeast, surf clams, and frogs (for review, see Murray and Hunt, 1993; Thomas, 1996). After cytokinesis, the cell enters a variable-length G1 phase. If environmental and cellular status is appropriate, the cell replicates its DNA (S phase). After replication of the entire genome, the cell enters G2, then undergoes mitosis and reenters G1. This simple picture, taught to all undergraduate biology students, masks a plethora of controlling signals.

Progression through the cell cycle is regulated primarily by fluctuating levels of kinase activity. The activity of these constitutively expressed kinases, cyclin-

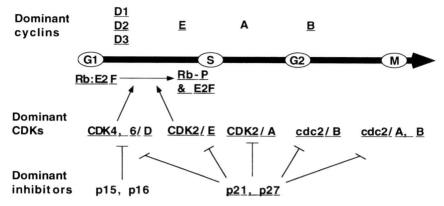

Fig. 1 Cell cycle overview. Genes or protein activities known to be regulated in RA-treated cells are underlined.

dependent kinases (CDKs), is controlled by their level of phosphorylation, by the concentration of regulatory subunits called cyclins, and by specific inhibitors. Cyclin levels rise continuously during specific cell cycle phases and then are precipitously destroyed to inactivate their associated CDK.

In single-cell organisms, such as yeast, the cycle is controlled primarily by nutrient and pheromone levels. In metazoans, a host of other variables cause striking spatial and temporal variability in the cell cycle. For example, the earliest embryonic cycles may be as short as 10 minutes in insect embryos and 30 minutes in amphibian embryos. These brief cycles are possible only in species whose large egg provides a rich store of cellular materials. However, even mammalian cells lacking a comparable maternal dowry can attain a 3-hour division time (Cunningham *et al.*, 1994). In contrast to rapid embryonic cycles, adult hepatocytes seldom divide, but, following liver injury, hepatocytes can reenter the cell cycle. At the far end of the spectrum, fully differentiated mammalian neurons or muscle cells apparently cannot divide. These extremes are controlled by at least 11 distinct cyclins, 7 CDKs, and a variety of inhibitory proteins (Thomas, 1996). Since most work on retinoids focuses on mammalian cells, Figure 1 provides a schematic overview of the primary mammalian cell cycle regulators. For more detailed reviews, consult Morgan, 1995; Sherr, 1994; Sherr and Roberts, 1995; Weinberg, 1995.

Cell division time is controlled largely by the length of the G1 phase; for example, embryonic and differentiated cells vary the most in G1 phase (Pardee, 1989). Serum-starved, quiescent (G0) fibroblasts are a useful model, since they can be induced to reenter the cycle during G1. This enables *in vitro* studies regarding G1 regulatory mechanisms. Such studies indicate that a multitude of oncogenes and tumor suppressors exert at least some of their function during G1.

Oncogenes often activate and tumor suppressors often inhibit the S-phase-promoting E2F transcription factors. The E2F factors regulate genes involved in DNA replication (e.g., DNA polymerase α, thymidylate synthase, thymidine kinase, ribonucleotide reductase) and cyclin-dependent kinase (CDK) activity (e.g., cyclins D, E, and A, CDC2, proliferating cell nuclear antigen). The proto-oncogenes c-*myc*, N-*myc*, and B-*myb* also contain E2F sites in their promoters (DeGregori *et al.*, 1995). The coordinate regulation of genes by E2F factors is required for the G1-to-S transition.

E2F factors are negatively regulated by binding hypophosphorylated members of the retinoblastoma tumor suppressor family. The retinoblastoma proteins, p105[rb], p107, and p130, are phosphorylated by various CDKs. P105[rb] is the best characterized member of this family. Once phosphorylated, p105[rb] cannot inhibit E2F activity and the cell proceeds into S phase. The G1 CDKs (*CDK2, 4*, and *6*) probably phosphorylate p105[rb]. CDK activity is controlled by the levels of the regulatory subunits (cyclins D1, D2, D3, and E) and by phosphorylation of specific activating or inactivating amino acids in the catalytic subunits. Kinase activity is also restrained by the CDK inhibitory proteins, p21[waf1/cip1/Sdi1/CDKN1], p27[kip1], p26[ink4A/MTS1/CDKN2], and p15[ink4B/MTS2]. Thus progression through G1 is controlled by a network of regulatory factors controlling kinase activity (for review, see Morgan, 1995; Peters *et al.*, 1996; Whyte, 1996). Retinoids might modulate any of these factors.

Interestingly, the decision to undergo apoptosis is influenced by many of the same genes regulating G1 arrest and progression. In some cases, cells undergoing apoptosis up regulate genes typically expressed during the G1-to-S transition (Pandey and Wang, 1995). P105[rb] and dominant-negative mutants of *CDC2*, *CDK2*, and *CDK3* can inhibit apoptosis (HaasKogan *et al.*, 1995; Meikrantz and Schlegel, 1996). P105[rb] overexpression and CDK inactivation may repress apoptosis by repressing E2F, since E2F-1 overexpression can induce apoptosis *in vitro* (Qin *et al.*, 1994; Shan and Lee, 1994; Wu and Levine, 1994) and its absence is tumorigenic (Field *et al.*, 1996; Yamasaki *et al.*, 1996). The tumor suppressor p53 and the oncogene c-*myc* appear to function upstream of the basic cell cycle machinery. Each plays a dual role in controlling proliferation and apoptosis. The path actually chosen by a cell depends on the activity of these internal factors along with the cell's environment. For a review regarding the coupling of proliferation and apoptosis, see Evan *et al.*, 1996 and refs. therein.

Although whether a cell stops proliferating or undergoes apoptosis is controlled largely by G1 regulators, other cell cycle proteins can play a role. For example, apoptosis caused by granzyme B activates *CDC2*, the G2 CDK catalyzing entry into mitosis (Shi *et al.*, 1994). The linkage between cell cycle and induction of apoptosis is specific to cell type and apoptotic stimulus. The context-dependent function of cell cycle/apoptosis regulatory proteins is reminiscent of the context-dependent effects of the retinoids. A particular retinoid can inhibit or stimulate proliferation or cause apoptosis, depending on the cell type.

B. Apoptosis

Apoptosis has two distinct phases. During the first phase, a cell becomes committed to undergoing apoptosis but does not exhibit any of the characteristic features of apoptosis. This phase is often linked to cell cycle regulation. Most retinoid studies concern commitment to undergo apoptosis. The actual mechanics of the death program occur during the second, "execution" phase. Retinoids may influence commitment or may directly activate the apoptotic machinery. Current knowledge regarding the apoptotic process has been exhaustively reviewed (Kroemer *et al.*, 1995; Schwartz and Osborne, 1995). The key difference between death by apoptosis and death by necrosis is regulation of the death process. Necrosis occurs as a result of gross cellular insult. The cell membrane loses its integrity and cellular disintegration occurs. Debris is removed by phagocytes during the inflammatory response. In contrast, apoptotic cells do not lose membrane integrity and do not cause inflammation. A set of characteristic morphological and molecular events occurs. Although the specific choreography varies from cell type to cell type, typically mitochondrial impairment, chromatin compaction, and shrinking and convolution of both cytoplasm and nucleus occur (Kroemer *et al.*, 1995). In many cases, although not all (Lockshin and Zakeri, 1996), rapid internucleosomal chromatin cleavage occurs. Ultimately the cell breaks into membrane-enclosed vesicles that are phagocytosed by surrounding cells. The process is rapid and causes little tissue disruption.

Although published descriptions of the influence of retinoids on the commitment to apoptosis are plentiful, little is known regarding the effects of retinoids on the execution phase. A notable exception is the observation that most apoptotic cells have high levels of the RA-inducible enzyme, tissue transglutaminase. This calcium-dependent enzyme forms irreversible protein crosslinks and may be partially responsible for the unique morphology of apoptotic cells.

III. Brief Review Regarding the Effects of Retinoids on Tissue Growth and Death

A. Retinoids and Morphogenesis

The retinoids are extremely teratogenic. In fact, Accutane (13-*cis*-RA), which is used for treatment of chronic cystic acne, is one of the most potent human teratogens. Affected children show malformations of the cardiovascular system, face, and central nervous system and absence of the thymus (Lammer *et al.*, 1985). Frequency of spontaneous abortion is also increased. Numerous animal models are also affected by excess retinoids in specific and reproducible ways. Vitamin A is a vital nutrient, however, and vitamin-A–deficient embryos exhibit a distinct spectrum of abnormalities. For a review of retinoid-induced teratogenesis, see Armstrong *et al.* (1994). Interestingly, the syndromes of malformations caused by retinoid deficiency or excess overlap. This suggests that retinoid levels must be precisely controlled in a subset of tissues.

The effects of retinoids on developments are caused by two mechanisms: modification of positional information, and direct effects on cell differentiation, proliferation, or death. A profusion of studies indicates that retinoids can alter the positional identity of cells within developing structures. This may occur because retinoids directly alter the expression of *hox* genes. *Hox* genes encode transcription factors expressed in overlapping patterns along the anterior–posterior axis of vertebrate embryos and developing limbs. The patterns of expression provide a code that might trigger a pattern of cell responses appropriate to that embryonic region. Retinoic acid treatment of mouse, chick, and frog embryos causes ectopic anterior *hox* gene expression and loss of anterior structures. Local application of RA to the anterior margin of the developing chick limb or ectopic expression of HOX 4.6 causes extra digits to develop in an exact mirror image of the first set (posterior to anterior) (Tabin, 1995). Retinoids also alter the proximal–distal axis of regenerating amphibian limbs (Hofmann and Eichele, 1994).

Since retinoids can transform cell positional identity, separating the direct effects of retinoids on cell behavior from effects due to a cell's perceiving an incorrect location is a challenge. Much of our understanding of the direct effects of retinoids on specific cells has been derived from studies of cells grown *in vitro*. Because retinoids can cause profound changes in differentiation, proliferation, and survival of a multitude of cell lines *in vitro*, it is likely that retinoids have similar consequences *in vivo*. Although *in vitro* work dominates the literature, there have been some studies regarding the influence of retinoids on cell survival in the embryo or in explanted tissues.

1. The Neural Crest

Neural crest cells are pluripotent cells, derived from the neural tube, that migrate throughout the embryo and generate ectomesenchymal derivatives (e.g., smooth muscle, bone, and cartilage), skin melanocytes, and most of the peripheral nervous system. Excess retinoids profoundly disrupt many neural-crest-derived tissues, e.g., the thymus, craniofacial structure, and cardiovasculature (e.g., Lammer *et al.*, 1985). Interestingly, vitamin A deficiency and null mutations of some retinoic acid receptors also affect neural-crest-derived tissues (Kastner *et al.*, 1995). The neural tube and other tissues containing neural crest cells contain endogenous retinoids and retinoid receptors, suggesting that RA signaling occurs in these tissues (see Ang *et al.*, 1996; Colbert *et al.*, 1995, and ref. therein).

The specific cellular mechanisms by which retinoids disrupt neural-crest-derived tissues are complex. Both *in vitro* and *in vivo* studies indicate that retinoid treatment causes alterations in neural crest cell migration (Lee *et al.*, 1995b; Leonard *et al.*, 1995; Smith-Thomas *et al.*, 1987; Thorogood *et al.*, 1982).

Retinoic acid also changes the differentiation of cultured neural crest cells (Dupin and Le Douarin, 1995; Henion and Weston, 1994; Ito and Morita, 1995), suggesting that neural crest cells may differentiate inappropriately in response to excess retinoid *in vivo*. Finally, as discussed next, neural crest cells may undergo aberrant growth or apoptosis if exposed to excess retinoids or if deprived of adequate retinoid levels.

Several observations suggest that some neural crest cells die in response to retinoid excess or deficiency. Early reports regarding the effects of retinol and 13-*cis* RA on cranial crest explants described rounded, blebbing cells (Smith-Thomas *et al.*, 1987; Thorogood *et al.*, 1982). This morphology is typical of apoptotic cells. Increased cell death was observed along the neural tube of RA-treated rodent embryos (Cunningham *et al.*, 1994; Sulik *et al.*, 1988). The death of neural crest cells destined to populate embryonic structures can explain retinoid-induced teratogenesis. However, the perplexing observation that similar defects occur in vitamin-A–deficient animals can be explained best by the hypothesis that some neural crest cells require retinoids for survival. This is supported by some experimental observations. For example, RA greatly increases colony formation by cranial neural crest cells but not by trunk neural crest cells (Ito and Morita, 1995). The rat neural crest line developed by Stemple and Anderson was selected in media containing 0.1 μM RA, a concentration that causes differentiation or apoptosis in most other cell lines (Stemple and Anderson, 1992). The paradox that both retinoid overdose and retinoid deficiency affect neural-crest-derived tissues may be due to the pluripotency of neural crest cells. Indeed, the response of the neural tube to RA is highly stage dependent, with distinct effects on embryos differing in age by only 12 hours (Cunningham *et al.*, 1994; Gale *et al.*, 1996; Leonard *et al.*, 1995; Wood *et al.*, 1994). Cells at different anterior–posterior levels or stages of development may require different retinoid levels for normal proliferation or survival.

2. The Developing Limb

The ability of retinoids to mimic the zone of polarizing activity (ZPA) in the developing chick limb is well known. Briefly, local application of retinoic acid to the anterior margin of a developing limb causes mirror-image duplication of the digit pattern (Tickle *et al.*, 1982). This is a position identity effect. Retinoids can also affect limb cell survival. The developing limbs are shaped by regions of apoptosis. In the limbs of embryos whose mothers are fed teratogenic amounts of all-*trans* RA, an increase in the number of apoptotic cells correlates with malformations (Kochhar and Agnish, 1977; Zakeri *et al.*, 1993).

It is not clear whether or not a retinoid is normally required to stimulate apoptosis in the mouse limb. There are, however, several genes involved in retinoid signal transduction or metabolism expressed in the interdigital regions and other apoptotic areas (Dollé *et al.*, 1990; Ruberte *et al.*, 1991, 1992). One is the retinoic acid receptor (RAR)β, a retinoid-activated transcription factor (Dollé *et al.*, 1989). Interestingly, the expression patterns of RARβ and another receptor, RARγ, exclude each other. RARβ is interdigital, and RARγ is expressed in the digit forming regions. RARα transcripts are evenly distributed in the developing limb. The cellular retinol binding protein (CRBP)I, and the cellular retinoic acid binding proteins (CRABP)I and II are expressed in the interdigital zone and may be involved in retinoid metabolism.

RARβ and the CRABPs, however, are not obligatory for cell death, because mice lacking RARβ or CRABPI and II are essentially normal (Kastner *et al.*, 1995; Lampron *et al.*, 1995). The only significant defect noted in mice lacking both CRABPI and II was a high frequency of one extra postaxial digit. This type of polydactyly is often noted in embryos exposed to treatments that alter cell death patterns in developing limbs; vitamin A deficiency and retinoid overdose, for example (Morriss-Kay and Sokolova, 1996). In contrast, RARα-deficient mice have fused digits, possibly due to decreased apoptosis (Lohnes *et al.*, 1994). Whether another receptor performs a function in mutant mice normally fulfilled in RARβ or whether RARβ is truly irrelevant will be difficult to sort out, because the RARs compensate readily for one another's activities (Kastner *et al.*, 1995; Taneja *et al.*, 1996).

Some observations indicate that RA is found in regions undergoing apoptosis in the limb. Several RA-inducible genes, e.g., RARβ, CRBPI, and CRABPII, are expressed in the apoptotic zones (Gudas *et al.*, 1994; Hofmann and Eichele, 1994). Likewise, tissue transglutaminase, an RA-inducible enzyme involved in the execution of apoptosis, is activated in the interdigital region (Davies *et al.*, 1992). Transgenic mice, which have RA-inducible β-galactosidase genes, express β-galactosidase in the interdigital zones (Mendelsohn *et al.*, 1991; Rossant *et al.*, 1991). The TGFβ-related growth factor, BMP-2, which is induced by RA in F9 embryonal carcinoma cells and in developing chick limbs (Francis *et al.*, 1994; Rogers *et al.*, 1992), colocalizes with BMP4 and -7 to the interdigital regions. One or more of these BMPs is required for cell death, because interference with BMP signaling prevents death (Zou and Niswander, 1996). Limb mesenchyme may be poised between apoptosis and chondrogenesis (Roach *et al.*, 1995). Since RA can inhibit chondrogenesis and promote apoptosis (Ballock *et al.*, 1994), RA may disrupt this delicate balance and cause malformations.

3. Palatogenesis

Excess RA can cause cleft palate in mice. Normally, the medial epithelium of the palate undergoes programmed cell death. A down-regulation of epidermal growth factor (EGF) receptors occurs prior to death. The resulting loss of an EGF signal to proliferate may cause death. RA prevents the normal developmental down regulation of EGF receptors in the palate (Abbott *et al.*, 1988). These cells continue to proliferate instead of dying. Expression of dominant-negative RARs that block RA signaling in transgenic mice also induces cleft palate (Damm *et al.*, 1993). Like the neural crest, the palate may be an RA-sensitive structure responsive to either an increase or a decrease in retinoid signaling.

4. Mutations with RA-Modulated Phenotypes

Several mutations have been identified that affect programmed cell death in mice. The phenotypes associated with some of these mutations are aggravated or alleviated in RA-treated mice. Many teratogens, including RA, affect neural tube closure by disrupting proliferation and/or apoptosis levels. *Pax3*, whose human homologue is associated with the human Waardenburg type I syndrome, is a transcription factor required for neural tube closure. *Splotch* is a murine mutation in *Pax3* (Epstein *et al.*, 1991; Vogan *et al.*, 1993). Depending on time of exposure, RA can increase or decrease the incidence of neural tube defects in *Splotch* embryos (Kapron-Bras and Trasler, 1985, 1988).

RA also prevents neural tube defects associated with the *curly tail (ct)* mutation. Just as RA-induced neural crest defects vary greatly with time of embryonic exposure in wild-type (Cunningham *et al.*, 1994) or *splotch* embryos, *curly tail* rescue was maximal within a 4-hour window (Chen *et al.*, 1994). This indicates that RA can complement the *ct* mutation only at a specific embryonic stage. Rescue by RA may involve restoration of normal receptor expression patterns, as RARβ and RARγ RNA levels are abnormally low in untreated *ct/ct* embryos and are partly normalized in RA-treated *ct/ct* embryos (Chen *et al.*, 1995b). Possibly, *ct* normally controls RAR expression and high levels of RA compensate for *ct* dysfunction.

RA also modulates the phenotype of two mutations that cause limb defects by increasing or decreasing the severity of malformation. The *legless* insertional mutant is hypersensitive to RA-induced limb malformations relative to wild-type embryos (Scott *et al.*, 1994). On the other hand, RA decreases the severity of the defects found in mice homozygous for the *polydactyly Nagoya* (*Pdn*) allele of *Gli3* (Schimmang *et al.*, 1994; Tamagawa *et al.*, 1995). Since RA changes apoptotic patterns (Kochhar and Agnish, 1977; Zakeri *et al.*, 1993) and polydactyly is often associated with aberrant apoptosis, the signaling pathways affected by these mutations and retinoids may intersect.

Defects in RA signaling can alter embryonic susceptibility to RA-induced malformations. For example, embryos with a disrupted RXRα are less sensitive to RA-induced limb defects. It has been suggested that the mechanisms controlling teratogenic responses differ subtly from those controlling normal morphogenesis (Kastner *et al.*, 1995). If so, then understanding how RA modulates mutant phenotypes may increase our understanding of how environmentally induced birth defects occur.

The effects of RA on many other spontaneous and induced mouse mutations affecting embryonic apoptosis and proliferation have not been determined (Informatics, 1996). Some of these genes have been shown to interact tangentially with RA. For example, the Wilms' tumor suppressor gene, *WT1*, may repress RARα expression (Goodyer *et al.*, 1995). A null mutation of the mouse homologue, *Wt1*, causes apoptosis of metanephric blastema cells, resulting in failure of kidney development (Kreidberg *et al.*, 1993). It would be interesting to see if vitamin A deficiency or RA overdose modulates the *Wt1* null mutation phenotype. Understanding the etiology of birth defects requires understanding interactions between genetic and environmental factors. It will be instructive to analyze systematically the effects of the well-characterized teratogen, RA, in various genetic backgrounds.

B. Effects on Cultured Cells and Tumors

Vitamin A deficiency (VAD) is clearly incompatible with normal differentiation and proliferation (Wolf, 1996). Vitamin A deficiency is associated with premalignant tissue changes and increased cancer risk, particularly in epithelial tissues (Moon *et al.*, 1994). The growth inhibition of many transformed cells by retinoids is of great clinical interest. Retinoic acid causes this inhibition by inducing differentiation and an accompanying decrease in proliferation rate or by stimulating apoptosis, depending on cell type. Rapidly entering the medical arsenal against a variety of cancers, retinoids have particular efficacy against acute promyelocytic leukemia and cancers of the aerodigestive tract. Reviews of many cancer therapy studies can be found in Lotan (1996).

In 1989, Amos and Lotan compiled a list of 102 articles describing retinoid-responsive cells (Amos and Lotan, 1990). A cursory search of the more recent Medline database identified over 500 publications describing the effects of retinoids on cell proliferation and over 100 describing effects on cell death. Out of 199 cell types listed by Amos and Lotan, 190 were growth inhibited by retinoids. It is interesting to note that in recent years, a small, but increasing, number of cell types have been found to be stimulated to proliferate by retinoids. See Table 1. A closer analysis of these exceptional cells is warranted. The idea that retinoids stimulate some transformed cells *in vivo* is supported by the observation that supplements of β-carotene, a major dietary source of vitamin A, may even increase the risk of lung cancer in smokers (Peterson, 1996). This suggests that retinoid treatment may be contraindicated for some cancer patients. Besides the obvious clinical relevance of knowing how a tumor will respond to retinoids, a comparison of these cell lines to inhibited lines may provide insight into the mechanisms by which retinoids regulate proliferation.

More relevant to our understanding of normal growth and differentiation is the fact that many of the stimulated cells are not transformed. This is consistent with the absolute requirement for vitamin A in normal growth and differentiation.

Table 1 Cells Induced to Proliferate or Survive by Retinoic Acid

Cell type and origin	Reference
Human melanoma, Hs294	Amos and Lotan, 1990
Hamster fibroblasts, DES-4 H-*ras* transformed	Amos and Lotan, 1990
Human lung adenocarcinoma lines (four lines)	Yang *et al.*, 1992
Human squamous carcinoma cells, 183A	Amos and Lotan, 1990
Rat fibroblasts, NRK-SA6, clone 536-3-9 transformed	Amos and Lotan, 1990
Rat fibroblasts, NRK-SA6, clone 536-3-1 untransformed	Amos and Lotan, 1990
Mouse C3H 10T1/2 fibroblasts	Mordan, 1989
Rat lung epithelial cells	P. Roberts *et al.*, 1990
Rat NB II bladder tumor	Amos and Lotan, 1990
Rabbit kidney epithelial cells, RK13	Argiles *et al.*, 1989
Guinea pig epidermal cells	Amos and Lotan, 1990
Human epidermal cells cultured on a collagen raft	Choi and Fuchs, 1990
Human normal adult skin	Fisher and Voorhees, 1996
TM4 Sertoli cells	Amos and Lotan, 1990
Bovine aortic endothelial cells	Amos and Lotan, 1990
Human fetal hematopoietic cells (CD34+)	Zauli *et al.*, 1995
Mouse cranial neural crest explant cultures	Ito and Morita, 1995
Rat neural crest cell line	Stemple and Anderson, 1992

Differentiating cells may proceed through stages with varying sensitivity to retinoids. As discussed in section III.A.1, some neural crest cells require RA for growth. Likewise, normal fetal hematopoietic stem cells require retinoids for proliferation, and retinoids prevent apoptosis due to serum deprivation (Zauli *et al.*, 1995). Similarly, human keratinocytes grown on collagen rafts and human skin *in vivo* are stimulated by retinoids (Choi and Fuchs, 1990; Fisher and Voorhees, 1996). In contrast, retinoids inhibit hundreds of transformed cells *in vitro* and acute promyelocytic leukemic cells *in vivo*.

Finally, a number of natural, bioactive retinoids have been investigated recently (Achkar *et al.*, 1996; Blumberg *et al.*, 1996; Buck *et al.*, 1991; Pijnappel *et al.*, 1993; Thaller and Eichele, 1990). Some cells, for example, B lymphocytes (Buck *et al.*, 1991) and Sertoli cells (Griswold *et al.*, 1989), specifically require retinol. Sertoli cells support and nurture developing spermatids within the semeniferous epithelium. Although many symptoms of vitamin A deficiency can be alleviated by retinoic acid supplements, testicular dysfunction is not, implying a critical role for another retinol metabolite. The role of other natural retinol metabolites in regulating cell growth and survival is just beginning to be explored.

IV. Transcription Factors

Retinoids exert many of their effects by altering transcription of specific genes. This section will review known interactions between retinoid-controlled transcription factors and proliferation and death. Since the mechanisms controlling retinoid-mediated transcription activation and repression are considerable, a brief review will be provided along with reference to several detailed reviews.

A. Retinoid Receptors

1. Review

Retinoic-acid-regulated gene expression is mediated by nuclear retinoid receptors, which act as ligand-dependent transcription factors (for review, see Chambon, 1996). These receptors are encoded by six different genes, RARα, β, and γ and RXRα, β, and γ. Numerous developmentally regulated isoforms arise from differential promoter usage and alternate splicing. Furthermore, the receptors act as homodimers and heterodimers, each with unique characteristics. The RXRs also bind other hydrophobic ligand receptors, such as the thyroid hormone, vitamin D, and peroxisome proliferator-activated receptors. Additional partners include orphan receptors that have no known ligand, for example, Nurr77/NGFI-B and LXR/RLD-1 (Leblanc and Stunnenberg, 1995). This plethora of receptors and gene pathways may begin to explain the multiple effects retinoids have on differentiation.

Different dimer combinations have different DNA-binding and ligand-binding affinities. RARs bind either all-*trans* RA or 9-*cis* RA, whereas RXRs can bind only 9-*cis* RA. Thus 9-*cis* RA is a panagonist. In heterodimers with RARs on some consensus sequences, RXRs cannot bind any ligand and are silent cofactors. As homodimers or in heterodimers with Nurr77 or LXR, RXR can bind 9-*cis* RA and transactivate genes (Leblanc and Stunnenberg, 1995; Zhang *et al.*, 1992). The interaction with Nurr77 will be discussed later, because Nurr77 is induced during the T-cell activation leading to *Fas*-induced apoptosis (Liu *et al.*, 1994; Woronicz *et al.*, 1994). Extensive effort has been devoted to the synthesis of retinoids that activate a subset of retinoid receptors (Dawson and Hobbs, 1994). Both receptor-selective agonists and receptor-selective antagonists have been identified. These retinoids have enormous potential clinical value. They are also useful research tools, for they can regulate a subset of the genes targeted by retinoids.

2. Receptor Mutants

The use of homologous recombination in ES cells to disrupt specific genes has revolutionized cell and developmental biology. However, there are many surprising examples of mice surviving null mutations of genes predicted to be essential to development, suggesting that there are many alternative pathways to normal development (Tautz, 1992). The retinoid receptors are no exception. For example, despite highly regulated and widespread embryonic expression, mice lacking RARβ function are apparently normal. Even RARα or γ null mutations are partially penetrant and display a subset of vitamin-A-deficiency–induced defects (Kastner et al., 1995). Unraveling the functions of each receptor may require a combination of genetic and pharmacological approaches (Boylan et al., 1995; Taneja et al., 1996). Careful analysis of the effects of selective retinoids in different genetic backgrounds may tease out the normal roles of each receptor.

3. The Function of Specific Receptors in Controlling Apoptosis

Receptor-selective retinoids were used to determine the relative contribution of RARs and RXRs to HL-60 leukemia cell differentiation and apoptosis (Nagy et al., 1995). All-trans RA induces granulocytic differentiation in HL-60 cells, which is followed by apoptosis. Since all-trans RA activates only RARs, one might initially assume that both differentiation and apoptosis require only RARs. However, all-trans RA can isomerize in cells to 9-cis RA (Heyman et al., 1992). Therefore, Nagy et al. used two synthetic retinoids that activated only RARs or RXRs (Nagy et al., 1995). The nonmetabolizable, RAR-selective retinoid, TTNPB, caused levels of differentiation equivalent to those observed with all-trans RA or 9-cis RA, but did not induce cell death. In contrast, the RXR-selective retinoid (AGN191701) alone induced neither differentiation nor apoptosis. Sequential treatment by the RAR-selective retinoid followed by the RXR-selective retinoid mimicked the natural retinoids. This work indicates that specific receptors play specific roles. In HL60 cells, RARs mediate differentiation and prime the cells to respond to a RXR-mediated death trigger.

Several observations show that RARα alone can mediate differentiation in HL60 cells: (1) RARα dominates in HL-60 cells (Nagy et al., 1995); (2) an RARα-selective retinoid causes differentiation (Apfel et al., 1992; Nagy et al., 1995); (3) an antagonist of RARα function prevents differentiation (Apfel et al., 1992); and (4) dominant-negative RARα mutations prevent differentiation (Robertson et al., 1992; Tsai et al., 1992). Along with differentiation, RARα may regulate genes required to commit a cell to undergo apoptosis following an RXR signal. The antiapoptotic gene bcl2 may be one target, since RAR-selective retinoids, but not RXR-selective retinoids, down-regulate bcl2 in HL-60 cells (Nagy et al., 1996b). The down regulation of bcl2 by RARα is necessary for apoptosis, because constitutive bcl2 expression prevents apoptosis (Naumovski and Cleary, 1994; Park et al., 1994).

Although RAR activation may be sufficient to stimulate differentiation in some HL-60 strains, extended RAR stimulation or a weak RXR signal may be required for complete differentiation. Brooks et al. failed to see differentiation with the same RAR-selective retinoid (TTNPB) used by Nagy *et al.*, although they did observe delayed differentiation after sequential treatment with TTNPB followed by RXR-selective retinoids (Brooks *et al.*, 1996). Both groups used NBT reduction to measure differentiation, and Nagy *et al.* used morphological criteria as well. One difference may be that Nagy *et al.* measured differentiation after 4 to 6 days of treatment, whereas Brooks *et al.* stopped at day 4. Also, Nagy *et al.* scored cells in the earliest stages of differentiation (metamyelocyte) as differentiated. *In vivo,* combined RAR and RXR signaling may optimize granulocytic differentiation in response to low concentrations of retinoids.

An RXR influences proliferation and apoptosis in F9 EC cells. Clifford and coworkers isolated F9 cells lacking functional RXRα (Clifford *et al.*, 1996). These cells are impaired in their ability to differentiate into parietal endoderm but not into visceral endoderm. Interestingly, in the absence of RA, the RXRα-null cells proliferated more rapidly than wild-type cells. This suggests a growth-inhibiting function for unliganded RXR. Certain strains of F9 cells undergo some apoptosis in response to RA, but the level of apoptosis was diminished in the RXR-deficient cells. In contrast, the level of apoptosis in RARγ-null cells resembled that of wild-type cells, although they were unable to differentiate (Boylan *et al.*, 1995; Clifford *et al.*, 1996). Thus, as in HL60 cells, induction of F9 differentiation may require an RAR signal, and induction of apoptosis may require an RXR signal.

In some cells, apoptosis may not require an RXR signal. RAR-selective retinoids induce apoptosis in nontumorigenic rat tracheal epithelial cells (SPOC-1) (Zhang *et al.*, 1995a). In these cells, an antagonist of RARα function and a dominant-negative RARα mutation inhibited apoptosis induced by the RAR-selective, nonmetabolizable retinoid, TTNPB (Zhang *et al.*, 1995a). Clearly, some apoptosis is induced by RARα signaling; however, inhibition by the antagonist was not complete, leaving open the possibility that other RARs or RXRs play a role.

Curiously, an RXR signal prevents apoptosis in activated T-cells. As will be discussed in Section VI.B, retinoids inhibit *Fas* ligand induction and subsequent apoptosis. These examples indicate that induction of apoptosis is not uniquely associated with any particular retinoid receptor.

4. Receptors and Neoplasia

Many tantalizing correlations between RAR function and growth suppression by retinoids have been observed. The most striking finding is that the chromosomal translocations causing acute promyelocytic leukemia involve RARα. These mutations will be discussed separately in section IV.B. RARβ is inducible directly by RA in many cells *in vivo* and *in vitro* (Gudas *et al.*, 1994). Low levels of RARβ or a lack of inducibility in response to RA treatment have been observed in many neoplastic cells (Seewaldt *et al.*, 1995, and refs. therein). Some correlation between RA inducibility of RARβ and growth suppression by RA has been observed. For example, RARβ is induced during the differentiation of embryonal carcinoma cells into benign, slow-growing cells (Hu and Gudas, 1990). Lotan *et al.* (1995) showed that normal oral mucosa expressed RARβ but that only 21 out of 52 premalignant lesions did. Even more interestingly, isotretinoin (13-*cis* RA) treatment induced RARβ in 29 out of 35 lesions and was associated with a positive clinical response. Finally, 58% of mice expressing an antisense RARβ transgene in the lungs developed lung tumors (Bérard *et al.*, 1996).

Another correlation was found in senescing cells. Nontransformed somatic cells have a finite division capacity *in vitro*. They will eventually enter a permanent quiescent state. RARβ is up-regulated in human senescent mammary epithelial cells and dermal fibroblasts (X. Lee *et al.*, 1995; Swisshelm *et al.*, 1994). Lee *et al.* isolated dermal fibroblasts from young and old donors. As expected, fibroblasts from older donors generally had decreased proliferative capacity. However, RA inducibility of RARβ correlated only with proliferative capacity and not with donor age. Essentially, young or old slow-growing cell lines induced RARβ, but, regardless of donor age, fast-growing cells did not.

Finally, mutant proteins that interfere with RAR function are oncogenic. For example, the transforming activity of v-*erb* A oncogene variants correlates with the ability to prevent RA-, thyroid-hormone-, and estrogen-induced transactivation (Privalsky, 1992). V-*erb* A cannot transactivate, but acts as a dominant-negative repressor of nuclear receptors. The precise mechanism may be to sequester RXRs (Barettino *et al.*, 1993; Chen and Privalsky, 1993; Hermann *et al.*, 1993). Likewise, the chimeric proteins created by the acute promyelocytic leukemia-associated translocations may act as a dominant-negative repressor. Together, there is much circumstantial evidence that loss of RAR function correlates with neoplasia.

Several groups have transfected RA-resistant cells with RARs and observed a gain in RA sensitivity. RARα or β expression in human breast cancer lines increased sensitivity to all-*trans* RA. The MDA-MB-231 line is normally unresponsive to RA. Transfection of RARα or β caused these cells to decrease their rate of proliferation in response to RA (Seewaldt *et al.*, 1995; van der Leede *et al.*, 1995). Normally RA halts the growth of MCF-7 breast cancer cells, but transfected RARβ causes apoptosis in response to RA (Seewaldt *et al.*, 1995). Transfection of RARγ into head and neck squamous carcinoma cells enhanced the growth inhibitory effect of RA in monolayer culture and abolished anchorage-independent growth (Oridate *et al.*, 1996). Thus loss of RAR function correlates with neoplasia, and gain of RAR function can restore RA responsiveness.

This rosy picture suggests that RARβ inducibility might be an excellent predictor of tumor response to retinoid treatment. Unfortunately, this is not always true. Kim *et al.* (1995) recently tested normal, premalignant, and malignant human bronchial epithelial cells for RA sensitivity and RARβ inducibility. There was an inverse correlation between basal RARβ expression and RA sensitivity.

Normal cells expressed low RARβ levels and were inhibited by RA. Malignant cells were refractory to RA treatment but expressed transcriptionally active RARβ. A similar situation was observed in non-small cell lung cancer biopsies. Overall, the growth response of a cell to RA cannot be simply correlated with receptor expression or function.

B. PML/RAR

All-*trans* RA was first used to treat acute promyelocytic leukemia (APL) in 1988 (Huang *et al.*, 1988). Since then, many clinical trials indicate that all-*trans* RA is a useful weapon against APL (Chomienne *et al.*, 1996). In APL patients, two specific chromosomal translocations, t(15;17) and, rarely, t(11:17), fuse the RARα gene with two previously uncharacterized genes, PML/*Myl* and PLZF, respectively (for review, see Zelent, 1994). The specific function of PML and PLZF are as yet unknown. However, PML is a nuclear-matrix-associated phosphoprotein (Chang *et al.*, 1995), and PLZF has sequence similarity to the *Krüpple* class of zinc finger transcription factors (Zelent, 1994). PML– or PLZF– RAR fusions create chimeric peptides that encode all but the most *N*-terminal transactivation domain of RARα. These chimeric proteins are invariably expressed in APL patients, but the reciprocal fusions are seldom expressed. A common feature of each fusion protein is the ability to repress wild-type RAR function. Like v-*erb* A, one mechanism of repression may be RXR sequestration (Perez *et al.*, 1993). The PML/RAR fusion also is activated more efficiently by RARα selective agonists than by all-*trans* RA (Gianni *et al.*, 1996).

Although disruption of RARα causes the APL phenotype, the PML and PLZF portions of the chimeric protein control patient response to RA. Leukemic cells containing the PML/RAR fusion differentiate in response to RA. PML/RAR-containing cells are particularly sensitive to RARα agonists (Gianni *et al.*, 1996). In contrast, those with the PLZF/RAR fusion are insensitive to RA (Licht *et al.*, 1995). Both PML and PLZF are nuclear proteins, but PML is localized to discrete nuclear bodies (Dyck *et al.*, 1994; Koken *et al.*, 1994; Weis *et al.*, 1994), whereas the PLZF pattern is "speckled" (Licht *et al.*, 1996). The PML/RAR fusion protein is abnormally localized in both cytoplasm and nucleus. Retinoic acid causes PML/RAR to assume the normal PML pattern (Dyck *et al.*, 1994; Koken *et al.*, 1994; Weis *et al.*, 1994). In contrast, the PLZF/RAR pattern is unaltered by RA treatment. Both aberrant PML localization and RAR function may contribute to oncogenesis. The APL story clearly indicates that disruption of RA signal transduction can cause cancer and that these cancers might be treatable by retinoids.

C. AP-1 Factors

Retinoids and several other hormones antagonize gene activation by the AP-1 transcription factors (Beato *et al.*, 1995). Since many growth factors induce AP-1 transactivation, this is an effective mechanism by which retinoids can inhibit proliferation. The AP-1 protein complex consists of heterodimers between members of the *jun* and *fos* families of proteins or *jun* homodimers. AP-1 activation follows stimulation of the MAP kinase or JNK signal transduction cascades by mitogenic stimuli such as phorbol esters. Inhibition of tumor promotion by phorbol esters was one of the earliest experiments providing evidence for retinoid-mediated AP-1 interference (Verma *et al.*, 1979). Recently published data suggest that the CREB-binding protein (CBP) integrates nuclear receptor, AP-1, and cAMP signal transduction (Kamei *et al.*, 1996). Apparently, ligand-bound RARs interfere with AP-1 signaling by competing for limiting concentrations of CBP. The fact that perturbations in one of these major signal transduction systems can be relayed to the others via CBP competition may explain the extreme cell-context dependence of retinoid responses.

Retinoids that repress AP-1 activity but do not induce RAR-mediated transcription have been isolated by several groups (J. Chen *et al.*, 1995; Fanjul *et al.*, 1994; Li *et al.*, 1996; Nagpal *et al.*, 1995). These retinoids may cause allosteric changes in the RARs that prevent transactivation without changing the affinity of the RAR for CBP. Such retinoids may be very useful clinically, since they may inhibit growth without causing differentiation-related side effects.

In addition to interfering directly with AP-1 transactivation, retinoids may also regulate *jun* and *fos* levels. For example, all-*trans* RA can inhibit polyoma virus transformation of rat fibroblasts. This appears to result from an inhibition of c-*fos* transcription (Talmage and Listerud, 1994). In contrast, RA induces c-*fos* and c-*jun* in F9 embryonal carcinoma cells. Overexpression of these oncogenes can induce differentiation and concomitant growth arrest in F9 cells, indicating a role distinct to that seen in fibroblasts (Müller and Wagner, 1984; Yamaguchi-Iwai *et al.*, 1990). Thus, RA may influence AP-1 activity by regulating *fos* or *jun* levels by unknown mechanisms or by interfering directly with AP-1 mediated transcription.

V. Direct Modulation of Proliferation and Apoptosis

Section IV discussed the known retinoid signaling molecules involved in transmitting a signal to stop growing or to undergo apoptosis. To understand growth and death control completely, it is necessary to identify which genes are regulated by the retinoids. Logical targets might be those genes involved directly in cell cycle and apoptosis control. A survey of how each gene responds in retinoid-treated cells might help integrate the retinoid signaling pathway with proliferation regulation. A beginning has been made in only a few cell types. Table 2 summarizes known effects on specific cell cycle molecules, with the exception of c-*myc*. C-*myc* levels have been determined in many RA-treated cells, and these are summarized in Table 3.

Table 2   Changes in Cell Cycle Gene Expression or Activity That Precede or Coincide with Growth Alterations in RA-Treated Cells

Gene product or activity	Increase (↑) or decrease (↓)	Cell type	Reference
CDK 4/cyclin D activity	↓	T-47D human breast cancer cells	Wilcken et al., 1996
CDK 4/cyclin D activity	↑	CGR8 mouse ES cells	Savatier et al., 1996
		P19 mouse EC cells	Kranenburg et al., 1995
CDK 6/cyclin D activity	↑	CGR8 mouse ES cells	Savatier et al., 1996
Cyclin D1 RNA & protein levels	↑	CGR8 mouse ES cells	Savatier et al., 1996
		P19 mouse EC cells	Kranenburg et al., 1995; Slack et al., 1995b
Cyclin D1 protein level	↑	Tera2 human EC cells	Kranenburg et al., 1995
Cyclin D2 RNA & protein levels	↑	CGR8 mouse ES cells	Savatier et al., 1996
		P19 mouse EC cells	Kranenburg et al., 1995; Slack et al., 1995b
Cyclin D1 protein level	↑	Tera2 human EC cells	Kranenburg et al., 1995
Cyclin D3 RNA & protein levels	↑	CGR8 mouse ES cells	Savatier et al., 1996
		P19 mouse EC cells	Slack et al., 1995b
CDK2 RNA	↓	HL-60 human leukemia cells	Burger et al., 1994
CDK 2/cyclin E activity	↑	CGR8 mouse ES cells	Savatier et al., 1996
CDK 2/cyclin E activity	↓	P19 mouse EC cells	Kranenburg et al., 1995
CDK 2 protein level	↓	P19 mouse EC cells	Kranenburg et al., 1995
Cyclin E RNA level	↑	CGR8 mouse ES cells	Savatier et al., 1996
	↓	HL-60 human leukemia cells	Burger et al., 1994
Cyclin E protein level	↓	P19 mouse EC cells	Kranenburg et al., 1995
Cyclin B RNA	↓	HL-60 human leukemia cells	Burger et al., 1994

(continued)

Table 2   *(Continued)*

Gene product or activity	Increase (↑) or decrease (↓)	Cell type	Reference
CDC2 (CDK1)/ Cyclin B activity	↓	P19 mouse EC cells	Kranenburg et al., 1995
CDC2 (CDK1) protein	↓	neuroblastoma cell line	Gaetano et al., 1991
p105rb RNA and protein level	↑	P19 mouse EC cells	Kranenburg et al., 1995; Slack et al., 1993
p105rb protein level	↓	Tera2 human EC cells	Kranenburg et al., 1995
p105rb phosphorylation level	↓	T-47D human breast cancer cells	Wilcken et al., 1996
		P19 mouse EC cells	Kranenburg et al., 1995; Slack et al., 1993
		Tera2 human EC cells	Kranenburg et al., 1995
		HL-60, U937, BE-13 human leukemic cell lines	Brooks et al., 1996
		neuroblastoma cell line	Gaetano et al., 1991
E2F binding activity	↓	P19 mouse EC cells	Kranenburg et al., 1995; Reichel, 1992
		F9 mouse EC cells	Bandara and La Thangue, 1991; Reichel et al., 1987
p21waf1/cip1 RNA*	↑	HL-60 mouse leukemic cells	Jiang et al., 1994
	↓	HL-60 mouse leukemic cells	Zhang et al., 1995b
p21waf1/cip1 protein	↓	HL-60 mouse leukemic cells	Schwaller et al., 1995
p27kip RNA	↑	HL-60 mouse leukemic cells	Roberts et al., 1994

*Opposing results.

A. Basic Cell Cycle Machinery

1. Inner-Cell-Mass–like Cells

The mouse blastocyst contains two cell types, inner cell mass (ICM) and trophectoderm. The trophectoderm will contribute to the placenta. All other extra-embryonic and embryonic tissues descend from the inner cell mass. Embryonic

Table 3 Regulation of C-*myc* in RA-Treated Cells

Up-regulated	Reference
P19 mouse EC cells	Slack *et al.*, 1995b
MCF-7 human breast cancer lines (transient early induction)	Sheikh *et al.*, 1993
N.1 ovarian adenocarcinoma	Krupitza *et al.*, 1995
Down-regulated	
F9 mouse EC cells	Dony *et al.*, 1985
Myeloid cells	Banavali *et al.*, 1993; Ferrari *et al.*, 1992; Ishida *et al.*, 1995; Larsson *et al.*, 1994; Yen *et al.*, 1992
HOC-7 ovarian carcinoma cells	Grunt *et al.*, 1992; Somay *et al.*, 1992
NIH:OVCAR3 human ovarian cancer cells	Saunders *et al.*, 1993
MCF-7 and T47-D human breast cancer lines	Saunders *et al.*, 1993
LA-N-5 cultured human neuroblastoma cells	Saunders *et al.*, 1995
NCI-H82 small-cell lung cancer cells	Kalemkerian *et al.*, 1994

stem (ES) cells are totipotent cells cultured from the inner cell mass. Following transplantation back into a blastocyst, they develop normally and may contribute to the germ line. Embryonal carcinoma (EC) cells are pluripotent stem cells derived from spontaneous teratocarcinomas or from teratocarcinomas induced by transplanting early embryos (blastocyst to egg cylinder stage) to nonuterine sites. Like ES cells, EC cells can differentiate normally following blastocyst injection. However, presumably due to their sojourn through a teratocarcinoma, EC cells rarely contribute to the germ line. Both cell types are highly malignant in extra-uterine locations. For general reviews, see Martin (1980); Robertson (1987).

The morphology, pattern of gene expression, and cell cycle of both ES and EC cells resemble those of the inner cell mass. Inner cell mass cells have division times of 9–12 hours, with a very short G1 phase (Gamow and Prescott, 1970). The generation times of EC and ES cells are similar, with a G1 phase of 0–1.5 hours (Rosenstraus *et al.*, 1982; Savatier *et al.*, 1994).

Analysis of known cell cycle regulators in ES cells and EC cells suggests that G1 control is distinct from that in more differentiated cells. In fibroblasts and non-EC tumor cells, the G1 cyclin-controlled kinases (*CDK2*, *4*, and *6*) regulate G1 progression and S-phase entry. In contrast, ES cells have little or no G1 CDK activity, due to an absence of the G1 cyclins D1, 2, 3 or E. CDK4 activity is also low in EC cells (Kranenburg *et al.*, 1995; Slack *et al.*, 1995b). This may be a general feature of early embryos, because G1 CDK complexes are absent prior to gastrulation. Cyclin D1 complexes with *CDK4* are strongly induced at gastrulation and the D2 and D3 complexes appear two days later (Savatier *et al.*, 1996).

Embryonic stem cells also lack the general CDK inhibitors $p21^{waf1}$ and $p27^{kip1}$. Thus, rapidly growing inner-cell-mass–like cells and pregastrulation embryos lack the primary G1 regulators found in other cells.

A major transition occurs upon RA-induced differentiation of EC and ES cells. Differentiation coincides with an increased doubling time, due largely to an elongated G1 phase (Rosenstraus *et al.*, 1982; Savatier *et al.*, 1996). In ES cells induced to differentiate by the addition of RA, G1 CDK activities are dramatically increased following the induction of cyclins D1, D2, D3, and E (Table 2). In fibroblasts, *CDK2* or *4* activity promotes G1 progression, so CDK activation seems contrary to the decrease in ES cell proliferation rate. However, CDK activity is negatively regulated by the CDK inhibitors, $p21^{waf1}$, $p27^{kip1}$, and $P16^{Ink4a}$. $P21^{waf1}$ and $p27^{kip1}$ are absent in undifferentiated cells. Although, the induction of $p21^{waf1}$ and $p27^{kip1}$ by RA in ES cells has not been reported, it does occur in ES cells induced to differentiate by aggregation. Also, undifferentiated ES cells are resistant to the antiproliferative effect of overexpressed $p16^{Ink4a}$, as expected for cells lacking cyclin D's. In contrast, $p16^{Ink4a}$ can inhibit the growth of RA-treated cells (Savatier *et al.*, 1996). It may be that rapid division of undifferentiated cells is possible because the cells lack complex G1 controls. RA may stimulate "normalization" of G1 progression by inducing both cyclins and CDK inhibitors.

The tumor suppressor $p105^{rb}$ and related proteins may be CDK targets. Hypophosphorylated $p105^{rb}$ causes growth arrest by binding and inactivating the transcription factor E2F. Phosphorylation of $p105^{rb}$ frees E2F to stimulate genes involved in S phase. Hypophosphorylated $p105^{rb}$ is very low to undetectable in early embryos and undifferentiated EC and ES cells (Bernards *et al.*, 1989; Bocco *et al.*, 1993; Savatier *et al.*, 1994; Slack *et al.*, 1993). This may account for the high level of E2F activity in undifferentiated EC cells (Kranenburg *et al.*, 1995; La Thangue and Rigby, 1987; Reichel *et al.*, 1987; Reichel, 1992). Within 24 hours of treating EC cells with RA, $p105^{rb}$ RNA and protein levels are upregulated, although the protein is hyperphosphorylated and presumably inactive (Kranenburg *et al.*, 1995; Slack *et al.*, 1993). $P105^{rb}$ induction may be a direct transcriptional response to RA (Slack *et al.*, 1993). Gradually, as the cells approach terminal neuronal differentiation, the proportion of phosphorylated protein declines (Kranenburg *et al.*, 1995; Slack *et al.*, 1993). The increase in hypophosphorylated $p105^{rb}$ correlates with reduced E2F DNA binding activity and decreased cell proliferation (Kranenburg *et al.*, 1995). Rampant E2F activity caused by a lack of $p105^{rb}$ control may contribute to the rapid growth of inner-cell-mass–like cells. Like the induction of cyclin D's and CDK inhibitors, $p105^{rb}$ activation may add another layer of complex G1 control, resulting in decreased proliferation rates.

The adenovirus E1A protein can bind to and inactivate $p105^{rb}$ and the related proteins p107 and p130. E1A also binds the unrelated p300 protein, which resembles the CREB-binding protein (CBP) that integrates AP-1, RAR, and

CREB function (Kamei *et al.*, 1996). Various E1A deletion mutants have been derived that bind either p300 or the retinoblastoma proteins. Undifferentiated P19 EC cells expressing E1A proteins that bind p300 die, and only rare, spontaneously differentiated cells survive (Slack *et al.*, 1995a). In contrast, P19 cells overexpressing mutant E1A proteins that bound only retinoblastoma proteins survive and proliferate. Wild-type P19 cells grown in suspension cultures in the presence of RA differentiate into neuronal and glial cells (Robertson, 1987). Following 3–4 days of RA treatment, however, the E1A-transfected cells died by apoptosis (Slack *et al.*, 1995b). Death must be triggered either by neuronal differentiation or by RA treatment, for it does not occur during mesenchymal differentiation induced by DMSO. Since neuronal apoptosis also occurs in embryos lacking p105^{rb}, death of the E1A-transfected cells is most likely a consequence of neurogenesis rather than RA treatment per se (Whyte, 1996).

Amphibian and insect embryos experience a "midblastula transition" that marks the onset of zygotic gene transcription and cell cycle normalization. Prior to the midblastula transition, these embryos lack the G1 and G2 phases. Although mammalian embryos continuously transcribe zygotic genes, the differentiation of early embryonic cells, EC cells, or ES cells is also accompanied by dramatic changes in transcriptional regulation, cell cycle control, and the onset of a normal G1 phase. Retinoic acid is a useful experimental trigger, enabling experimental study of the mechanism underlying this critical embryonic event.

2. Breast Cancer Cell Lines

Retinoids may influence breast cancer cell growth by many mechanisms. Retinoic acid induces insulin-like growth factor (IGF) binding proteins, which prevent growth stimulation by sequestering IGF (Gucev *et al.*, 1996; Martin *et al.*, 1995). Retinoic acid may also inhibit growth by down-regulating the progesterone receptor (Clarke *et al.*, 1991). Since both RA and antiestrogens block G1 progression, an inquiry was made into which cell cycle genes are affected by these compounds in T-47D human breast cancer cells (Wilcken *et al.*, 1996). The antiestrogen ICI 164384 causes down-regulation of c-*myc* and cyclin D1, but RA did not. Retinoic acid also did not repress cyclin D3, E, or A, CDK2, or cdk 4 RNA or protein levels or repress CDK2 activity until after cell proliferation slowed. Retinoic acid did cause a decrease in cdk4 activity that correlated with the decrease in S-phase cells. Although the timing of inhibition and the genes regulated by RA and ICI 164384 differed, dephosphorylation of p105^{rb} occurred with both agents. Since this effect preceded the decrease in growth rate, it may be a causative factor. Despite the facts that each chemical acts via ligand-activated transcription factors on G1 progression, it appears that each acts by a different mechanism that ultimately modulates p105^{rb} phosphorylation.

3. Neuroblastoma Cells

Retinoids inhibit the growth of most neuroblastoma cells. One cell cycle player down-regulated by RA is the cyclin-dependent kinase, *CDC2* (*CDK1*). This CDK is best known as the kinase subunit of mitosis-promoting factor, MPF. During RA-induced neuroblastoma cell differentiation, *CDC2* protein levels decline 75-fold (Gaetano *et al.*, 1991). Surprisingly, *CDC2* RNA levels did not parallel the protein decrease, and the *CDC2* RNA half-life actually doubled. This suggests that RA does not directly regulate *CDC2* in neuroblastoma cells. In addition, since RA partially inhibited proliferation in a differentiation-defective cell line without affecting *CDC2* protein levels, other genes must be involved.

4. Myeloid Cells

HL-60 leukemic cells are a premier model of RA-induced differentiation. HL-60 cells can be induced to differentiate into granulocytes with either RA or DMSO and into macrophages with TPA. It should be possible to distinguish cell cycle changes due to RA treatment and those due to differentiation lineage by comparing cell cycle gene expression during different treatments. Retinoic acid, DMSO, or TPA lengthen the G1 phase, probably by decreasing Rb phosphorylation (Brooks *et al.*, 1996). A comparison of cell cycle gene expression in HL-60 cells showed that the mechanism by which G1 increased differs with both lineage and inducing agent. Distinct patterns of cyclin and CDK expression were observed in TPA-treated cells relative to RA- or DMSO-treated cells. For example, cyclin A RNA levels were unaltered in RA-treated cells, but plummeted following TPA treatment (Burger *et al.*, 1994). Retinoic acid, but not TPA, induced p27[kip] RNA levels (Roberts *et al.*, 1994). This might have been predicted, since these drugs induce different cell types. Unexpectedly, the patterns of cell cycle gene expression in granulocytes induced by RA and by DMSO were also dissimilar. For example, RA strongly inhibited cyclin E RNA levels, but DMSO did not. These studies, although an important first step, allow limited interpretation because only RNA levels were measured. Protein abundance and CDK activity should be measured.

5. C-*myc*

The proto-ongocene c-*myc* is a transcription factor of the basic helix-loop leucine zipper class. Intriguingly, c-*myc* overexpression can induce proliferation or apoptosis, depending on growth factor levels (Evan and Littlewood, 1993; Evan *et al.*, 1992). The CDK-activating phosphatase, *cdc25A*, recently has been shown to be an essential c-*myc* target for inducing both proliferation or apoptosis (Galaktionov *et al.*, 1996). Of all the major cell cycle players, c-*myc* levels have been measured in the most different retinoid-treated cells. A sampling of studies is listed in Table 3. Down regulation by retinoic acid is common. This is consistent with its role in promoting proliferation. However, the decrease in expression is often modest, and in several cell lines c-*myc* levels rise. It has been suggested that decreased E2F activity causes the c-*myc* decrease in HL-60 cells (Ishida *et al.*, 1995). This cannot be true in all cells, because E2F activity declines in P19 EC cells (Kranenburg *et al.*, 1995; Reichel, 1992), but the abundance of c-*myc* RNA increases (Slack *et al.*, 1995b). Though c-*myc* regulation may contribute to growth inhibition by RA, it cannot be the sole retinoid target.

6. Summary

A clear picture of the effect of RA on cell cycle genes has not yet emerged. As always, the effects of RA depend on which cell type is analyzed. There may be two general cell cycle effects of RA. In pluripotent inner-cell-mass–like cells (EC and ES cells), RA may induce the protein machinery required for the measured G1 progression found in lineage-committed but not terminally differentiated cells. In committed cells expressing G1 cyclins, e.g., HL60 cells, RA promotes the exit from this normal G1 phase during terminal differentiation. This may involve both inducing inhibitors and repressing regulators like cyclins. Can the same genes be involved? The data are too incomplete to be sure, but some may be. For example, the induction of the CDK inhibitors, e.g., p27^{kip}, would slow any cell also expressing cyclins. Identification of direct targets of RA will require a complete description of how each cell cycle gene responds to RA in various cell types.

B. Death Machinery

1. *Bcl2* and Related Proteins

Once a cell has become committed to cell death by cell cycle perturbation or other stimuli, it can often be rescued by the proto-oncogene *bcl2*. *Bcl2* and its *C. elegans* homologue *ced-9* dampen the activity of the ICE cysteine proteases. *Bcl2* interacts with many other proteins via two *bcl2* homology domains (BH-1 and BH-2). The number of known proteins that contain BH domains or bind to members of the *bcl2* family is now quite large (Farrow and Brown, 1996). Whereas *bcl2* protects against apoptosis, other members of the family, such as *bax*, promote apoptosis. Whether or not *bcl2* or *bax* dimers dominate determines the propensity of a cell to die. It is plausible that retinoids might repress *bcl2* or other protective proteins or induce *bax* or other death-promoting proteins.

The normal differentiation of hematopoietic cells is followed by spontaneous apoptosis. Likewise, HL-60 leukemic cells, induced by all-*trans* RA to differentiate into granulocytes, eventually die by apoptosis. *Bcl2* is down-regulated during differentiation induced by DMSO, TPA, or all-*trans* RA (Delia *et al.*, 1992; Nagy *et al.*, 1996b). If *bcl2* levels are forced to remain elevated in RA-treated cells, differentiation proceeds normally but apoptosis is abrogated (Naumovski and Cleary, 1994; Park *et al.*, 1994). This indicates that *bcl2* repression is necessary, but not sufficient, to prime HL-60 cells for death. As mentioned previously, Nagy *et al.* showed that HL-60 differentiation is largely RAR controlled. Commitment to apoptosis is gained during differentiation, but actual death requires an RXR-mediated signal (Nagy *et al.*, 1995). An RAR-selective retinoid, but not a RXR-selective retinoid, can suppress *bcl2*, but not as efficiently as all-*trans* or 9-*cis* RA (Nagy *et al.*, 1996b). It may be that low levels of RXR activation are required to sufficiently suppress *bcl2* (Roy *et al.*, 1995). Alternatively, an additional priming event may occur in response to RAR-activating ligands.

Bcl2 can also prevent RA-induced apoptosis in P19 EC cells. Okazawa *et al.* (1996) characterized apoptosis in P19 cultures induced to differentiate into neurons and glial cells by all-*trans* RA. They found that overexpression of *bcl2* inhibited apoptosis and increased the number of surviving neurons. Considering the extremely low levels of *bcl2* RNA found in undifferentiated cells, other factors may normally play a protective role in undifferentiated cells. Interestingly, the apoptosis observed in the cells described in this paper occurred more rapidly (12 hours) than that observed by others in wild-type P19 cells (24 hours) (Glozak and Rogers, 1996; Horn *et al.*, 1996) or E1A-transfected P19 cells (3–4 days) (Slack *et al.*, 1995b). Although strain variability may explain some of the differences in wild-type cells, the mechanism causing early apoptosis in response to RA may be distinct from that occurring in E1A-transfected cells arresting during terminal neuronal differentiation. In cells lacking p105[rb] function due to E1A repression, apoptosis is probably caused by the lack of p105[rb] during neurogenesis, rather than directly by RA (Whyte, 1996).

2. Tissue Transglutaminase (Transglutaminase Type II)

Activation of tissue transglutaminase is associated with apoptosis in a wide variety of cells (Piacentini *et al.*, 1994). Activated tissue transglutaminase catalyzes stable protein crosslinks that cannot be destroyed without degradation of the amino acid chain. These crosslinks may help maintain the membrane integrity of apoptotic cells. The presence of tissue transglutaminase protein is insufficient to cause apoptosis. Several cell types, such as smooth muscle, endothelial, and mesangial cells, express tissue transglutaminase constitutively but do not undergo apoptosis (Piacentini *et al.*, 1994). Furthermore, tissue transglutaminase overexpressing fibroblasts can proliferate, although they show some features of apoptotic cells (Gentile *et al.*, 1992). Since tissue transglutaminase is inactive at normal cellular calcium concentrations, it is probably activated by the increase in calcium concentration associated with apoptosis (Piacentini *et al.*, 1994).

Tissue transglutaminase is RA inducible in many cells (Davies *et al.*, 1992). This is a direct transcriptional response involving RARα (Benedetti *et al.*, 1996; Zhang *et al.*, 1995a). Interestingly, Nagy *et al.* (1996b) showed that RXR-selective retinoids could also induce transglutaminase activity in HL-60 leukemia cells, but Benedetti *et al.* (1996) failed to observe activity following RXR stimulation in NB4 leukemic cells. The discrepancy may be due to the use of different RXR agonists (AGN191701 vs. SR11217). A more interesting explanation would relate to the fact that NB4 cells have the PML/RAR translocation and HL-60 cells do not. As discussed earlier, PML/RAR-mediated transactivation has unique characteristics. In any case, the observation that transglutaminase activity was induced but that apoptosis was not further confirms that multiple triggers are required to induce apoptosis. Since neither *bcl2* repression nor tissue transglutaminase activation are individually sufficient to cause death, the next logical step will be to down-regulate *bcl2* artificially, and simultaneously to activate transglutaminase in the absence of retinoic acid. This will determine if only these two events are sufficient to induce apoptosis.

The retinoic-acid-responsive element (RARE) for the mouse tissue transglutaminase gene has been identified. It is a complex element that overlaps an IL-6-responsive element (Nagy *et al.*, 1996a). The RARE has been used to drive a β-galactosidase reporter gene in transgenic mice. Since β-galactosidase expression coincides with programmed cell death in these embryos (P. J. A. Davies, personal communication), this strain will be a useful tool for elucidating how apoptosis is regulated *in vivo*.

VI. External Signals Regulating Proliferation and Death

A. Growth Factors

Many peptide growth factors are induced or repressed by retinoids. In addition, the response of cells to a particular growth factor may change in the presence of a retinoid. Synergistic and antagonistic effects have been observed (Gudas *et al.*, 1994). Understanding the interactions between growth factor and retinoid signaling pathways will go a long way to explaining the pleiotropic effects of retinoids. The proliferative and apoptosis-inducing effects of retinoid-influenced growth factors will be emphasized, rather than effects on differentiation.

1. Insulin-like Growth Factors (IGF-1 and IGF-2) and Platelet-Derived Growth Factor (PDGF)

Insulin-like growth factors (IGF-1 and IGF-2), insulin, and platelet-derived growth factor (PDGF) regulate cell proliferation from the earliest stages of embryogenesis through the adult animal (Harvey and Kaye, 1992). Thus it is not surprising that retinoids have been observed to affect these growth factors in some cells. Retinoic acid can inhibit IGF-1 secretion in rat glioma cells (Lowe, Jr., *et al.*, 1992). The IGF binding proteins-3 and -6, which inhibit IGF activity, are up regulated in RA-treated breast cancer cells (Gucev *et al.*, 1996; Martin *et al.*, 1995). Simultaneous retinoid and dibutyryl cAMP or forskolin treatment induces the binding proteins six- to twelve-fold (Martin *et al.*, 1995). Thus, in at least some cells, retinoids may decrease sensitivity to proliferation-stimulating growth factors.

IGF-1 and PDGF are required for G1 progression in fibroblasts (Pardee, 1989). Retinyl acetate, all-*trans* retinoic acid, and 4-hydroxyphenyl-retinamide (4-HPR) inhibit serum-stimulated mitogenesis in many types of fibroblasts. Retinoids may interfere with the PDGF-mediated transition from G0 to G1 (Desbois *et al.*, 1991; Mordan, 1989). Since retinoids are known to interfere with AP-1 activation, their effect in fibroblasts may result from repression of AP-1–stimulated mitogenic genes. *Ras* activation blocks RA's inhibition (Scita *et al.*, 1996). RA-resistant tumor cells may circumvent growth repression by stimulation of alternative growth-promoting signaling paths (Hill and Treisman, 1995).

2. Epidermal Growth Factor (EGF) and Transforming Growth Factor (TGF-α)

EGF and TGF-α activate the same receptor and are expressed in a wide variety of tissues (Carpenter and Wahl, 1990; Wiley *et al.*, 1995). Retinoic acid induces or represses the number of surface EGF receptors in many interesting developmental systems. Since many other factors besides RA must control receptor expression, EGF's or TGF-α's precise role in regulating proliferation in response to RA is unclear.

3. Transforming Growth Factor (TGF-β's)

Like retinoids, members of the TGFβ superfamily have diverse effects on a motley assortment of cell types. Retinoids and TGFβs often interact to alter proliferation, differentiation, and apoptosis (Gudas *et al.*, 1994). TGFβ1, β2, and β3 are potent inhibitors of proliferation in many cells. Inhibition is due to regulation of G1 progression via c-*myc*, G1 cyclins, or CDK inhibitors (for review, see Alexandrow and Moses, 1995). Retinoids have perplexing effects on TGFβ expression. Retinoic acid treatment decreased TGFβ1 and β2 RNA, but not β3, levels in midgestation mouse embryos in many tissues (Mahmood *et al.*, 1992). Paradoxically, inhibition of retinol-binding protein (RBP) synthesis, which causes RA deficiency, down-regulates embryonic TGFβ1 levels (Båvik *et al.*, 1996). Injection of low concentrations of RA restored normal TGFβ1 expression. There are conflicting reports regarding the induction of TGFβ2 in keratinocytes. Glick *et al.* (1989) showed that TGFβ2 was posttranscriptionally induced in keratinocytes grown on plastic and in mouse epidermis, but Choi and Fuchs (1990) found that keratinocytes grown on feeder layers did not synthesize detectable levels of TGFβ2. The same researchers also observed different responses to retinoids, depending on culture conditions. This illustrates the critical need to understand the environmental context of a cell's response to retinoids.

The bone morphogenetic proteins (BMPs) form a large and evolutionarily conserved subgroup within the TGFβ superfamily (for review, see Hogan, 1996). The BMP2 and 6 RNAs are induced and BMP4 RNA is repressed by all-*trans* RA in F9 EC cells (Lyons *et al.*, 1989; Rogers *et al.*, 1992). Retinoic acid also induces BMP2 RNA in limb mesenchyme (Francis *et al.*, 1994). Transcriptional induction of BMP2 is mediated largely by RAR α and γ (Boylan *et al.*, 1995; Rogers, 1996). As is typical for TGFβs, exogenous BMP2 decreased the proliferation of F9 EC cells. This occurred only in the presence of all-*trans* RA (Rogers *et al.*, 1992). A more striking synergism occurs in pluripotent P19 EC cells. Retinoic acid and BMP2 or 4 synergize to cause extensive cell death in P19 EC cells (Glozak and Rogers, 1996). Although high concentrations of RA induce some apoptosis in P19 cells, several orders of magnitude less RA is required to induce the same level of apoptosis when BMP2 or 4 is present. BMPs alone cause very little apoptosis.

This observation has interesting *in vivo* implications regarding the induction of embryonic apoptosis. Bone morphogenetic proteins are linked to apoptosis in several embryonic tissues. BMP2 and BMP4 cause apoptosis of neural crest cells located in rhombomeres 3 and 5 (Graham *et al.*, 1994) and of limb bud cells (Gañan *et al.*, 1996). Disruption of BMP signaling by a dominant-negative BMP receptor prevents interdigital apoptosis in chick (Zou and Niswander, 1996). Since the BMP-sensitive neural crest and limb cells are located in regions known to contain endogenous retinoids (Hofmann and Eichele, 1994), it is possible that RA collaborates with the BMPs to induce apoptosis *in vivo*. Thus, malformations caused by exogenous retinoids may be due to alterations in BMP levels or increased sensitivity to BMP-induced apoptosis.

Another TGFβ, activin, also affects cell death in P19 EC cells, but in this case it is a survival factor (Schubert *et al.*, 1990). If grown in N2 synthetic, serum-free media with retinoic acid, P19 cells differentiate into neurons. Without RA, P19 cells die. Activin can rescue these cells. Although the mechanism of this effect is not yet known, others have published data supporting the role of activin as a survival factor *in vivo* (Andreasson and Worley, 1995). Some modest changes in activin and activin-receptor levels have been reported, but only the activin-binding protein, follistatin, is regulated significantly in RA-treated P19 cells grown in serum-containing media (Hashimoto *et al.*, 1992; van den Eijnden-van Raaij *et al.*, 1992). The significance of follistatin induction to activin's role as a survival factor is unclear.

4. Fibroblast Growth Factors (FGF)

FGF4 (kFGF) is strongly down regulated by RA in EC cells and in ES cells (Schofield *et al.*, 1991; Schoorlemmer and Kruijer, 1991; Velcich *et al.*, 1989).

FGF4-deficient mice die shortly after implantation due to impaired proliferation of the inner cell mass. The normal rapid pace of ICM growth is restored when FGF4 null mutant embryos are cultured in FGF4-supplemented media (Feldman *et al.*, 1995). These observations suggest that RA may inhibit growth of ICM-like cells by down regulating the autocrine stimulation of FGF4.

B. *Fas*-Mediated Cell Death

Stimulation of the *Fas* (*Apo-1*) cell surface receptor causes apoptosis (Trauth *et al.*, 1989). Although *Fas* and its ligand, *FasL*, have been detected in a wide variety of tissues, they have been best characterized in hematopoietic cells (Schulze-Osthoff, 1994). T-cell receptor activation on T cells, thymocytes, and T-cell hybridomas causes rapid apoptosis by inducing *Fas* and *FasL* expression. Receptor and ligand then interact, causing apoptosis (Brunner *et al.*, 1995; Dhein *et al.*, 1995; Ju *et al.*, 1995). Retinoids, particularly 9-*cis* RA, antagonize activation-induced apoptosis (Yang *et al.*, 1993). This may be clinically useful, since human immunodeficiency virus (HIV)-induced T-cell depletion involves *Fas* activation. A recent study indicated that the HIV-associated apoptosis in T cells isolated from patients was alleviated by retinoic acid treatment (Yang *et al.*, 1995a).

Recently, several laboratories have shown that retinoic acid prevents apoptosis by preventing T-cell-receptor activation-induced expression of the *Fas* ligand (Bissonette *et al.*, 1995; Cui *et al.*, 1996; Yang *et al.*, 1995b). Several observations indicate that both RARs and RXRs are required for full *FasL* repression. 9-*cis* RA, an RAR/RXR panagonist, is more active than all-*trans* RA, which activates only RARs. Likewise, a mixture of RAR and RXR-selective retinoids is required to fully mimic 9-*cis* RA (Bissonnette *et al.*, 1995; Yang *et al.*, 1995b). Finally, expression of a dominant-negative RXRβ prevented 9-*cis* RA rescue (Yang *et al.*, 1995b).

This simple mechanism hints at a new mode of retinoid function. Direct transcriptional repression by retinoid receptors is relatively rare, although retinoids do interfere with AP-1-mediated gene expression. An AP-1-regulated event during T-cell receptor activation is IL-2 induction. Since retinoids do not prevent IL-2 production, it is unlikely that *FasL* repression is due entirely to AP-1 interference (Bissonnette *et al.*, 1995; Yang *et al.*, 1993). The observation that an RXR must be stimulated suggests that alternative heterodimer partners of RXR might be involved. The orphan receptor *Nurr77/NGFI-B* may be one. *Nurr77* is rapidly induced by activation and is required for death *in vitro* (Liu *et al.*, 1994; Woronicz *et al.*, 1994). *Nurr77* binds efficiently to RXRs, changing the RXR function in this heterodimer relative to that in an RAR/RXR heterodimer (Forman *et al.*, 1995; Perlmann and Jansson, 1995). One problem with this hypothesis is that mice containing *Nurr77* null mutants undergo normal T-cell apoptosis *in vivo* (S. Lee, 1995). It may be that binding of ligand to RXR causes formation of RXR homodimers (Zhang *et al.*, 1992). In this case, the function of *Nurr77* or any potential complementing receptor is irrelevant to retinoid function. 9-*cis* RA may simply disrupt the normal RXR hetero-homodimer balance.

The role of *Fas* signaling in other tissues is still unclear. *Fas* and *FasL* are expressed in other tissues, including testis, which normally undergoes a high level of apoptosis (Schulze-Osthoff, 1994; Suda *et al.*, 1993). Although known genetic defects in *Fas* (*lpr* and *lpr*^cg^) and *FasL* (*gld*) cause mouse immune disorders (Schulze-Osthoff, 1994), obvious testicular abnormalities are not seen. However, these mutations do not completely abolish *Fas* or *FasL* expression, so some function may remain. Clearly, if *Fas* signaling plays a role in embryonic programmed cell death, then retinoids might cause perturbations by interfering with *FasL* production in other tissues.

C. Extracellular Matrix (ECM)

Extracellular matrix (ECM) plays a pivotal role in controlling cell proliferation and differentiation. More recently, ECM was found to promote the survival of cells. Mammary epithelial cells deposit a thick ECM, which is required for differentiation and milk production. When lactation ceases, the ECM is actively degraded and the secretory epithelium undergoes apoptosis (Lund *et al.*, 1996). One ECM function is to suppress the apoptosis-inducing enzyme ICE and to stave off death (Boudreau *et al.*, 1996). The ECM also protects the embryonic ectoderm during the formulation of the proamniotic cavity in rodent embryos (Coucouvanis and Martin, 1995). In this situation, the surrounding visceral endoderm secretes a death-promoting signal that causes apoptosis in all ectoderm cells except those contacting the ECM.

It has been long known that retinoids modulate the deposition of extracellular matrix. Many extracellular matrix proteins, for example, laminins A, B1, and B2, and collagen IV RNAs are induced in RA-treated cells (Gudas *et al.*, 1994). Induction is often transcriptional and requires an RAR. Retinoic acid also modulates ECM *in vivo* (Morriss-Kay and Mahmood, 1992). Retinoids repress ECM-degrading enzymes, such as metalloproteinases and collagenase (Pan *et al.*, 1995; Schüle *et al.*, 1991). This is due to the ability of retinoids to interfere with AP-1 transcriptional activation. Although a direct link has not yet been drawn between retinoid-mediated regulation of ECM synthesis and apoptosis, it is quite likely that one (or more) will be found.

Retinoid-mediated suppression of ECM-degrading enzymes may interest those with sun-ravaged skin. Only 15 minutes of exposure to UVB triples AP-1 activity in skin (Fisher *et al.*, 1996). Subsequently, dramatic increases in the RNA abundance and activity of three AP-1–regulated ECM-degrading enzymes, interstitial collagenase, stromelysin, and 92K gelatinase, occur. Topical treatment of skin with all-*trans* RA reduces UV-induced AP-1 and proteinase activity by 50–80%. Since ECM degradation may contribute to photo-aging (wrinkles), perhaps future sun creams will contain retinoids. On a more serious note, ectopic stromelysin expression in the mammary glands of transgenic mice is associated with mammary gland hyperproliferation and a predisposition to tumor development (Sympson *et al.*, 1994). If repeated stromelysin induction and ECM degradation promotes skin cancer, then the identification of safe AP-1–repressing retinoids will be very beneficial.

VII. Retinoic Acid-Independent Responses

As mentioned earlier, the correlation between receptor activity and sensitivity to RA is good but not perfect. Those studies focused on all-*trans* RA. Many new natural and synthetic retinoids are being tested for efficacy against tumor cells. N-(4-hydroxphenyl)retinamide (4-HPR) is in clinical trials as a chemopreventive agent for breast cancer. 4-HPR may not act via normal retinoid signaling mechanisms (Formelli *et al.*, 1996). For example, an RA-resistant subclone of MCF-7 breast cancer cells is sensitive to 4-HPR. Likewise, a 4-HPR-resistant subclone is sensitive to RA (Sheikh *et al.*, 1995). This lack of cross-resistance indicates that different mechanism are involved in RA– or 4-HPR–induced growth inhibition. 4-HPR also minimally binds RARs α, β, and γ and does not activate RAR- or RXR-binding enhancers. Finally, 4-HPR does not interfere with AP-1 activity. The 4-HPR studies suggests that receptor-independent mechanisms may mediate some retinoid responses.

Retinoic acid is not the only natural retinoid that regulates cell growth. Takatsuka and coworkers (1996) recently analyzed RA metabolism in 15 cell lines and found a perfect correlation between extensive metabolism and RA sensitivity. Nine RA-responsive cell lines, including NIH 3T3 cells, keratinocytes, leukemia cells, adenocarcinoma cells, normal breast epithelial, breast ductal carcinoma, and breast carcinoma cells, metabolized 90% of the added RA within 48 hours. In contrast, six resistant lines (*ras*-transformed NIH 3T3 cells, malignant melanoma, human bronchial tumor, breast adenocarcinoma, breast carcinoma, and colon adenocarcinoma) metabolized only 25% of the added RA in 48 hours. The authors suggested that an RA metabolite, not RA, causes the decreased proliferation of these cells. The observation that the RA-metabolizing enzymes, RA-4- and 18-hydroxylase, are induced in an RA-sensitive breast cancer line (T-47D) is consistent with this idea (Han and Choi, 1996).

Regulation of the enzymes controlling retinol or RA metabolism and signaling adds another layer of complexity to retinoid biology. The following proteins or enzymatic activities have been shown to be regulated in RA-treated tissues: cellular retinol-binding protein (CRBP), cellular retinoic-acid-binding protein II (CRABP II), lecithin:retinol acyltransferase (LRAT), alcohol dehydrogenase

(ADH), and RARβ (Ang *et al.*, 1996; Duester *et al.*, 1991; Fisher and Voorhees, 1996; Napoli, 1996). Recent studies also indicate that endogenous retinoids vary with species and developmental stage (Blumberg *et al.*, 1996; Creech Kraft *et al.*, 1995). Novel, biologically active metabolites have been discovered (Achkar *et al.*, 1996; Blumberg *et al.*, 1996; Pijnappel *et al.*, 1993; Thaller and Eichele, 1990) and shown to be required by certain cells (e.g., Buck *et al.*, 1991). Although universally accepted that vitamin A is required for vertebrate embryogenesis and adult health, the identity and function of all active derivatives remains an enigma.

VIII. Future Directions

Clearly, the story of how retinoids influence life-and-death decisions is incomplete. As always, the influence of environmental agents, such as growth factors and extracellular matrix, confounds the simple models. Retinoids can stimulate or repress distant events in a complex web of regulation. We are, however, in a position to begin integrating retinoid signaling with other signaling paths controlling proliferation and death. Some future experiments will be a matter of characterizing the effects of retinoids on the newly identified cell cycle and apoptosis genes. Which genes, if any, are direct targets for retinoid-regulated gene expression? Since many cell cycle genes are transcriptionally regulated, those studying cell cycle gene regulatory elements should examine retinoid responsiveness.

One newly emerging area of research will address the question "Which retinoid?" Because there are thousands of synthetic retinoids that can bind and activate various receptors and binding proteins, the ligand-binding pockets appear to be rather flexible. Therefore, the newly discovered endogenous retinoids may be capable of activating several different signaling paths. The ability of a retinoid to cause a biological response does not prove that it does so *in vivo*. All our experimental tricks will be required to sort out which ligands Mother Nature uses in which cellular context.

Finally, assessing the influence of retinoids on various mutations affecting proliferation and death will use the power of genetics to lead us in unexpected directions. Since the majority of birth defects have no known cause, it will be helpful to understand how one well-characterized teratogen acts in different genetic backgrounds. In particular, the observations that retinoid teratogenesis is exceedingly stage-dependent, particularly during neural development, suggests that additional basic embryological principles will be revealed by retinoid studies.

Acknowledgments

The author appreciates the helpful comments regarding this review provided by Drs. W. J. Pledger, P. J. A. Davies, W. D. Cress, and members of her laboratory, particularly Drs. M. A. Glozak and L. C.

Heller. The review was enhanced by information provided by Drs. Davies, J. Clifford, and R. Taneja before publication. The research of this laboratory is supported by a USF Presidential Young Faculty Award and NIH grant HD31117-03.

References

Abbott, B., Adamson, E., and Pratt, R. (1988). Retinoic acid alters EGF receptor expression during palatogenesis. *Development* **102**, 853–867.

Achkar, C. C., Derguini, F., Blumberg, B., Langston, A., Levin, A. A., Speck, J., Evans, R M., Bolado, J., Jr., Nakanishi, K., Buck, J., and Gudas, L. J. (1996). 4-Oxoretinol, a new natural ligand and transactivator of the retinoic acid receptors. *Proc. Natl. Acad. Sci. U.S.A.* **93**, 4879–4884.

Alexandrow, M. G., and Moses, H. L. (1995). Transforming growth factor beta and cell cycle regulation. *Cancer Res.* **55**, 1452–1457.

Amos, B., and Lotan, R. (1990). Retinoid-sensitive cells and cell lines. *Meth. Enzymol.* **190**, 217–225.

Andreasson, K., and Worley, P. F. (1995). Induction of beta-A activin expression by synaptic activity and during neocortical development. *Neuroscience* **69**, 781–796.

Ang, H., Deltour, L. Hayamizu, T., Zgombic-Knight, M., and Duester, G. (1996). Retinoic acid synthesis in mouse embryos during gastrulation and craniofacial development linked to class IV alcohol dehydrogenase gene expression. *J. Biol. Chem.* **271**, 9526–9534.

Apfel, C., Bauer, F., Crettaz, M., Forni, L., Kamber, M., Kaufmann, F., LeMotte, P., Pirson, W., and Klaus, M. (1992). A retinoic acid receptor alpha antagonist selectively counteracts retinoic acid effects. *Proc. Natl. Acad. Sci. U.S.A.* **89**, 7129–7133.

Argiles, A., Kraft, N. E., Hutchinson, P., Senes-Ferrari, S., and Atkins, R. C. (1989). Retinoic acid affects the cell cycle and increases total protein content in epithelial cells. *Kidney Int.* **36**, 954–959.

Armstrong, R., Ashenfelter, K., Eckhoff, C., Levin, A., and Shapiro, S. (1994). General and reproductive toxicology of retinoids. *In* "The Retinoids: Biology, Chemistry, and Medicine (M. Sporn, A. Roberts, and D. Goodman, eds.), pp. 545–572. Raven Press, New York.

Ballock, R. T., Heydemann, A., Wakefield, L. M., Flanders, K. C., Roberts, A. B., and Sporn, M. B. (1994). Inhibition of the chondrocyte phenotype by retinoic acid involves upregulation of metalloprotease genes independent of TGF-beta. *J. Cell Physiol.* **159**, 340–346.

Banavali, S. D., Pancoast, J. R., Tricot, G., Larson, R., Goldberg, J., Raza, A., Bismayer, J. A., and Preisler, H. D. (1993). The serial study of c-*myc* expression in bone marrow biopsy specimens during treatment for acute myelogenous leukaemia. *Eur. J. Cancer* **29**, 1162–1167.

Bandara, L. R., and La Thangue, N. B. (1991). Adenovirus Ela prevents the retinoblastoma gene product from complexing with a cellular transcription factor. *Nature* **351**, 494–497.

Barettino, D., Bugge, T. H., Bartunek, P., Vivanco Ruiz, M. D., Sonntag-Buck, V., Beug, H., Zenke, M., and Stunnenberg, H. G. (1993). Unliganded T3R, but not its oncogenic variant, v-erbA, suppresses RAR-dependent transactivation by titrating out RXR. *EMBO J.* **12**, 1343–1354.

Båvik, C., Ward, S., and Chambon, P. (1996). Developmental abnormalities in cultured mouse embryos deprived of retinoic acid by inhibition of yolk-sac retinol binding protein synthesis. *Proc. Natl. Acad. Sci. U.S.A.* **93**, 3110–3114.

Beato, M., Herrlich, P., and Schutz, G. (1995). Steroid hormone receptors: Many actors in search of a plot. *Cell* **83**, 851–857.

Benedetti, L., Grignani, F., Scicchitano, B. M., Jetten, A. M., Diverio, D., Lo Coco, F., Avvisati, G., Gambacorti-Passerini, C., Adamo, S., Levin, A. A., Pelicci, P. G., and Nervi, C. (1996).

Retinoid-induced differentiation of acute promyelocytic leukemia involves PML-RARalpha-mediated increase of type II transglutaminase. *Blood* **87**, 1939–1950.

Bérard, J., Laboune, F., Mukuna, M., Massé, S., Kothary, R., and Bradley, W. (1996). Lung tumors in mice expressing an antisense RARb2 transgene. *FASEB J.* **10**, 1091–1097.

Bernards, R., Schacleford, G., and Gerber, M. (1989). Structure and expression of the murine retinoblastoma gene and characterization of its encoded protein. *Proc. Natl. Acad. Sci. U.S.A.* **86**, 6474–6478.

Bissonnette, R. P., Brunner, T., Lazarchik, S. B., Yoo, N. J., Boehm, M. F., Green, D. R., and Heyman, R. A. (1995). 9-Cis retinoic acid inhibition of activation-induced apoptosis is mediated via regulation of fas ligand and requires retinoic acid receptor and retinoid X receptor activation. *Mol. Cell. Biol.* **15**, 5576–5585.

Blumberg, B., Bolado, J., Derguini, F., Craig, A. G., Moreno, T. A., Chakravarti, D., Heyman, R. A., Buck, J., and Evans, R. M. (1996). Novel retinoic acid receptor ligands in Xenopus embryos. *Proc. Natl. Acad. Sci. U.S.A.* **93**, 4873–4878.

Bocco, J. L., Reimund, B., Chatton, B., and Kedinger, C. (1993). Rb may act as a transcriptional co-activator in undifferentiated F9 cells. *Oncogene* **8**, 2977–2986.

Boudreau, N., Werb, Z., and Bissell, M. (1996). Suppression of apoptosis by basement membrane requires three-dimensional tissue organization and withdrawal from the cell cycle. *Proc. Natl. Acad. Sci. U.S.A.* **93**, 3509–3513.

Boylan, J., Lufkin, T., Achkar, C., Taneja, R., Chambon, P., and Gudas, L. (1995). Targeted disruption of retinoic acid receptor a (RARa) and RARg results in receptor-specific alterations in retinoic acid-mediated differentiation and retinoic acid metabolism. *Mol. Cell. Biol* **15**, 843–851.

Brooks, S. C., Kazmer, S., Levin, A. A., and Yen, A. (1996). Myeloid differentiation and retinoblastoma phosphorylation changes in HL-60 cells induced by retinoic acid receptor– and retinoid X receptor–selective retinoic acid analogs. *Blood* **87**, 227–237.

Brunner, T., Mogil, R., LaFace, D., Yoo, J., Mahboubi, A., Echeverri, F., Martin, S., Force, W., Lynch, D., Ware, C., and Green, D. (1995). Cell-autonomous Fas (CD95)/Fas-ligand interaction mediates activation-induced apoptosis in T-cell hybridomas. *Nature* **373**, 441–444.

Buck, J., Myc, A., Garbe, A., and Cathomas, G. (1991). Differences in the action and metabolism between retinol and retinoic acid in B lymphocytes. *J. Cell Biol.* **115**, 851–859.

Burger, C., Wick, M., and Muller, R. (1994). Lineage-specific regulation of cell cycle gene expression in differentiating myeloid cells. *J. Cell Sci.* **107**, 2047–2054.

Carpenter, G., and Wahl, M. (1990). The epidermal growth factor family. *In* "Peptide Growth Factors and Their Receptors I" (M. Sporn and A. Roberts, eds.), pp. 69–172. Springer-Verlag, Berlin.

Chambon, P. (1996). A decade of molecular biology of retinoic acid receptors. *FASEB J.* **10**, 940–954.

Chang, K. S., Fan, Y. H., Andreeff, M., Liu, J., and Mu, Z. M. (1995). The PML gene encodes a phosophoprotein associated with the nuclear matrix. *Blood* **85**, 3646–3653.

Chen, H. W., and Privalsky, M. L. (1993). The erbA oncogene represses the actions of both retinoid X and retinoid A receptors but does so by distinct mechanisms. *Mol. Cell Biol.* **13**, 5970–5980.

Chen, J. Y., Penco, S., Ostrowski, J., Balaguer, P., Pons, M., Starrett, J. E., Reczek, P., Chambon, P., and Gronemeyer, H. (1995). RAR-specific agonist/antagonists which dissociate transactivation and AP1 transrepression inhibit anchorage-independent cell proliferation. *EMBO J.* **14**, 1187–1197.

Chen, W. H., Morriss-Kay, G. M., and Copp, A. J. (1994). Prevention of spinal neural tube defects in the curly tail mouse mutant by a specific effect of retinoic acid. *Dev. Dynamics* **199**, 93–102.

Chen, W. H., Morriss-Kay, G. M., and Copp, A. J. (1995). Genesis and prevention of spinal neural tube defects in the curly tail mutant mouse: involvement of retinoic acid and its nuclear receptors RAR-beta and RAR-gamma. *Development* **121**, 681–691.

Choi, Y., and Fuchs, E. (1990). TGF-beta and retinoic acid: Regulators of growth and modifiers of differentiation in human epidermal cells. *Cell Regul.* **1**, 791–809.

Chomienne, C., Fenaux, P., and Degos, L. (1996). Retinoid differentiation therapy in promyelocytic leukemia. *FASEB J.* **10**, 1025–1030.

Clarke, C. L., Graham, J., Roman, S. D., and Sutherland, R. L. (1991). Direct transcriptional regulation of the progesterone receptor by retinoic acid diminishes progestin responsiveness in the breast cancer cell line T-47D. *J. Biol. Chem.* **266**, 18969–18975.

Clifford, J., Chiba, H., Sobieszczuk, D., Metzger, D., and Chambon, P. (1996). RXRa-null F9 embryonal carcinoma cells are resistant to the differentiation, anti-proliferative and apoptotic effects of retinoids. *EMBO J.* **15**, 4142–4155.

Colbert, M., Rubin, W., Linney, E., and LaMantia, A. (1995). Retinoid signaling and the generation of regional and cellular diversity in the embryonic mouse spinal cord. *Dev. Dynamics* **204**, 1–12.

Coucouvanis, E., and Martin, G. (1995). Signals for death and survival: A two-step mechanism for cavitation in the vertebrate embryo. *Cell* **83**, 279–287.

Creech Kraft, J., Kimelman, D., and Juchau, M. R. (1995). Xenopus laevis: A model system for the study of embryonic retinoid metabolism. II. Embryonic metabolism of all-*trans*-3,4-didehydroretinol to all-*trans*-3,4-didehydroretinoic acid. *Drug Metab. Dispos.* **23**, 83–89.

Cui, H., Sherr, D. H., el-Khatib, M., Matsui, K., Panka, D. J., Marshak-Rothstein, A., and Ju, S. T. (1996). Regulation of T-cell death genes: Selective inhibition of FasL- but not Fas-mediated function. *Cell. Immunol.* **167**, 276–284.

Cunningham, M. L., Mac Auley, A., and Mirkes, P. E. (1994). From gastrulation to neurulation: Transition in retinoic acid sensitivity identifies distinct stages of neural patterning in the rat. *Dev. Dynamics* **200**, 227–241.

Damm, K., Heyman, R. A., Umesono, K., and Evans, R. M. (1993). Functional inhibition of retinoic acid response by dominant negative retinoic acid receptor mutants. *Proc. Natl. Acad. Sci. U.S.A.* **90**, 2989–2993.

Davies, P., Stein, J., Chiocca, E., Basilion, J., Gentile, V., Thomazy, V., and Fesus, L. (1992). Retinoid-regulated expression of transglutaminases: Links to the biochemistry of programmed cell death. *In* "Retinoids in Normal Development and Teratogenesis" (G. Morriss-Kay, ed.), p. 249. Oxford University Press, New York.

Dawson, M., and Hobbs, P. (1994). The synthetic chemistry of retinoids. *In* "The Retinoids: Biology, Chemistry, and Medicine" (M. Sporn, A. Roberts, and D. Goodman, eds.), pp. 5–178. Raven Press, New York.

DeGregori, J., Kowalik, T., and Nevins, J. R. (1995). Cellular targets for activation by the E2F1 transcription factor include DNA synthesis- and G1/S-regulatory genes. *Mol. Cell Biol.* **15**, 4215–4224.

Delia, D., Aiello, A., Soligo, D., Fontanella, E., Melani, C., Pezzella, F., Pierotti, M. A., and Della Porta, G. (1992). Bcl-2 proto-oncogene expression in normal and neoplastic human myeloid cells. *Blood* **79**, 1291–1298.

Desbois, C., Aubert, D., Legrand, C., Pain, B., and Samarut, J. (1991). A novel mechanism of action for v-Erb-A: Abrogation of the inactivation of transcription factor AP-1 by retinoic acid and thyroid hormone receptors. *Cell* **67**, 731–740.

Dhein, J., Walczak, H., Baumler, C., Debatin, K., and Krammer, P. (1995). Autocrine T-cell suicide mediated by APO-1 (Fas/CD95). *Nature* **373**, 438–441.

Dollé, P., Ruberte, E., Kastner, P., Petkovich, M., Stoner, C. M., Gudas, L. J., and Chambon, P. (1989). Differential expression of genes encoding α, β, and γ retinoic acid receptors and CRABP in the developing limbs of the mouse. *Nature* **342**, 702–705.

Dollé, P., Ruberte, E., Leroy, P., Morriss-Kay, G., and Chambon, P. (1990). Retinoic acid receptors and cellular retinoid binding proteins. I. A systematic study of their differential pattern of transcription during mouse organogenesis. *Development* **110,** 1133–1151.

Dony, C., Kessel, M., and Gruss, P. (1985). Post-transcriptional control of *myc* and p53 expression during differentiation of the embryonal carcinoma cell line F9. *Nature* **317,** 636–639.

Duester, G., Shean, M., McBride, M., and Stewart, M. (1991). Retinoic acid response element in the human alcohol dehydrogenase gene ADH3: Implications for regulation of retinoic acid synthesis. *Mol. Cell Biol.* **11,** 1638–1646.

Dupin, E., and Le Douarin, N. M. (1995). Retinoic acid promotes the differentiation of adrenergic cells and melanocytes in quail neural crest cultures. *Dev. Biol.* **168,** 529–548.

Dyck, J., Maul, G., Miller, W., Chen, J., Kakizuka, A., and Evans, R. (1994). A novel macromolecular structure is a target of the promyelocyte-retinoic acid receptor oncoprotein. *Cell* **76,** 333–343.

Epstein, D. J., Vekemans, M., and Gros, P. (1991). Splotch (Sp2H), a mutation affecting development of the mouse neural tube, shows a deletion within the paired homeodomain of Pax-3. *Cell* **67,** 767–774.

Evan, G., and Littlewood, T. (1993). The role of c-*myc* in cell growth. *Curr. Opin. Genet. Dev.* **3,** 44–49.

Evan, G., Wyllie, A., Gilbert, C., Littlewood, T., Land, H., Brooks, M., Waters, C., Penn, L., and Hancock, D. (1992). Induction of apoptosis in fibroblasts by c-*myc* protein. *Cell* **63,** 119–125.

Evan, G., Harrington, E., McCarthy, N., Gilbert, C., Benedict, M., and Nuñez, G. (1996). Integrated control of cell proliferation and apoptosis by oncogenes. *In* "Apoptosis and Cell Cycle Control" (N. Thomas, ed.), pp. 109–129. BIOS Scientific Publishers, Oxford.

Fanjul, A., Dawson, M. I., Hobbs, P. D., Jong, L., Cameron, J. F., Harlev, E., Graupner, G., Lu, X. P., and Pfahl, M. (1994). A new class of retinoids with selective inhibition of AP-1 inhibits proliferation. *Nature* **372,** 107–111.

Farrow, S., and Brown, R. (1996). New members of the Bcl-2 family and their protein partners. *Curr. Opin. Genet. Dev.* **6,** 45–49.

Feldman, B., Poueymirou, W., Papaioannou, V. E., DeChiara, T. M., and Goldfarb, M. (1995). Requirement of FGF-4 for postimplantation mouse development. *Science* **267,** 246–249.

Ferrari, S., Tagliafico, E., Manfredini, R., Grande, A., Rossi, E., Zucchini, P., Torelli, G., and Torelli, U. (1992). Abundance of the primary transcript and its processed product of growth-related genes in normal and leukemic cells during proliferation and differentiation. *Cancer Res.* **52,** 11–16.

Field, S., Tsai, J., Kuo, F., Zubiaga, A., Jr., W. K., Livingston, D., Orkin, S., and Greenberg, M. (1996). E2F-1 functions in mice to promote apoptosis and suppress proliferation. *Cell* **85,** 549–561.

Fisher, G., and Voorhees, J. (1996). Molecular mechanisms of retinoid actions in skin. *FASEB J.* **10,** 1002–1013.

Fisher, G., Datta, S., Talwar, H., Wang, Z., Varani, J., Kang, S., and Voorhees, J. (1996). Molecular basis of sun-induced premature skin aging and retinoid antagonism. *Nature* **379,** 335–339.

Forman, B., Umesono, K., Chen, J., and Evans, R. (1995). Unique response pathways are established by allosteric interactions among nuclear hormone receptors. *Cell* **81,** 541–550.

Formelli, F., Barua, A., and Olson, J. (1996). Bioactivities of *N*-(4-hydroxyphenyl) retinamide and retinoyl *b*-glucuronide. *FASEB J.* **10,** 1014–1024.

Francis, P. H., Richardson, M. K., Brickell, P. M., and Tickle, C. (1994). Bone morphogenetic proteins and a signalling pathway that controls patterning in the developing chick limb. *Development* **120,** 209–218.

Gaetano, C., Matsumoto, K., and Thiele, C. J. (1991). Retinoic acid negatively regulates p34cdc2 expression during human neuroblastoma differentiation. *Cell Growth Differ.* **2,** 487–493.

Galaktionov, K., Chen, X., and Beach, D. (1996). Cdc25 cell-cycle phosphatase as a target of c-*myc*. *Nature* **382,** 511–517.

Gale, E., Prince, V., Lumsden, A., Clarke, J., Holder, N., and Maden, M. (1996). Late effects of retinoic acid on neural crest and aspects of rhombomere identity. *Development* **122**, 783–793.

Gamow, E., and Prescott, D. (1970). The cell life cycle during early embryogenesis of the mouse. *Exp. Cell Res.* **59**, 117–123.

Gañan, Y., Macias, D., Duterque-Coquillaud, M., Ros, M., and Hurle, J. (1996). Role of TGFbs and BMPs as signals controlling the position of the digits and the areas of interdigital cell death in the developing chick limb autopod. *Development* **122**, 2349–2357.

Gentile, V., Thomazy, V., Piacentini, M., Fesus, L., and Davies, P. (1992). Expression of tissue transglutaminase in Balb-C 3T3 fibroblasts: Effects on cellular morphology and adhesion. *J. Cell Biol.* **119**, 463–474.

Gianni, M., Li Calzi, M., Terao, M., Guiso, G., Caccia, S., Barbui, T., Rambaldi, A., and Garattini, E. (1996). AM580, a stable benzoic derivative of retinoic acid, has powerful and selective cyto-differentiating effects on acute promyelocytic leukemia cells. *Blood* **87**, 1520–1531.

Glick, A., Flanders, K., Danelpour, D., Yuspa, S., and Sporn, M. (1989). Retinoic acid induces transforming growth factor-beta2 in cultured keratinocytes and mouse epidermis. *Cell Reg.* **1**, 87–97.

Glozak, M., and Rogers, M. (1996). Specific induction of apoptosis in P19 embryonal carcinoma cells by retinoic acid and BMP2 or BMP4. *Dev. Biol.* 458–470.

Goodyer, P., Dehbi, M., Torban, E., Bruening, W., and Pelletier, J. (1995). Repression of the retinoic acid receptor-alpha gene by the Wilms' tumor suppressor gene product, wt1. *Oncogene* **10**, 1125–1129.

Graham, A., Francis-West, P., Brickell, P., and Lumsden, A. (1994). The signalling molecule BMP4 mediates apoptosis in the rhombencephalic neural crest. *Nature* **372**, 684–686.

Griswold, M. D., Bishop, P. D., Kim, K.-H., Ping, R., Siiteri, J. E., and Morales, C. (1989). Function of vitamin A in normal and synchronized semeniferous tubules. *Ann. N.Y. Acad. Sci.* **564**, 154–172.

Grunt, T. W., Somay, C., Oeller, H., Dittrich, E., and Dittrich, C. (1992). Comparative analysis of the effects of dimethyl sulfoxide and retinoic acid on the antigenic pattern of human ovarian adenocarcinoma cells. *J. Cell Sci.* **103**(Pt 2), 501–509.

Gucev, Z. S., Oh, Y., Kelley, K. M., and Rosenfeld, R. G. (1996). Insulin-like growth factor binding protein 3 mediates retinoic acid- and transforming growth factor beta2-induced growth inhibition in human breast cancer cells. *Cancer Res.* **56**, 1545–1550.

Gudas, L., Sporn, M., and Roberts, A. (1994). Cellular biology and biochemistry of the retinoids. *In* "The Retinoids: Biology, Chemistry, and Medicine" (M. Sporn, A. Roberts, and D. Goodman, eds.), pp. 443–520. Raven Press, New York.

HaasKogan, D., Kogan, S., Levi, D., Dazin, P., Tang, A., Fung, Y., and Israel, M. (1995). Inhibition of apoptosis by the retinoblastoma gene product. *EMBO J.* **14**, 461–472.

Han, I., and Choi, J. (1996). Highly specific cytochrome P450-like enzymes for all-*trans*-retinoic acid in T47D human breast cancer cells. *J. Clinical Endocrin. Metab.* **81**, 2069–2075.

Harvey, M. B., and Kaye, P. L. (1992). Insulin-like growth factor-1 stimulates growth of mouse preimplantation embryos *in vitro*. *Molec. Reproduc. Dev.* **31**, 195–199.

Hashimoto, M., Nakamura, T., Inoue, S., Kondo, T., Yamada, R., Eto, Y., Sugino, H., and Muramatsu, M. (1992). Follistatin is a developmentally regulated cytokine in neural differentiation. *J. Biol. Chem.* **267**, 7203–7206.

Henion, P. D., and Weston, J. A. (1994). Retinoic acid selectively promotes the survival and proliferation of neurogenic precursors in cultured neural crest cell populations. *Dev. Biol.* **16**, 243–250.

Hermann, T., Hoffmann, B., Piedrafita, F. J., Zhang, X. K., and Pfahl, M. (1993). V-erbA requires auxiliary proteins for dominant negative activity. *Oncogene* **8**, 55–65.

Heyman, R., Mangelsdorf, D., Dyck, J., Stein, R., Eichele, G., Evans, R., and Thaller, C. (1992). 9-*Cis* retinoic acid is a high affinity ligand for the retinoid X receptor. *Cell* **68**, 397–406.

Hill, C. S., and Treisman, R. (1995). Transcriptional regulation by extracellular signals: Mechanisms and specificity. *Cell* **80,** 199–211.

Hofmann, C., and Eichele, G. (1994). Retinoids in development. *In* "The Retinoids: Biology, Chemistry, and Medicine" (M. B. Sporn, A. B. Roberts, and D. S. Goodman, eds.), pp. 387–441. Raven Press, New York.

Hogan, B. (1996). Bone morphogenetic proteins: multifunctional regulators of vertebrate development. *Genes Dev.* **10,** 1580–1594.

Horn, V., Minucci, S., Ogryzko, V., Adamson, D., Howard, B., Levin, A., and Ozato, K. (1996). RAR and RXR selective ligands cooperatively induce apoptosis and neuronal differentiation in P19 embryonal carcinoma cells. *FASEB J* **10,** 1071–1077.

Hu, L., and Gudas, L. J. (1990). Cyclic AMP analogs and retinoic acid influence the expression of retinoic acid receptor α, β, and γ mRNAs in F9 teratocarcinoma cells. *Mol. Cell. Biol.* **10,** 391–396.

Huang, M., Ye, Y., Chen, S., Rong, C., Ren, C., Xiang, L., Lin, Z., Jun, G., and Yi, W. (1988). Use of all-*trans* retinoic acid in the treatment of acute promyelocytic leukemia. *Blood* **72,** 567–572.

Informatics, M. G. (1996). Mouse Genome Database (MGD) (URL: http://www.informatics.jax.org/). The Jackson Laboratory, Bar Harbor, Maine.

Ishida, S., Shudo, K., Takada, S., and Koike, K. (1995). A direct role of transcription factor E2F in c-*myc* gene expression during granulocytic and macrophage-like differentiation of HL60 cells. *Cell Growth Differ.* **6,** 229–237.

Ito, K., and Morita, T. (1995). Role of retinoic acid in mouse neural crest cell development in vitro. *Dev. Dynamics* **204,** 211–218.

Jiang, H., Lin, J., Su, Z. Z., Collart, F. R., Huberman, E., and Fisher, P. B. (1994). Induction of differentiation in human promyelocytic HL-60 leukemia cells activates p21, WAF1/CIP1, expression in the absence of p53. *Oncogene* **9,** 3397–3406.

Ju, S., Panka, D., Cui, H., Ettinger, R., El-Khatib, M., Sherr, D., Stanger, B., and Marshak-Rothstein, A. (1995). Fas(CD95)/FasL interactions required for programmed cell death after T-cell activation. *Nature* **373,** 444–448.

Kalemkerian, G. P., Jasti, R. K., Celano, P., Nelkin, B. D., and Mabry, M. (1994). All-*trans*-retinoic acid alters *myc* gene expression and inhibits *in vitro* progression in small cell lung cancer. *Cell Growth Differ.* **5,** 55–60.

Kamei, Y., Zu, L., Heinzel, T., Torchia, J., Kurokawa, R., Gloss, B., Lin, S.-C, Heyman, R., Rose, D., Glass, C., and Rosenfeld, M. (1996). A CBP integrator complex mediates transcriptional activation and AP-1 inhibition by nuclear receptors. *Cell* **85,** 402–414.

Kapron-Bras, C., and Trasler, D. (1985). Reduction in the frequency of neural tube defects in splotch mice by retinoic acid. *Teratology* **32,** 87–92.

Kapron-Bras, C. M., and Trasler, D. G. (1988). Interaction between the splotch mutation and retinoic acid in mouse neural tube defects *in vitro*. *Teratology* **38,** 165–173.

Kastner, P., Mark, M., and Chambon, P. (1995). Nonsteroid nuclear receptors: What are genetic studies telling us about their role in real life? *Cell* **83,** 859–869.

Kim, Y. H., Dohi, D. F., Han, G. R., Zou, C. P., Oridate, N., Walsh, G. L., Nesbitt, J. C., Xu, X. C., Hong, W. K., Lotan, R., and et al. (1995). Retinoid refractoriness occurs during lung carcinogenesis despite functional retinoid receptors. *Cancer Res.* **55,** 5603–5610.

Kochhar, D. M., and Agnish, N. D. (1977). "Chemical surgery" as an approach to study morphogenetic events in embryonic mouse limb. *Dev. Biol.* **61,** 388–394.

Koken, M. H., Puvion-Dutilleul, F., Guillemin, M. C., Viron, A., Linares-Cruz, G., Stuurman, N., de Jong, L., Szostecki, C., Calvo, F., Chomienne, C., et al. (1994). The t(15;17) translocation alters a nuclear body in a retinoic acid–reversible fashion. *EMBO J.* **13,** 1073–1083.

Kranenburg, O., DeGroot, R., VanDerEb, A., and Zantema, A. (1995). Differentiation of P19 EC cells leads to differential modulation of cyclin-dependent kinase activities and to changes in the cell cycle profile. *Oncogene* **10,** 87–95.

Kreidberg, J. A., Sariola, H., Loring, J. M., Maeda, M., Pelletier, J., Housman, D., and Jaenisch, R. (1993). WT-1 is required for early kidney development. *Cell* **74,** 679–691.

Kroemer, G., Petit, P., Zamzami, N., Vayssiére, J., and Mignotte, B. (1995). The biochemistry of programmed cell death. *FASEB J.* **9,** 1277–1287.

Krupitza, G., Hulla, W., Harant, H., Dittrich, E., Kallay, E., Huber, H., Grunt, T., and Dittrich, C. (1995). Retinoic acid induced death of ovarian carcinoma cells correlates with c-*myc* stimulation. *Int. J. Cancer* **61,** 649–657.

La Thangue, N. B., and Rigby, P. W. (1987). An adenovirus E1A-like transcription factor is regulated during the differentiation of murine embryonal carcinoma stem cells. *Cell* **49,** 507–513.

Lammer, E. J., Chen, D. T., Hoar, R. M., Agnish, N. D., Benke, P. J., Braun, J. T., Curry, C. J., Fernoff, P. M., Grix, A. W., Lott, I. T., Richard, J. M., and Sun, S. C. (1985). Retinoic Acid Embryopathy. *N. Engl. J. Med.* **313,** 837–841.

Lampron, C., Rochette-Egly, C., Gorry, P., Dolle, P., Mark, M., Lufkin, T., LeMeur, M., and Chambon, P. (1995). Mice deficient in cellular retinoic acid binding protein II (CRABPII) or in both CRABPI and CRABPII are essentially normal. *Development* **121,** 539–548.

Larsson, L. G., Pettersson, M., Oberg, F., Nilsson, K., and Luscher, B. (1994). Expression of mad, mxi1, max and c-*myc* during induced differentiation of hematopoietic cells: Opposite regulation of mad and c-*myc*. *Oncogene* **9,** 1247–1252.

Leblanc, B., and Stunnenberg, H. (1995). 9-*Cis* retinoic acid signaling: changing partners causes some excitement. *Genes Dev.* **9,** 1811–1816.

Lee, S. L., Wesselschmidt, R. L., Linette, G. P., Kanagawa, O., Russell, J. H., and Milbrandt, J. (1995). Unimpaired thymic and peripheral T cell death lacking the nuclear receptor NGFI-B (Nur77). *Science* **269,** 532–535.

Lee, X., Si, S. P., Tsou, H. C., and Peacocke, M. (1995). Cellular aging and transformation suppression: A role for retinoic acid receptor beta 2. *Exp. Cell Res.* **218,** 296–304.

Lee, Y. M., Osumi-Yamashita, N., Ninomiya, Y., Moon, C. K., Eriksson, U., and Eto, K. (1995). Retinoic acid stage dependently alters the migration pattern and identity of hindbrain neural crest cells. *Development* **121,** 825–837.

Leonard, L., Horton, C., Maden, M., and Pizzey, J. A. (1995). Anteriorization of CRABP-I expression by retinoic acid in the developing mouse central nervous system and its relationship to teratogenesis. *Dev. Biol.* **168,** 514–528.

Li, J. J., Dong, Z., Dawson, M. I., and Colburn, N. H. (1996). Inhibition of tumor promoter induced transformation by retinoids that transrepress AP-1 without transactivating retinoic acid response element. *Cancer Res.* **56,** 483–489.

Licht, J. D., Chomienne, C., Goy, A., Chen, A., Scott, A. A., Head, D. R., Michaux, J. L., Wu, Y., DeBlasio, A., Miller, W. H., Jr., and et al. (1995). Clinical and molecular characterization of a rare syndrome of acute promyelocytic leukemia associated with translocation (11;17). *Blood* **85,** 1083–1094.

Licht, J., Shaknovich, R., English, M., Melnick, A., Li, J.-Y., Reddy, J., Ong, S., Chen, S.-J., Zelent, A., and Waxman, S. (1996). Reduced and altered DNA-binding and transcriptional properties of the PLZF-retinoic acid receptor-a chimera generated in t(11:17)-associated acute promyelocytic leukemia. *Oncogene* **12,** 323–336.

Liu, Z., Smith, S., McLaughlin, K., Schwartz, L., and Osborne, B. (1994). Apoptotic signals delivered through the T-cell receptor of a T-cell hybrid require the immediate-early gene nur77. *Nature* **367,** 281–284.

Lockshin, R., and Zakeri, Z. (1996). The biology of cell death and its relationship to aging. *In* "Cellular Aging and Cell Death" (N. Holbrook, G. Martin, and R. Lockshin, eds.), pp. 167–180. Wiley-Liss, New York.

Lohnes, D., Mark, M., Mendelsohn, C., Dolle, P., Dierich, A., Gorry, P., Gansmuller, A., and
    Chambon, P. (1994). Function of the retinoic acid receptors (RARs) during development (I).
    Craniofacial and skeletal abnormalities in RAR double mutants. *Development* **120**, 2723–2748.
Lotan, R. (1996). Retinoids in cancer prevention. *FASEB J.* **10**, 1031–1039.
Lotan, R., Xu, S., Lippman, S., Ro, J., Lee, J., Lee, J., and Hong, W. (1995). Suppression of
    retinoic acid receptor-beta in premalignant oral lesions and its up-regulation of isotretinoin.
    N. Engl. J. Med. **332**, 1405–1410.
Lowe, Jr., W., Meyer, T., Karpen, C., and Lorentzen, L. (1992). Regulation of insulin-like growth
    factor I production in ra C6 glioma cells: Possible role as an autocrine/paracrine growth factor.
    Endocrinology **130**, 2683–2691.
Lund, L. Rømer, J., Thomasset, N., Solberg, H., Pyke, C., Bissell, M., Dane, K., and Werb, Z.
    (1996). Two distinct phases of apoptosis in mammary gland involution: Proteinase-independent
    and -dependent pathways. *Development* **122**, 181–193.
Lyons, K., Graycar, J. L., Lee, A., Hashmi, S., Lindquist, P. B., Chen, E. Y., Hogan, B. L. M.,
    and Derynck, R. (1989). *Vgr-1*, a mammalian gene related to *Xenopus Vg-1*, is a member of
    the transforming growth factor β gene super family. *Proc. Natl. Acad. Sci. USA* **86**, 4554–
    4558.
Mahmood, R., Flanders, D., and Morriss-Kay, G. (1992). Interactions between retinoids and TGF
    betas in mouse morphogenesis. *Development* **115**, 67–74.
Martin, G. R. (1980). Teratocarcinomas and mammalian embryogenesis. *Science* **209**, 768–776.
Martin, J. L., Coverley, J. A., Pattison, S. T., and Baxter, R. C. (1995). Insulin-like growth factor-
    binding protein-3 production by MCF-7 breast cancer cells: Stimulation by retinoic acid and
    cyclic adenosine monophosphate and differential effects of estradiol. *Endocrinology* **136**,
    1219–1226.
Meikrantz, W., and Schlegel, R. (1996). Suppression of apoptosis by dominant-negative mutants
    of cyclin-dependent protein kinases. *J. Biol. Chem.* **271**, 10205–10209.
Mendelsohn, C., Ruberte, E., LeMeur, M., Morriss-Kay, G., and Chambon, P. (1991). Develop-
    mental analysis of the retinoic acid-inducible RAR-b2 promoter in transgenic animals. *Devel-
    opment* **113**, 723–734.
Milligan, C., and Schwartz, L. (1996). Programmed cell death during development of animals. *In*
    "Cellular Aging and Cell Death" (N. Holbrook, G. Martin, and R. Lockshin, eds.), pp. 181–
    208. Wiley-Liss, New York.
Moon, R., Mehta, R., and Rao, K. (1994). Retinoids and cancer in experimental animals. *In* "The
    Retinoids: Biology, Chemistry and Medicine" (M. Sporn, A. Roberts, and D. Goodman, eds.),
    pp. 573–595. Raven Press, New York.
Mordan, L. (1989). Inhibition by retinoids of platelet growth factor-dependent stimulation of
    DNA synthesis and cell1 division in density-arrested C3H 10T1/2 fibroblasts. *Cancer Res.* **49**,
    906–909.
Morgan, D. O. (1995). Principles of CDK regulation. *Nature* **374**, 131–134.
Morriss-Kay, G., and Mahmood, R. (1992). Morphogenesis-related changes in extracellular matrix
    induced by retinoic acid. *In* "Retinoids in Normal Development and Teratogenesis" (G. Mor-
    riss-kay, ed.), pp. 165–180. Oxford University Press, New York.
Morriss-Kay, G., and Sokolova, N. (1996). Embryonic development and pattern formation. *FASEB
    J.* **10**, 961–968.
Müller, R., and Wagner, E. F. (1984). Differentiation of F9 teratocarcinoma stem cells after trans-
    fer of c-*fos* proto-oncogenes. *Nature* **311**, 438–442.
Murray, A., and Hunt, T. (1993). The Cell Cycle: An Introduction. W. H. Freeman, New York.
Nagpal, S., Athanikar, J., and Chandraratna, R. A. (1995). Separation of transactivation and AP1
    antagonism functions of retinoic acid receptor alpha. *J. Biol. Chem.* **270**, 923–927.
Nagy, L., Thomazy, V. A., Shipley, G. L., Fesus, L., Lamph, W., Heyman, R. A., Chandraratna,

R. A., and Davies, P. J. (1995). Activation of retinoid X receptors induces apoptosis in HL-60 cell lines. *Mol. Cell Biol.* **15**, 3540–3551.

Nagy, L., Saydak, M., Shipley, N., Lu, S., Basilion, J. P., Yan, Z. H., Syka P., Chandraratna, R. A., Stein, J. P., Heyman, R. A., and Davies, P. J. (1996a). Identification and characterization of a versatile retinoid response element (retinoic acid receptor response element–retinoid X receptor response element) in the mouse tissue transglutaminase gene promoter. *J. Biol. Chem.* **271**, 4355–4365.

Nagy, L., Thomázy, V., Chandraratna, R., Heyman, R., and Davies, P. (1996b). Retinoid-regulated expression of BCL-2 and tissue transglutaminase during the differentiation and apoptosis of human myeloid leukemia (HL-60) cells. *Leukemia Res.* 499–505.

Napoli, J. (1996). Retinoic acid biosynthesis and metabolism. *FASEB J.* **10**, 993–1001.

Naumovski, L., and Cleary, M. L. (1994). *Bcl2* inhibits apoptosis associated with terminal differentiation of HL-60 myeloid leukemia cells. *Blood* **83**, 2261–2267.

Okazawa, H., Shimizu, J., Kamei, M., Imafuku, I., Hamada, H., and Kanazawa, I. (1996). Bcl-2 inhibits retinoic acid-induced apoptosis during the neural differentiation of embryonal stem cells. *J. Cell Biol.* **132**, 955–968.

Oridate, N., Esumi, N., Lotan, D., Hing, W. K., Rochette-Egly, C., Chambon, P., and Lotan, R. (1996). Implication of retinoic acid receptor gamma in squamous differentiation and response to retinoic acid in head and neck SqCC/Y1 squamous carcinoma cells. *Oncogene* **12**, 2019–2028.

Pan, L., Eckhoff, C., and Brinckerhoff, C. (1995). Suppression of collagenase gene expression by all-*trans* and 9-*cis* retinoic acid is ligand dependent and requires both RARs and RXRs. *J. Cell. Biochem.* **57**, 575–589.

Pandey, S., and Wang, E. (1995). Cells en route to apoptosis are characterized by the upregulation of c-*fos*, c-*myc*, c-*jun*, cdc2, and RB phosphorylation, resembling events of early cell-cycle traverse. *J. Cell. Biochem.* **58**, 135–150.

Pardee, A. (1989). G1 events and regulation of cell proliferation. *Science* **246**, 602–608.

Park, J. R., Robertson, K., Hickstein, D. D., Tsai, S., Hockenbery, D. M., and Collins, S. J. (1994). Dysregulated bcl-2 expression inhibits apoptosis but not differentiation of retinoic acid–induced HL-60 granulocytes. *Blood* **84**, 440–445.

Perez, A., Kastner, P., Sethi, S., Lutz, Y., Reibel, C., and Chambon, P. (1993). PMLRAR homodimers: Distinct DNA binding properties and heteromeric interactions with RXR. *EMBO J.* **12**, 3171–3182.

Perlmann, T., and Jansson, L. (1995). A novel pathway for vitamin A signaling mediated by RXR heterodimerization with NGFI-B and NURR1. *Genes Dev.* **9**, 769–782.

Peters, G., Bates, S., and Parry, D. (1996). The D-cyclins, their kinases and their inhibitors. *In* "Apoptosis and Cell Cycle Control" (N. Thomas, ed.), pp. 77–92. BIOS Scientific Publishers, Oxford.

Peterson, K. (1996). "Natural" cancer prevention trial halted. *Science* **271**, 441.

Piacentini, M., Davies, P., and Fesus, L. (1994). Tissue transglutaminase in cells undergoing apoptosis. *In* "Apoptosis II: The Molecular Basis of Apoptosis in Disease" (L. D. Tomei and F. O. Cope, eds.), pp. 143–163. CSH Laboratory Press, New York.

Pijnappel, W. W., Hendriks, H. F., Folkers, G. E., van den Brink, C. E., Dekker, E. J., Edelenbosch, C., van der Saag, P. T., and Durston, A. J. (1993). The retinoid ligand 4-oxo-retinoic acid is a highly active modulator of positional specification. *Nature* **366**, 340–344.

Privalsky, M. (1992). Retinoid and thyroid hormone receptors: Ligand-regulated transcription factors as proto-oncogenes. *Sem. Cell Biol.* **3**, 66–106.

Qin, X., Livingston, D., Kaelin, W., and Adams, P. (1994). Deregulated transcription factor E2F-1 expression leads to S-phase entry and p53-mediated apoptosis. *Proc. Natl. Acad. Sci., USA* **91**, 10918–10922.

Reichel, R. R. (1992). Regulation of E2F/cyclin A–containing complex upon retinoic acid–induced differentiation of teratocarcinoma cells. *Gene Expr.* **2**, 259–271.

Reichel, R., Kovesdi, I., and Nevins, J. R. (1987). Developmental control of a promoter-specific factor that is also regulated by the E1A gene product. *Cell* **48**, 501–506.

Roach, H. I., Erenpreisa, J., and Aigner, T. (1995). Osteogenic differentiation of hypertrophic chondrocytes involves asymmetric cell divisions and apoptosis. *J. Cell Biol.* **131**, 483–494.

Roberts, J., Koff, A., Polyak, K., Firpo, E., Collins, S., Ohtsubo, M., and Massagué, J. (1994). Cyclins, cdks, and cyclin kinase inhibitors. *CSH Symp. Quant. Biol.* **LIX**, 31–38.

Roberts, P. E., Phillips, D. M., and Mather, J. P. (1990). A novel epithelial cell from neonatal rat lung: Isolation and differentiated phenotype. *Am. J. Physiol.* **259**, L415–L425.

Robertson, E. J. (1987). "Teratocarcinomas and Embryonic Stem Cells." IRL Press, Oxford.

Robertson, K., Emami, B., and Collins, S. (1992). Retinoic acid–resistant HL-60R cells harbor a point mutation in the retinoic acid receptor ligand binding domain that confers dominant negative activity. *Blood* **80**, 1885–1889.

Rogers, M. (1996). Receptor-selective retinoids implicate RAR alpha and gamma in the regulation of *bmp-2* and *bmp-4* in F9 embryonal carcinoma cells. *Cell Growth Diff.* **7**, 115–122.

Rogers, M. B., Rosen, V., Wozney, J. M., and Gudas, L. J. (1992). Bone morphogenetic proteins-2 and 4 are involved in the retinoic acid–induced differentiation of embryonal carcinoma cells. *Molec. Biol. Cell* **3**, 189–196.

Rosenstraus, M., Sundell, C., and Liskay, R. (1982). Cell-cycle characteristics of undifferentiated and differentiating embryonal carcinoma cells. *Dev. Biol.* **89**, 516–520.

Rossant, J., Zirngibl, R., Cado, D., Shago, M., and Giguére, V. (1991). Expression of a retinoic acid response element—*hsplacZ* transgene defines specific domains of transcriptional activity during mouse embryogenesis. *Genes Dev.* **5**, 1333–1344.

Roy, B., Taneja, R., and Chambon, P. (1995). Synergistic activation of retinoic acid (RA)–responsive genes and induction of embryonal carcinoma cell differentiation by an RA receptor alpha (RAR alpha)-, RAR beta-, or RAR gamma-selective ligand in combination with a retinoid X receptor-specific ligand. *Mol. Cell Biol.* **15**, 6481–6487.

Ruberte, E., Dolle, P., Chambon, P., and Morriss-Kay, G. (1991). Retinoic acid receptors and cellular retinoid binding proteins. II. Their differential pattern of transcription during early morphogenesis in mouse embryos. *Development* **111**, 45–60.

Ruberte, E., Friederich, V., Morriss-Kay, G., and Chambon, P. (1992). Differential distribution patterns of CRABP I and CRABP II transcripts during mouse embryogenesis. *Development* **115**, 973–987.

Saunders, D. E., Christensen, C., Wappler, N. L., Schultz, J. F., Lawrence, W. D., Malviya, V. K., Malone, J. M., and Deppe, G. (1993). Inhibition of c-*myc* in breast and ovarian carcinoma cells by 1,25- dihydroxyvitamin D3, retinoic acid and dexamethasone. *Anticancer Drugs* **4**, 201–208.

Saunders, D. E., Zajac, C. S., and Wappler, N. L. (1995). Alcohol inhibits neurite extension and increases *N-myc* and c-*myc* proteins. *Alcohol* **12**, 475–483.

Savatier, P., Huang, S., Szekely, L., Wiman, K. G., and Samarut, J. (1994). Contrasting patterns of retinoblastoma protein expression in mouse embryonic stem cells and embryonic fibroblasts. *Oncogene* **9**, 809–818.

Savatier, P., Lapillonne, H., VanGrunsven, L., Rudkin, B., and Samarut, J. (1996). Withdrawal of differentiation inhibitory activity/leukemia inhibitory factor up-regulates D-type cyclins and cyclin-dependent kinase inhibitors in mouse embryonic stem cells. *Oncogene* **12**, 309–322.

Schimmang, T., Oda, S. I., and Ruther, U. (1994). The mouse mutant *Polydactyly Nagoya* (Pdn) defines a novel allele of the zinc finger gene *Gli3. Mamm. Genome* **5**, 384–386.

Schofield, P. N., Ekstrom, T. J., Granerus, M., and Engstrom, W. (1991). Differentiation associated modulation of K-FGF expression in a human teratocarcinoma cell line and in primary germ cell tumors. *FEBS Lett.* **280**, 8–10.

Schoorlemmer, J., and Kruijer, W. (1991). Octamer-dependent regulation of the *kFGF* gene in embryonal carcinoma and embryonic stem cells. *Mech. Dev.* **36,** 75–86.

Schubert, D., Kimura, H., LaCorbiere, M., Vaughan, J., Karr, D., and Fischer, W. H. (1990). Activin is a nerve cell survival molecule. *Nature* **344,** 868–870.

Schüle, R., Rangarajan, P., Yang, N., Kliewer, S., Ransone, L. J., Bolado, J., Verma, I., and Evans, R. M. (1991). Retinoic acid is a negative regulator of AP-1 responsive genes. *Proc. Natl. Acad. Sci. USA* **88,** 6092–6096.

Schulze-Osthoff, K. (1994). The Fas/APO-1 receptor and its deadly ligand. *Trends Cell Biol.* **4,** 421–426.

Schwaller, J., Koeffler, H. P., Niklaus, G., Loetscher, P., Nagel, S., Fey, M. F., and Tobler, A. (1995). Posttranscriptional stabilization underlies p53-independent induction of p21WAF1/CIP1/SDI1 in differentiating human leukemic cells. *J. Clin. Invest.* **95,** 973–979.

Schwartz, L., and Osborne, B. (1995). Cell Death. *In* "Methods in Cell Biology" (L. Wilson and P. Matsudaira, eds.), p. 459. Academic Press, San Diego.

Scita, G., Darwiche, N., Greenwald, E., Rosenberg, M., Politi, K., and De Luca, L. M. (1996). Retinoic acid down-regulation of fibronectin and retinoic acid receptor alpha proteins in NIH-3T3 cells. Blocks of this response by *ras* transformation. *J. Biol. Chem.* **271,** 6502–6508.

Scott, W. J., Collins, M. D., Ernst, A. N., Supp, D. M., and Potter, S. S. (1994). Enhanced expression of limb malformations and axial skeleton alterations in legless mutants by transplacental exposure to retinoic acid. *Dev. Biol.* **164,** 277–289.

Seewaldt, V. L., Johnson, B. S., Parker, M. B., Collins, S. J., and Swisshelm, K. (1995). Expression of retinoic acid receptor beta mediates retinoic acid–induced growth arrest and apoptosis in breast cancer cells. *Cell Growth Differ.* **6,** 1077–1088.

Shan, B., and Lee, W. (1994). Deregulated expression of E2F-1 induces *S*-phase entry and leads to apoptosis. *Mol. Cell. Biol.* **14,** 8166–8173.

Sheikh, M. S., Shao, Z. M., Chen, J. C., Ordonez, J. V., and Fontana, J. A. (1993). Retinoid modulation of c-*myc* and *max* gene expression in human breast carcinoma. *Anticancer Res.* **13,** 1387–1392.

Sheikh, M. S., Shao, Z. M., Li, X. S., Ordonez, J. V., Conley, B. A., Wu, S., Dawson, M. L., Han, Q. X., Chao, W. R., Quick, T., et al. (1995). *N*-(4-hydroxyphenyl)retinamide (4-HPR)-mediated biological actions involve retinoic receptor–independent pathways in human breast carcinoma. *Carcinogenesis* **16,** 2477–2486.

Sherr, C. J. (1994). G1 phase progressions: Cycling on cue. *Cell* **79,** 551–555.

Sherr, C. J., and Roberts, J. M. (1995). Inhibitors of mammalian G1 cyclin-dependent kinases. *Genes Dev.* **9275,** 1149–1163.

Shi, L., Nishioka, W. K., Th'ng, J., Bradbury, E. M., Litchfield, D. W., and Greenberg, A. H. (1994). Premature p34cdc2 activation required for apoptosis. *Science* **263,** 1143–1145.

Slack, R. S., Hamel, P. A., Bladon, T. S., Gill, R. M., and McBurney, M. W. (1993). Regulated expression of the retinoblastoma gene in differentiating embryonal carcinoma cells. *Oncogene* **8,** 1585–1591.

Slack, R. S., Craig, J., Costa, S., and McBurney, M. W. (1995a). Adenovirus 5 E1A–induced differentiation of P19 embryonal carcinoma cells requires binding to p300. *Oncogene* **10,** 19–25.

Slack, R. S., Skerjanc, I. S., Lach, B., Craig, J., Jardine, K., and McBurney, M. W. (1995b). Cells differentiating into neuroectoderm undergo apoptosis in the absence of functional retinoblastoma family proteins. *J. Cell Biol.* **129,** 779–788.

Smith-Thomas, L., Lott, I., and Bronner-Fraser, M. (1987). Effects of isotretinoin on the behavior of neural crest cells *in vitro*. *Dev. Biol.* **123,** 276–281.

Somay, C., Grunt, T. W., Mannhalter, C., and Dittrich, C. (1992). Relationship of *myc* protein expression to the phenotype and to the growth potential of HOC-7 ovarian cancer cells. *Br. J. Cancer* **66,** 93–98.

Sporn, M., and Roberts, A. (1994). Introduction. *In* "The Retinoids: Biology, Chemistry, and Medicine" (M. Sporn, A. Roberts, and D. Goodman, eds.), pp. 1–3. Raven Press, New York.

Stemple, D. L., and Anderson, D. J. (1992). Isolation of a stem cell for neurons and glia from the mammalian neural crest. *Cell* **71**, 973–985.

Suda, T., Takahashi, T., Golstein, P., and Nagata, S. (1993). Molecular cloning and expression of the Fas ligand, a novel member of the tumor necrosis factor family. *Cell* **75**, 1169–1178.

Sulik, K., Cook, C., and Webster, W. (1988). Teratogens and craniofacial malformations: Relationships to cell death. *Development* **103** suppl., 213–232.

Swisshelm, K., Ryan, K., Lee, X., Tsou, H. C., Peacocke, M., and Sager, R. (1994). Down-regulation of retinoic acid receptor beta in mammary carcinoma cell lines and its up-regulation in senescing normal mammary epithelial cells. *Cell Growth Differ.* **5**, 133–141.

Sympson, C. J., Talhouk, R. S., Alexander, C. M., Chin, J. R., Clift, S. M., Bissell, M. J., and Werb, Z. (1994). Targeted expression of stromelysin-1 in mammary gland provides evidence for a role of proteinases in branching morphogenesis and the requirement for an intact basement membrane for tissue-specific gene expression. *J. Cell Biol.* **125**, 681–693.

Tabin, C. (1995). The initiation of the limb bud: Growth factors, *Hox* genes, and retinoids. *Cell* **80**, 671–674.

Takatsuka, J., Takahashi, N., and DeLuca, L. (1996). Retinoic acid metabolism and inhibition of cell proliferation: An unexpected liaison. *Cancer Res.* **56**, 675–678.

Talmage, D., and Listerud, M. (1994). Retinoic acid suppresses polyoma virus transformation by inhibiting transcription of the c-*fos* proto-oncogene. *Oncogene* **9**, 3557–3563.

Tamagawa, M., Morita, J., and Naruse, I. (1995). Effects of all-*trans*-retinoic acid on limb development in the genetic polydactyly mouse. *J. Toxicol. Sci.* **20**, 383–393.

Taneja, R., Roy, B., Plassat, J., Zusi, C., Ostrowski, J., Reczek, P., and Chambon, P. (1996). Cell-type and promoter-context dependent retinoic acid receptor (RAR) redundancies fr *RARb2* and *Hoxa-1* activation in F9 and P19 cells can be artefactually generated by gene knockouts. *Proc. Natl. Acad. Sci. USA* **93**, 6197–6202.

Tautz, D. (1992). Redundancies, development and the flow of information. *Bioessays* **14**, 263–266.

Thaller, C., and Eichele, G. (1990). Isolation of 3,4-didehydroretinoic acid, a novel morphogenetic signal in the chick wing bud. *Nature* **345**, 815–819.

Thomas, N. (1996). "Apoptosis and Cell Cycle Control." UCL Molecular Pathology Series (D. Latchman, ed.). BIOS Scientific Publisher, Oxford.

Thorogood, P., Smith, L., Nicol, A., McGinty, R., and Garrod, D. (1982). Effects of vitamin A on the behavior of migratory neural crest cells *in vitro*. *J. Cell Sci.* **57**, 331–350.

Tickle, C., Alberts, B., Wolpert, L., and Lee, J. (1982). Local application of retinoic acid to the limb bond mimics the action of the polarizing region. *Nature* **296**, 564–566.

Trauth, B. C., Klas, C., Peters, A. M., Matzku, S., Moller, P., Falk, W., Debatin, K. M., and Krammer, P. H. (1989). Monoclonal antibody–mediated tumor regression by induction of apoptosis. *Science* **245**, 301–305.

Tsai, S., Bartelmez, S., Heyman, R., Damm, K., Evans, R., and Collins, S. (1992). A mutated retinoic acid receptor-a exhibiting dominant–negative activity alters the lineage development of a multipotent hematopoietic cell line. *Genes Dev.* **6**, 2258–2269.

van den Eijnden-van Raaij, A. J., Feijen, A., Lawson, K. A., and Mummery, C. L. (1992). Differential expression of inhibin subunits and follistatin, but not of activin receptor type II, during early murine embryonic development. *Dev. Biol.* **154**, 356–365.

van der Leede, B. J., Folkers, G. E., van den Brink, C. E., van der Saag, P. T., and van der Burg, B. (1995). Retinoic acid receptor alpha I isoform is induced by estradiol and confers retinoic acid sensitivity in human breast cancer cells. *Mol. Cell Endocrinol.* **109**, 77–86.

Velcich, A., Delli-Bovi, P., Mansukhani, A., Ziff, E. B., and Basilico, C. (1989). Expression of the K-*fgf* protooncogene is repressed during differentiation of F9 cells. *Oncogene Res.* **5**, 31–37.

Verma, A. K., Shapas, B. G., Rice, H. M., and Boutwell, R. K. (1979). Correlation of the inhibition by retinoids of tumor promoter–induced mouse epidermal ornithine decarboxylase activity and of skin tumor promotion. *Cancer Res.* **39,** 419–425.

Vogan, K. J., Epstein, D. J., Trasler, D. G., and Gros, P. (1993). The splotch-delayed (*Spd*) mouse mutant carries a point mutation within the paired box of the *Pax-3* gene. *Genomics* **17,** 364–369.

Weinberg, R. A. (1995). The retinoblastoma protein and cell cycle control. *Cell* **81,** 323–330.

Weis, K., Rambaud, S., Lavau, C., Jansen, J., Carvalho, T., Carmo-Fonseca, M., Lamond, A., and Dejean, A. (1994). Retinoic acid regulates aberrant nuclear localization of PML-RAR alpha in acute promyelocytic leukemia cells. *Cell* **76,** 345–356.

Whyte, P. (1996). The retinoblastoma family of proteins. *In* "Apoptosis and Cell Cycle Control" (N. Thomas, ed.), pp. 77–92. BIOS Scientific Publishers, Oxford.

Wilcken, N., Sarcevic, B., Musgrove, E., and Sutherland, R. (1996). Differential effects of retinoids and antiestrogenes on cell cycle progression and cell cycle regulatory genes in human breast cancer cells. *Cell Growth Diff.* **7,** 65–74.

Wiley, L., Adamson, E., and Tsark, E. (1995). Epidermal growth factor receptor function in early mammalian development. *BioEssays* **17,** 839–846.

Wolf, G. (1996). A History of Vitamin A and Retinoids. *FASEB J.* **10,** 1102–1107.

Wood, H., Pall, G., and Morriss-Kay, G. (1994). Exposure to retinoic acid before or after the onset of somitogenesis reveals separate effects on rhombomeric segmentation and *3' HoxB* gene expression domains. *Development* **120,** 2279–2285.

Woronicz, J., Calnan, B., Ngo, V., and Winoto, A. (1994). Requirement for the orphan steroid receptor Nur77 in apoptosis of T cell hybridomas. *Nature* **367,** 277–281.

Wu, X., and Levine, A. (1994). p53 and E2F-1 expression leads to *S*-phase entry and p53-mediated apoptosis. *Proc. Natl. Acad. Sci. USA* **91,** 3602–3606.

Yamaguchi-Iwai, Y., Satake, M., Murakami, Y., Sakai, M., Muramatsu, M., and Ito, Y. (1990). Differentiation of F9 embryonal carcinoma cells induced by the c-*jun* and activated c-*Ha-ras* oncogenes. *Proc. Natl. Acad. Sci. USA* **87,** 8670–8674.

Yamasaki, L., Jacks, T., Bronson, R., Goillot, E., Harlow, E., and Dyson, N. (1996). Tumor induction and tissue atrophy in mice lacking E2F-1. *Cell* **85,** 537–548.

Yang, P. C., Luh, K. T., Wu, R., and Wu, C. W. (1992). Characterization of the mucin differentiation in human lung adenocarcinoma cell lines. *Am. J. Respir. Cell Mol. Biol.* **7,** 161–171.

Yang, Y., Vacchio, M. S., and Ashwell, J. D. (1993). 9-*Cis*-retinoic acid inhibits activation-driven T-cell apoptosis: Implications for retinoid X receptor involvement in thymocyte development. *Proc. Natl. Acad. Sci. USA* **90,** 6170–6174.

Yang, Y., Bailey, J., Vacchio, M. S., Yarchoan, R., and Ashwell, J. D. (1995a). Retinoic acid inhibition of ex vivo human immunodeficiency virus–associated apoptosis of peripheral blood cells. *Proc. Natl. Acad. Sci. USA* **92,** 3051–3055.

Yang, Y., Minucci, S., Ozato, K., Heyman, R. A., and Ashwell, J. D. (1995b). Efficient inhibition of activation-induced Fas ligand up-regulation and T cell apoptosis by retinoids requires occupancy of both retinoid X receptors and retinoic acid receptors. *J. Biol. Chem.* **270,** 18672–18677.

Yen, A., Chandler, S., Forbes, M. E., Fung, Y. K., T'Ang, A., and Pearson, R. (1992). Coupled down-regulation of the RB retinoblastoma and c-*myc* genes antecedes cell differentiation: Possible role of RB as a "status quo" gene. *Eur. J. Cell Biol.* **57,** 210–221.

Zakeri, Z., Quaglino, D., Latham, T., and Lockshin, R. (1993). Delayed internucleosomal DNA fragmentation in programmed cell death. *FASEB J.* **7,** 470–478.

Zauli, G., Visani, G., Vitale, M., Gibellini, D., Bertolaso, L., and Capitani, S. (1995). All-*trans* retinoic acid shows multiple effects on the survival, proliferation and differentiation of human fetal CD34+ haemopoietic progenitor cells. *Br. J. Haematol.* **90,** 274–282.

Zelent, A. (1994). Translocation of the RAR alpha locus to the *PML* or *PLZF* gene in acute promyelocytic leukaemia. *Br. J. Haematol.* **86,** 451–460.

Zhang, X., Lehmann, J., Hoffmann, B., Dawson, M., Cameron, J., Graupner, G., Hermann, T., Tran, P., and Pfahl, M. (1992). Homodimer formation of retinoid X receptor induced by 9-*cis* retinoic acid. *Nature* **358**, 587–591.

Zhang, L. X., Mills, K. J., Dawson, M. I., Collins, S. J., and Jetten, A. M. (1995a). Evidence for the involvement of retinoic acid receptor RAR alpha–dependent signaling pathway in the induction of tissue transglutaminase and apoptosis by retinoids. *J. Biol. Chem.* **270**, 6022–6029.

Zhang, W., Grasso, L., McClain, C. D., Gambel, A. M., Cha, Y., Travali, S., Deisseroth, A. B., and Mercer, W. E. (1995b). p53-independent induction of WAF1/CIP1 in human leukemia cells is correlated with growth arrest accompanying monocyte/macrophage differentiation. *Cancer Res.* **55**, 668–674.

Zou, H., and Niswander, L. (1996). Requirement for BMP signaling in interdigital apoptosis and scale formation. *Science* **272**, 738–741.

2

Developmental Modulation of the Nuclear Envelope

Jun Liu,[1] Jacqueline M. Lopez,[1] and Mariana F. Wolfner
Sections of Genetics and Development and [1]Biochemistry, Molecular, and Cell Biology
Cornell University, Ithaca, NY 14853-2703

I. Introduction

Developmental processes often require or exploit variations in normal cellular mechanisms or organelles. Understanding the nature of these variations can simultaneously inform understanding of the developmental process and the organelle or cellular mechanism. This review considers developmentally-regulated changes in the nuclear envelope. To set the stage, we begin with a brief general consideration of the nuclear envelope's structure, components, and function, providing references to numerous recent reviews for readers to consult for more in depth discussion of the nuclear envelope per se. We then turn to our major focus: a consideration of processes in which changes in the nuclear envelope have been implicated or are expected, and consider the reasons and molecular correlates for those changes. Among the examples we cite, we have a bias towards *Drosophia*, both because of our own familiarity with this organism and because this organism has permitted a genetic analysis of the roles of particular nuclear envelope components in development.

---

[1]These authors contributed equally to the ideas in this paper and are listed in alphabetical order. Present address: JL: Dept. of Embryology, Carnegie Institute of Washington, Baltimore, MD. JML: Dept. of Biology, MIT and Whitehead Institute, Cambridge, MA.

Current Topics in Developmental Biology, Vol. 35

II. The Nuclear Envelope

A. Structure and Composition

The nuclear envelope is a defining feature of eukaryotic cells. It compartmentalizes a cell by separating the chromosomes from the cytoplasm, and plays important roles in mitosis, DNA replication, protein transport, and regulated gene expression. The nuclear envelope consists of three major structural components: the nuclear membranes, the nuclear pore complexes, and the nuclear lamina (summarized in Figure 1; reviewed in Gerace, 1986; Gerace and Burke, 1988; Nigg, 1989, 1992; Burke, 1990; Dessev, 1992; Dingwall and Laskey, 1992; Moir and Goldman, 1993; Hutchison et al., 1994; Georgatos et al., 1994; Moir et al., 1995; Goldberg and Allen, 1995).

1. Nuclear Membranes

There are two bilayer membranes surrounding the nucleus. The inner and outer nuclear membranes are biochemically and functionally distinct. The outer membrane is continuous with and functionally similar to the endoplasmic reticulum (ER). The inner nuclear membrane instead contains integral membrane proteins that are not found in the peripheral ER (for recent reviews, see Georgatos et al., 1994; Gerace and Foisner, 1994; Cowin and Burke, 1996). These include, in vertebrates, the lamin B receptor (LBR) (Worman et al., 1988), lamin-associated proteins (LAPs) 1A, 1B, 1C, and 2 (Senior and Gerace, 1988; Foisner and Gerace, 1993; and Furukawa et al., 1995), p54 (Bailer et al., 1991) and p55 (Powell and Burke, 1990), and, in Drosophila, the otefin protein (Harel et al., 1989; Padan et al., 1990). Several of these proteins have been shown to interact with lamins and/or chromosomes, and are thus hypothesized to be involved in attaching the nuclear lamina and/or chromosomes to the nuclear membranes.

2. Nuclear Pore Complexes

Nuclear pore complexes (NPCs) are supramolecular structures perforating the nuclear membranes. Filaments are attached to NPCs and extend into the cytoplasm, where they may be attached to the cytoskeleton (Goldberg and Allen, 1995). On the nucleoplasmic side of NPCs, the filaments extend to form a basketlike structure. Nuclear pores are eight-sided channels with molecular weights of about 125×10^6 Da. They allow passive diffusion of small molecules and ATP-dependent, signal-mediated translocation of proteins and RNAs (for recent reviews, see Pante and Aebi, 1994, 1995; Fabre and Hurt, 1994; Gorlich and Mattaj, 1996).

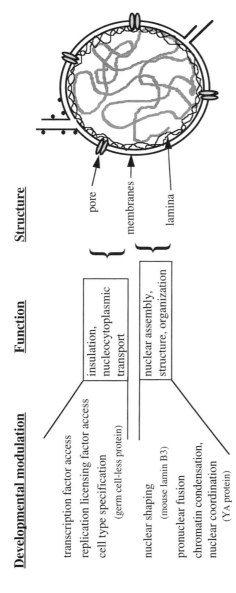

Developmental modulation   **Function**   **Structure**

transcription factor access
replication licensing factor access
cell type specification
(germ cell-less protein)

insulation,
nucleocytoplasmic
transport

nuclear shaping
(mouse lamin B3)
pronuclear fusion
chromatin condensation,
nuclear coordination
(YA protein)

nuclear assembly,
structure, organization

pore

membranes

lamina

Fig. 1. This figure summarizes developmental processes that require, or are likely to require, modulation of the nuclear envelope. The center of the figure lists major functions of the nuclear envelope. To the right of those functions is a schematic drawing of nuclear envelope structures, modified from Gerace and Burke (1986). The nuclear membranes, the lamina, and one of the pore complexes are labeled. Their relationship to particular nuclear envelope functions is indicated. The squiggly line inside the nucleus symbolizes the chromosomes, including their regions of association with the nuclear envelope. The small dots on the outer nuclear membrane are ribosomes, included only to indicate that this membrane is continuous with the rough endoplasmic reticulum.

To the left of the list of general functions of the nuclear envelope is a list of ways in which those functions are developmentally modulated. Examples of developmentally regulated nuclear envelope proteins are also included in this list, written in parentheses. If those proteins have been assigned a function, they are listed underneath that function, indented.

A considerable inventory of vertebrate and yeast NPC proteins is known; these are called nucleoporins, or NUPS (reviewed in Bastos *et al.*, 1995; Pante and Aebi, 1994, 1995). Many of these proteins are O-linked glycoproteins, such as gp210 (Gerace *et al.*, 1982; known as gp188 in *Drosophila*; Harel, *et al.*, 1989; Greber *et al.*, 1990; Berrios *et al.*, 1995), which is thought to play a role in anchoring the pores to the nuclear membranes (reviewed in Forbes, 1992). In addition, a myosin-heavy-chain-like protein has been seen associated with nuclear pores in *Drosophila* (Berrios *et al.*, 1991), as has the germ cell–less protein, which is important in germ cell specification (Jongens *et al.*, 1992, 1994).

3. Nuclear Lamina

The nuclear lamina is a proteinaceous network underlying the nucleoplasmic face of the inner nuclear membrane. It is found in animals, from *C. elegans* to humans, and in plants (see Moir *et al.*, 1995, for a recent review). The major components of the lamina are intermediate filament-like proteins called lamins (for reviews, see Gerace, 1986; Newport, 1987; Moir *et al.*, 1995). In vertebrates, lamins are classified into subtypes A, C, and B. Lamins A and C are differential splice products derived from a single gene; they differ at their C-termini. B-type lamins are found associated with membranes during the cell cycle, possibly functioning in the attachment of the nuclear lamina to the nuclear membranes. Lamins A and B contain a CaaX (Cysteine–two aliphatic residues–any residue) motif at their C-termini, a target for farnesylation and carboxymethylation. This motif is required for targeting lamins to the nuclear envelope (Holtz *et al.*, 1989; Krohne *et al.*, 1989; Kitten and Nigg, 1991; Lutz *et al.*, 1992; Firmbach-Kraft and Stick, 1993; Hennekes and Nigg, 1994). In *Drosophila*, a single gene for a B-type lamin, Dm_0, and another, encoding a lamin C–like protein, have been identified (Gruenbaum *et al.*, 1988; Bossie and Sanders, 1993; Riemer and Weber, 1994).

Though lamins are the most abundant proteins in the nuclear lamina, nonlamin proteins are present in the lamina as well. For example, nuclear laminae in *Drosophila* eggs and embryos contain a developmentally regulated nonlamin protein called YA (Young Arrest; H. Lin and Wolfner, 1991; Lopez *et al.*, 1994; Liu *et al.*, 1995). This protein is present in the nuclear envelopes of the four female meiotic products, the male pronucleus, and all nuclei in cleavage-stage embryos before cellularization. In the fertilized egg, the YA protein's function is required during the transition from meiosis to the mitotic divisions of cleavage. Examples of vertebrate nonlamin proteins in the nuclear lamina are the three human polypeptides known as MAN antigens, whose molecular identity and function await further study (Paulin-Levasseur *et al.*, 1996).

B. Dynamics During the Cell Cycle

The nuclear envelope goes through dramatic changes during the cell cycle (reviewed in Gerace and Burke, 1988). In many organisms, the nuclear envelope disassembles completely at metaphase. Lamins A/C become soluble and disperse in the cell, whereas lamin B remains associated with vesicles derived from fragmented nuclear membranes and with vesicles associated with chromosomes (reviewed in McKeon, 1991). Nuclear pore complexes also gradually disappear from the nuclear envelope at this time (reviewed in Forbes, 1992). At the end of mitosis, the nuclear envelope reassembles. In *Drosophila*, disassembly of the nuclear envelope is incomplete. In this organism, most of the lamin and all the NPCs dissociate from the nuclear envelope at the end of metaphase. However, nuclear membranes remain visible, though "fenestrated," open at the spindle poles (Stafstrom and Staehelin, 1984; Harel *et al.*, 1989; Paddy *et al.*, 1996), and surrounded by another layer of membranes in a structure known as the "spindle envelope" (Stafstrom and Staehelin, 1984).

Lamins have been implicated in the assembly and disassembly of the nuclear envelope during the cell cycle. Using cell-free nuclear assembly systems, lamins were found to be essential for nuclear envelope assembly (Burke and Gerace, 1986, for mammalian cells; Dabauvalle *et al.*, 1991, for *Xenopus*; and Ulitzur *et al.*, 1992, for *Drosophila*). When antibodies against lamins were added to the extracts, nuclear envelopes failed to assemble. However, a lamin-independent pathway for nuclear envelope assembly was proposed based on studies of nuclear envelope assembly using *Xenopus* egg extracts depleted of lamin L_{III} (Newport *et al.*, 1990; Meier *et al.*, 1991; Jenkins *et al.*, 1993; also termed lamin B3: Moir *et al.*, 1995). Nuclei assembled in such extracts, but were more fragile than normal, suggesting that lamin L_{III} is important for maintenance of nuclear structural integrity. Lourim and Krohne (1993) suggest that formation of the nuclear envelope in the absence of L_{III} is probably due to lamin L_{II} that is still remaining in these extracts (reviewed in Lourim and Krohne, 1994; Moir *et al.*, 1995).

Disassembly of the nuclear envelope in vertebrates has been shown to be due to phosphorylation of lamins by mitosis-promoting factor (MPF), and possibly other kinases (Peter *et al.*, 1990; Heald and McKeon, 1990; Ward and Kirschner, 1990; Peter *et al.*, 1991, 1992; Hass and Jost, 1993). Both *in vitro* and *in vivo* studies have identified mitosis-specific phosphorylation sites in lamins. That these phosphorylation sites are required for nuclear envelope disassembly is shown by the prevention of nuclear disassembly by lamins mutant at the putative cdc2 phosphorylation sites. In other organisms, such as *Drosophila* (D. Smith *et al.*, 1987; Stuurman *et al.*, 1995; D. Smith and Fisher, 1989; L. Lin and Fisher, 1990), cell-cycle-stage–dependent changes in phosphorylation states of lamin Dm_0 have also been documented. For example, phosphorylation states differ between two interphase isoforms of this lamin, Dm_1 and Dm_2, and also between them and the meiotic/mitotic isoform Dm_{mit}.

C. Functions

1. Separating Chromosomes from Cytoplasm and Controlling Nucleocytoplasmic Transport

The presence of nuclear membranes insulates the chromosomes from the cytoplasm. In order for transcription and replication factors to access chromatin, or for RNA to reach the translational machinery, regulated nucleocytoplasmic transport must occur. Pore complexes in the nuclear envelope allow such transport of proteins and RNAs between the cytoplasm and the nucleus. The importance of pore complexes in nucleocytoplasmic transport has been shown through a combination of both *in vivo* and *in vitro* studies (see Fabre and Hurt, 1994, for review). For example, a number of yeast mutants lacking functional pore proteins show abnormal nuclear protein import (Nehrbass *et al.*, 1993; Schlenstedt *et al.*, 1993). These mutants are either nonviable or temperature-sensitive, suggesting the necessity of nucleocytoplasmic transport in essential cellular processes (Nehrbass *et al.*, 1990; Davis and Fink, 1990; Wimmer *et al.*, 1992; Wente and Blobel, 1993).

The mechanisms of protein import and export through the pore complexes are being revealed through studies using the *Xenopus in vitro* reconstitution system (reviewed in Gorlich and Mattaj, 1996). This system has identified several proteins, including the NLS receptor importin (importin-δ and importin-β) and the small GTPase Ran/TC4, that play roles in protein import into the nucleus (reviewed in Sweet and Gerace, 1995).

2. DNA Replication

a. The Nuclear Envelope Is Required for DNA Replication.   Definition of the role of the nuclear envelope in DNA replication comes primarily from studies using *Xenopus* or *Drosophila* cell-free systems. When demembranated sperm chromatin or naked DNA are incubated in *Xenopus* egg extracts or *Drosophila* embryo extracts, nuclear structures can form that have intact nuclear envelopes (Blow and Laskey, 1986; Newport, 1987; Sheehan *et al.*, 1988; Blow and Sleeman, 1990; Crevel and Cotterill, 1991; Ulitzur *et al.*, 1992). These nuclei can replicate their DNA. There is a correlation between the efficiency of nuclear assembly and DNA replication: Only assembled nuclei with intact nuclear envelopes can replicate their DNA, and DNA replication does not occur if formation of the nuclear envelope is prevented (Blow and Laskey, 1986; Newport, 1987; Sheehan *et al.*, 1988; Blow and Sleeman, 1990; Crevel and Cotterill, 1991).

The nuclear lamina in particular has been shown to be required for DNA replication; evidence comes from studies of immunodepletion experiments

(Newport *et al.*, 1990; Meier *et al.*, 1991; Jenkins *et al.*, 1993, 1995; Hutchison *et al.*, 1994; Goldberg *et al.*, 1995). Nuclei assembled in *Xenopus* extracts depleted of lamin L_{III} failed to grow and could not assemble preinitiation complexes, as indicated by their abnormal PCNA staining. They did not replicate DNA (Meier *et al.*, 1991; Jenkins *et al.*, 1993). When purified lamin L_{III} was added back to the depleted extracts, both lamina assembly and DNA replication were rescued (Jenkins *et al.*, 1995; Goldberg *et al.*, 1995). The observation that, in mouse 3T3 cells, lamin B is localized to foci in the nucleoplasm that correspond to sites of DNA replication further supports the hypothesis that lamins (at least lamin B) are involved in DNA replications (Moir *et al.*, 1994).

Studies by Leno and Laskey (1991) suggest that the nuclear envelope also defines the unit of replication. When demembranated erythrocyte nuclei are added to cell-free *Xenopus* egg extracts, many of these nuclei become enclosed by a common nuclear membrane. Nuclei in such "multinuclear aggregates" lack individual envelopes. Whereas nuclei excluded from the multinuclear aggregates replicate their DNA asynchronously, nuclei within an aggregate replicate their DNA synchronously; different aggregates replicate out of synchrony from one another. These results suggest that the nuclear envelope allows coordinated replication of the DNA it encloses.

b. The Nuclear Envelope Is Needed to Coordinate Rounds of Replication with Mitosis.   Blow and Laskey (1988) proposed that the nuclear envelope is necessary for the mechanism that prevents reinitiation of DNA replication within a single cell cycle. They had found that nuclei formed from demembranated sperm in *Xenopus* egg extracts can only replicate their DNA once; rereplication required permeabilization or breakdown of the nuclear envelope. Nuclear envelope breakdown was thus needed to "license" chromatin for each subsequent round of replication. Blow and Laskey hypothesized that only upon nuclear envelope breakdown could replication-initiating molecules present in the cytoplasm gain access to their chromosomal targets. Sufficient amounts of these "licensing factor" molecules were retained in the nucleus to permit a single round of replication, but the licensing factor was destroyed or inactivated upon DNA replication. Thus, rereplication is prevented until the chromatin again has access to cytoplasmic licensing factor, upon nuclear envelope breakdown. Subsequent studies (e.g., Coverly *et al.*, 1993; Blow, 1993) have supported this hypothesis, and recent work suggests that likely candidates for components of licensing factor are the MCM protein family members (reviewed in Tye, 1994; see also Kubota *et al.*, 1995; Chong *et al.*, 1995; Madine *et al.*, 1995).

3. Chromatin Organization of Interphase Nuclei

The nuclear envelope has been suggested to play a role in organizing chromatin structure in interphase nuclei. Association of interphase chromosomes and the nuclear envelope has been seen in a number of systems (reviewed in Gerace and Burke, 1988). For example, cells in a range of animals and plants show the "Rabl" orientation of chromosomes, in which telomeres and centromeres are located at opposite poles of the nucleus (similar to chromosome orientation in telophase), associated with the nuclear envelope (Rabl, 1885; Comings, 1980; Cremer *et al.*, 1982; Fussell, 1992). This organization has been suggested to facilitate smooth transitions into nuclear divisions, by minimizing chromosomal tangling (Rabl, 1885). Another example of chromatin interaction with the nuclear envelope is seen in polytene chromosomes in *Drosophila*, where specific loci are seen associated with the nuclear envelope (Agard and Sedat, 1983; Mathog *et al.*, 1984; Hochstrasser *et al.*, 1986; Hochstrasser and Sedat, 1987). Attachment of interphase chromosomes to the nuclear envelope through these loci is hypothesized to be important for organizing the chromatin structure within the nucleus.

Several nuclear envelope components have been suggested to bind chromosomes, via either direct or indirect interactions with DNA. Using high-resolution microscopy techniques, Belmont *et al.* (1993) observed a close juxtaposition of lamin B and chromatin in mammalian and *Drosophila* cells, suggesting that the nuclear lamina interacted directly with chromosomes. Consistent with a tight association of the lamina and chromatin, both A/C and B types of lamins bind to chromatin or to specific DNA fragments *in vitro*; their domains conferring this binding have been identified (J. Glass and Gerace, 1990; Hoger *et al.*, 1991; Yuan *et al.*, 1991; Luderus *et al.*, 1992; C. Glass *et al.*, 1993; Taniura *et al.*, 1995; Zhao *et al.*, 1996). *In vitro*, purified rat lamin B_1 and *Drosophila* interphase lamin Dm_0 have been shown to bind to DNA, specifically to AT-rich matrix-attachment regions (MARs) (Luderus *et al.*, 1992; Zhao *et al.*, 1996). The *in vivo* significance of these observations is not fully clear, since at least two MARs to which *Drosophila* lamin Dm_0 binds, those in the *ftz* and histone genes, are not localized to the nuclear envelope *in vivo* (Marshall *et al.*, 1996); also, MARs do not bind to *Drosophila* type II scaffolds, which lack the internal nuclear matrix and are composed primarily of lamina proteins (Izaurralde *et al.*, 1988). The *Drosophila* YA nuclear lamina protein also has the capacity to bind chromatin as seen *in vitro* and upon ectopic expression in polytene cells (Lopez and Wolfner, 1997), but its chromatin-binding domain has not yet been identified (Liu, 1996; Liu and Wolfner, submitted). Taken together, these results suggest that the nuclear lamina may function to organize the nucleus through binding to chromosomes or directly to DNA. Some proteins in the inner nuclear membrane, such as the lamin B receptor and LAP2 (Ye and Worman, 1994; Foisner and Gerace, 1993) and nuclear pore proteins nup153 and RanBP2s (Sukegawa and Blobel, 1993; Yokoyama *et al.*, 1995), are likely to associate also with chromatin, as they have DNA binding motifs or demonstrated DNA binding capacities.

4. Mitotic Functions

As already described, the nuclear lamina and its lamin constituents are necessary for assembly and re-formation of functional nuclear envelopes during the cell cycle. In addition, another role for the nuclear envelope in mitosis is suggested from studies of the transmembrane protein NDC1, a component of the yeast nuclear envelope (Winey *et al.*, 1993). Mutations in the *ndc1* gene cause defects in duplication of the yeast cell's spindle pole body (SPB), a centrosomelike organelle that is embedded in the nuclear envelope (reviewed by Winey and Byers, 1992). This has led Winey *et al.* (1993) to suggest that the nuclear envelope has a function in mitotic spindle formation.

III. Developmental Variations of the Nuclear Envelope and Their Possible Functions

A number of developmental events could be expected to be directly influenced by the nuclear envelope (summarized in Figure 1).

A. Developmentally Modulated Nuclear Transport

Developmentally modulated nuclear transport can alter the expression of genes or the replication of DNA. We first discuss two examples of the importance of the separation of nucleus and cytoplasm in development, even though an active role for the nuclear envelope has not been shown in either examples: In one case the cell circumvents a process normally tied to the presence of a nuclear envelope; in the other, the cell exploits the presence of an insulating nuclear envelope. We then discuss a developmentally regulated protein that is associated with nuclear pores and thus could modulate nucleocytoplasmic transpost function.

In the first example, some nuclei, such as yolk nuclei in *Drosophila* embryos, undergo endoreplication cycles in which S phases continue in the absence of mitosis (Smith and Orr-Weaver, 1991; Orr-Weaver, 1994). Similarly, *Drosophila* polytene cells replicate their DNA in the absence of mitosis. This means that these cells must have found a way to circumvent the need for licensing factor or to transport it into their nuclei without nuclear envelope breakdown.

An example in which the insulation of nucleus from cytoplasm is exploited to control developmental processes is illustrated by a crucial step in dorso–ventral patterning control in *Drosophila* embryos (for a recent review, see Morisato and Anderson, 1995). Ventral development is specified by transcription of a class of genes regulated by the DORSAL transcription factor in the nucleus. If this transcription factor is not in the nucleus, a different set of genes is transcribed, and dorsal development results. An intermediate level of DORSAL protein in the nucleus is correlated with lateral development. The DORSAL transcription factor is normally sequestered in the cytoplasm by interaction with the CACTUS protein. Upon receipt of a signal from the ventral side of the egg, DORSAL protein is released from this sequestration and is now able to enter the nucleus. This only occurs to a full extent on the ventral side of the embryo, resulting in the transcriptional changes that specify ventral development. On the dorsal side of the embryo, the DORSAL protein remains cytoplasmic, unable to access its target genes, and dorsal development ensues. In between the dorsal and ventral sides of the embryo, some DORSAL protein enters the nucleus, triggering lateral development.

Although in the case of dorso–ventral patterning in *Drosophila* the nuclear envelope plays a crucial role, it is passive. One could, however, imagine an active role of the nuclear envelope if there were developmental modulation of nuclear pores or their function. The *Drosophila* nuclear pore–associated protein germ cell–less (GCL) may provide an example of this (Jongens *et al.*, 1992, 1994). This protein associates with nuclear pores only in the germ cell precursor cells located at the posterior pole of *Drosophila* embryos. Genetic studies indicate that GCL is necessary to specify germ cell fate: Embryos containing reduced levels of GCL develop into animals lacking germ cells. Moreover, ectopic expression of GCL at the anterior pole of the embryos causes cells that would normally be somatic to adopt characteristics of germ cell precursors. These results suggest that the presence of GCL in the nuclear envelope might alter nuclear trafficking in a way that allowed chromosomal access to a regulatory molecule that specifies germ cell fate; restriction of GCL to germ line precursor cell nuclei would allow a potentially ubiquitous determinant to access only the chromosomes of those nuclei. A converse model could have the GCL protein required for transport to the cytoplasm of an RNA molecule whose translation was needed for germ cell fate.

B. Structural Changes in the Nuclear Envelope

Structural changes in the nuclear envelope play a role in several critical developmental events, as will be discussed shortly as specific cases. Nuclear shape changes in spermatogenesis and nuclear envelope reorganization in fertilized eggs are among the processes that require modulation of nuclear envelope function. Consistent with the importance of and need for these changes, developmental variations in the protein composition of the nuclear lamina have been observed in many animals. In vertebrates and *Drosophila*, the general rule seems to be that lamin B is expressed in almost all cell types, whereas lamins A and C are induced as cells differentiate (Stewart and Burke, 1987; Rober *et al.*, 1989; Riemer *et al.*, 1995; Benavente *et al.*, 1985; Stick and Hausen, 1985; Gruenbaum *et al.*, 1988; Liu, *et al.*, 1997). Since ectopic expression of lamin A does not induce differentiation of lamin A–negative embryonic carcinoma cells (Peter and Nigg, 1991), it appears that expression of lamin A is not sufficient to induce a developmental program; rather, its expression in differentiated cells might be required to maintain the structural organization of the nucleus required for specific developmental programs. In addition to variation in lamin forms during development, there are also variations of other nuclear lamina components. For example, the YA nuclear lamina protein, which has been shown genetically to be essential for embryos to complete the transition from meiosis to mitosis, is found in laminae only at these particular stages (H. Lin and Wolfner, 1991; Lopez *et al.*, 1994; Liu *et al.*, 1995).

1. Alterations of the Nuclear Lamina and Lamina Components During Gametogenesis

Changes in the composition of the nuclear lamina occur during oogenesis and spermatogenesis (earlier reviews include Krohne and Benavente, 1986; Stricker *et al.*, 1989). For example, during oogenesis in *Xenopus*, an oocyte-specific lamin, L_{III}, surrounds nuclei (Stick and Hausen, 1985; Stick, 1988). In this organism, somatic lamins, L_I and L_{II}, were reported to become undetectable at the pachytene stage of meiosis, when the nuclear lamina structure disappears (Stick and Schwarz, 1983), though subsequent studies have shown that there are low levels of L_{II} in oocytes and early embryos (Lourim and Krohne, 1993). Lamin L_{III} appears at early diplotene stage as the nuclear lamina reassembles. It becomes the major lamin of the mature oocyte. The reason for oogenesis-specific lamins is not clear. Perhaps special lamins are needed to provide longterm stability to the lamina during the prolonged oogenesis period (several months in *Xenopus*), or perhaps they exert a special function in nuclear structure or organization during meiosis or its arrest.

2. Spermatogenesis

Dramatic changes of cell shape and nuclear organization occur during spermatogenesis, resulting in the formation of sperm with highly compact chromatin [for reviews see: Lindsley and Tokuyasu, 1980, and Fuller, 1993 (*Drosophila*); Pundey, 1995 (vertebrates); Roth and Allis, 1992 (sea urchin)]. Germ line stem cells in the testis divide and give rise to spermatocytes. Spermatocytes then undergo two meiotic divisions to become spermatids, which elongate and differentiate to mature sperm with highly compact chromatin. A number of studies suggest that the nuclear lamina is absent, or present in an altered form, during spermatogenesis and that this is accompanied by changes in lamina composition, including, in some cases, the presence of germ line–specific lamins.

The first piece of evidence came from ultrastructural studies. Fawcett (1966) found that although a lamina structure is present in many cell types, it is absent in cat spermatocytes. Subsequently, ultrastructural and immunofluorescence studies showed that a nuclear lamina is also absent in nuclei of chicken spermatocytes and spermatids (Stick and Schwarz; 1982). This was confirmed in biochemical studies by Lehner *et al.* (1987), who showed on Western blots that there are no detectable lamins A, B1, and B2 in male germ cells after the pachytene stage of meiosis.

Studies on *Xenopus*, mouse, and rat suggest that a nuclear lamina is present, but probably in an altered form, during spermatogenesis in these species. In

Xenopus, somatic forms of lamin, L_I and L_{II}, are not present in late stages of spermatogenesis. Instead, a male germ line–specific lamin, L_{IV}, is present (Krohne and Benavente, 1986). Immunoelectron microscopic studies using monoclonal antibodies against L_{IV} showed that L_{IV} is present in patches in the nuclear envelopes of spermatids and sperm (Benavente and Krohne, 1985). These results suggest that a nuclear lamina in spermatids and sperm is present in an altered form.

Immunofluorescence studies of lamina composition during mouse spermatogenesis showed that although lamin A/C epitopes are absent in germ cells at any stage of spermatogenesis, lamin B epitopes are present throughout spermatogenesis (Schatten *et al.*, 1985; Maul *et al.*, 1986; Moss *et al.*, 1987; Moss *et al.*, 1993). Moreover, in mature sperm, lamin B epitopes are present in patches along the sperm. These studies used antibodies against somatic lamins, so they could not determine whether the lamin staining observed reflected the presence of normal somatic lamins or spermatogenesis-specific lamin variants. Subsequent molecular cloning of germ cell–specific lamins from mouse indicates that, as in *Xenopus*, there are special forms of lamin present and functional during mouse spermatogenesis. A mouse spermatocyte–specific lamin, lamin B3, was cloned and shown to be present only during meiosis; it is absent from mature sperm (Furukawa and Hotta, 1993). Moreover, ectopic expression of this lamin in cultured somatic cells causes those cells to adopt a hook-shaped nuclear morphology. This suggests that lamin B3 plays a role in organizing the nuclear shape and structures during spermatogenesis. Another cloned-mouse male germ cell–specific A/C-type lamin, lamin C2, is found in mouse pachytene-stage spermatocytes (Furukawa *et al.*, 1994). Interestingly, lamin C2 and lamin B3 share striking similarity in their structural organization: both B3 and C2 lack the head domain and part of the rod domain that are present in all somatic lamins. Therefore, the possibility exists that, like lamin B3, lamin C2 might also function in organizing nuclear structures during mouse spermatogenesis.

Male germ cell–specific lamin forms have also been identified in rats. Biochemical studies showed that a 60-kDa protein related to somatic lamin B is present in the lamina of rat spermatogonia, and becomes a component of synaptonemal complexes as germ cells enter meiosis (Sudhakar and Rao, 1990). Following meiosis, it relocalizes to the lamina and is present in the lamina in round spermatids and mature sperm. Immunofluorescence studies further show that this lamin form is present in the nuclear envelope in both pre- and postmeiotic germ cells. Moreover, it is also present at the same stage of spermatogenesis in grasshopper, rooster, and frog, as well as in plant meiocytes (Sudhakar and Rao, 1990; Sudhakar *et al.*, 1992). The presence of this lamin in synaptonemal complexes suggests that it might play important roles during meiotic prophase, particularly during homologous chromosome pairing and recombination, though this hypothesis still needs further experimental tests. In addition to this 60-kDa lamin form, Smith and Benavente (1992) reported the identification of a 52-kDa novel lamin form selectively expressed during meiotic stages of spermatogenesis, and Vester *et al.* (1993) also reported the presence of a protein closely related to lamin B1 in rat pachytene spermatocytes. None of these germ cell–specific lamin forms in rats has been cloned. It is possible that some of them might be the same protein [for example, the 60-kDa protein identified by Sudhakar and Rao (1990) and the lamin B1–like protein identified by Vester *et al.* (1993)] and that they may also be homologues of the male germ cell–specific lamin forms identified in mouse. Once cloned, the possible functions of these male germ cell–specific lamin forms can also be tested.

Compared to studies of the nuclear lamina during vertebrate spermatogenesis, there are fewer studies of changes of lamina structure during spermatogenesis of invertebrates. Immunofluorescence studies by Schatten *et al.* (1985) showed that both lamin A/C and B epitopes are present in sea urchin sperm, but only localized to acrosomal and centriolar fossae, suggesting that, as with vertebrate sperm, sea urchin sperm might also contain an altered nuclear lamina. Support for this notion comes from Collas *et al.* (1995). Using several antilamin antibodies, Collas *et al.* (1995) found that although lamin epitopes are present over the whole periphery of sperm nuclei, "lateral lamin labeling" is absent in very low concentrations of detergent (0.1% Triton X-100, rather than the 0.5% that is normally used to purify the lamina-matrix fraction). This suggests a weaker association of lamin epitopes with the lateral surface of the sperm nucleus as compared to their strong association toward the acrosomal and centriolar fossae. Therefore, the structure of the nuclear lamina in sea urchin sperm might also be different from that in somatic cells. Since only one B-type lamin has been cloned in sea urchin so far (Holy *et al.*, 1995), it is not possible to determine whether the cross-reactive lamin epitopes in sea urchin sperm indicate the presence of a ubiquitous lamin or of a sperm-specific lamin that is cross-reactive with the somatic one.

In *Drosophila*, lamin Dm_0 and its phospho-derivatives are expressed throughout development, and present in all cell types tested except for sperm (cited in Riemer *et al.*, 1995; Liu *et al.*, 1997). *Drosophila* lamin C is also not found in sperm (Riemer *et al.*, 1995). When the lamina protein YA is ectopically expressed in testes, it is present in nuclear envelopes of spermatocytes but not in mature sperm (Liu *et al.*, 1997). These results suggest that the nuclear lamina in *Drosophila* sperm is either absent or atypical. It is also formally possible that, as in the vertebrate systems described earlier, there are as-yet-unidentified sperm-specific lamin proteins in *Drosophila* that do not cross-react with available antilamin protein antibodies.

Therefore, it appears that the nuclear lamina is either absent, or present in an altered form, in sperm, and that there are variations in lamina protein compositions during spermatogenesis. But what are the functional significances of these variations? During spermatogenesis, nuclear and chromosomal structures go through dramatic changes, including nuclear shape changes and compaction of sperm chromatin. The presence of germ cell–specific lamins appears to play a structural role in the nuclear shaping, as seen in the case of mouse lamin B3 (Furukawa and Hotta, 1993). Given that chromatin condensation appears to initiate at the nuclear periphery (Hiraoka *et al.*, 1989), the absence or an altered lamina might also be required for the proper condensation of chromatin during spermatogenesis. In addition, as will be discussed later, the sperm nucleus goes through a series of changes upon fertilization, including breakdown and reformation of its nuclear envelope and decondensation of its DNA. It is possible that the absence of an intact lamina facilitates breakdown of the sperm nuclear envelope. This could be important for rapid formation of a male pronucleus with a new nuclear envelope with similar composition to that of the female pronucleus, which in turn allows the occurrence of the first mitotic division. In *Drosophila*, the whole process from sperm entry to the end of the first mitotic division takes only about 17 minutes (reviewed in Foe *et al.*, 1993). Thus rapid breakdown of the nuclear envelope of the sperm nucleus might be a requirement for the coordination of pronuclear activities required for the first mitotic division.

3. Nuclear Envelope Reorganization During Fertilization and Pronuclear Formation

Upon fertilization, the sperm nucleus reorganizes and becomes the male pronucleus. Both *in vivo* and *in vitro* studies have indicated that formation of the male pronucleus includes the following steps: (1) breakdown of the sperm nuclear envelope, (2) dispersion of the highly condensed sperm chromatin, (3) formation of a new nuclear envelope, called the pronuclear envelope, surrounding the dispersed chromatin, and (4) further chromatin decondensation and DNA replication of the pronucleus (Longo, 1985; Yamashita *et al.*, 1990; Cothren and Poccia, 1993; Longo *et al.*, 1994; Cameron and Poccia, 1994; Collas and Poccia, 1995). Once formed, the male pronucleus, in conjunction with the female pronucleus, will coordinately go through the first mitotic division to form two zygotic nuclei.

Studies of the gynogenetic crucian carp ginbuna, *Carassius auratus langsdorfii* (Yamashita *et al.*, 1990), suggest that breakdown of the sperm nuclear envelope is required for male pronuclear formation and successful fertilization. In the wild, sperm from a related subspecies can inseminate ginbuna eggs, but the sperm nucleus remains condensed and does not form the male pronucleus, thus making no contribution the zygote's genome. Yamashita *et al.* (1990) found that the failure of the sperm nucleus to become the male pronucleus is due to the failure of the breakdown of the sperm nuclear envelope. When they injected sperm without plasma membranes and nuclear envelopes into eggs, the sperm nuclei could transform into male pronuclei.

Decondensation of the sperm nucleus follows the breakdown of its own nuclear envelope. *In vitro* experiments studying the formation of the male pronucleus, using cell-free *Xenopus* and sea urchin egg extracts, showed that there are two steps involved in the decondensation of the sperm nucleus: a membrane-independent chromatin decondensation and a nuclear envelope–dependent nuclear swelling, or further decondensation (Lohka and Masui, 1984; Wilson and Newport, 1988; Collas and Poccia, 1995). Moreover, partially decondensed sperm nuclei fail to go through further decondensation if lamin L_{III} is depleted in the *Xenopus* cell-free extracts (Newport *et al.*, 1990; Meier *et al.*, 1991; Jenkins *et al.*, 1993, 1995; Hutchison *et al.*, 1994; Goldberg *et al.*, 1995). These studies suggest that reformation of the nuclear envelope with an intact lamina around the partially decondensed sperm nucleus is a prerequisite for the formation of a functional male pronucleus.

Therefore, reorganization of the nuclear envelope plays important roles in the formation of the male pronucleus, which leads to successful fertilization. Studies on both vertebrates and invertebrates showed that, accompanying the structural reorganization of the male pronucleus, the composition of its nuclear envelope also changes. Lamins that have been vastly decreased during spermatogenesis are restored upon formation of the male pronucleus, as seen in mouse, sea urchin, and *Xenopus* (for reviews see Schatten and Schatten, 1987; Stricker *et al.*, 1989).

In *Drosophila*, reorganization of the male pronucleus involves recruitment of both lamin Dm and YA into its nuclear envelope (Lopez *et al.*, 1994; Liu *et al.*, 1997), thus resulting in its being surrounded by a nuclear envelope apparently like that of the four female meiotic products. The maternally provided YA protein is absolutely essential at this particular early developmental stage. In the absence of YA function, embryos cannot advance into the cleavage divisions: Nuclei in YA-deficient unfertilized eggs and early zygotes show abnormal chromatin condensation and fusion of the meiotic products (Liu *et al.*, 1995), and the coordination of their behaviors is abnormal (Lopez, 1996). This suggests that YA is needed to specify the identities of the female meiotic products and the male pronucleus, allowing them to coordinate their associations and condensations during the transition from meiosis to syngamy and the first mitotic divisions. YA provides the first example of a developmentally regulated nuclear lamina protein that plays a critical role *in vivo* in early embryonic development.

4. Variations of the Nuclear Envelope Components at Other Developmental Stages

In addition to dramatic variations of the nuclear envelope components during gametogenesis and fertilization, variations in nuclear envelope components also occur at other developmental stages. For example, in *Xenopus* only L_{III} and low levels of L_{II}, contributed by the maternal pool, are present in early embryos. Expression of the somatic lamin L_I starts at the midblastula transition; L_I and L_{II} become the predominant somatic lamin forms in swimming tadpoles (reviewed in Krohne and Benavente, 1986; see also Lourim and Krohne, 1993). Changes in expression of different lamin epitopes during development also have been detected in mouse, chicken, sea urchin, and clam, mostly by immunofluorescence studies (see Stricker *et al.*, 1989, for review, and also Schatten *et al.*, 1985; Maul *et al.*, 1987; Lehner *et al.*, 1987; Prather and Schatten, 1992; Holy *et al.*, 1995). However, the functional relevance of these differential expression patterns of lamin forms is still unknown.

IV. Conclusions

There are developmental variations in the composition and consequent structure and function of the nuclear envelope. Mitotic state, transcription pattern, condensation state, and shape of a nucleus also undergo modulation during development. Changes in the nuclear envelope can, and in some cases are known to, play essential roles in these developmental processes. Many nuclear envelope proteins are known from studies of tissue culture cells, mixed cell populations, and yeast, and much is known about their biochemistry and sequence. However, the developmental modulation of their presence and function is not well understood, nor is the spectrum of developmentally regulated components of the nuclear envelope. We think that the next few years will see the identification and cloning of additional developmentally regulated nuclear envelope proteins, as well as the cloning of ones identified thus far only by cross-reactivity, such as the vimentin cross-reactive epitope seen at nuclear peripheries in young *Drosophila* embryos (Riparbelli and Callaini, 1992; Callaini and Riparbelli, 1996).

Molecular reagents will allow determination of the times and processes in which these developmentally regulated proteins function. We will also learn with which nuclear components they interact and how such interactions modify nuclear behavior and structure. Finally, ectopic expression and knockout experiments in genetically manipulatable multicellular systems, such as *Drosophila*, mouse, and *C. elegans*, and antisense experiments in systems offering other developmental and cell biological advantages, such as sea urchin and *Xenopus*, will define which of these developmentally regulated nuclear envelope proteins are essential for which developmental process, and will determine why this is so. In addition to their developmental significance, these results will contribute to a more complete understanding of fundamental nuclear envelope structure/function, since they will reveal which aspects of the nuclear envelope require flexibility and modulation in different cell types or stages.

Acknowledgements

We thank Drs. B. Brown, K. Kemphues, and J. Lis for comments on an early version of this manuscript. Our work on the nuclear envelope is supported by an NIH grant to M.F.W., who is also grateful to the American Cancer Society for support from a Faculty Research Award that helped initiate our studies into the nuclear envelope and its developmental significance.

References

Agard, D. A., and Sedat, J. W. (1983). Three-dimensional architecture of a polytene nucleus. *Nature* **302**, 676–681.

Bailer, S. M., Eppenberger, H. M., Griffiths, G., and Nigg, E. A. (1991). Characterization of a 54-kD protein of the inner nuclear membrane: Evidence for cell cycle–dependent interaction with the nuclear lamina. *J. Cell Biol* **114**, 389–400.

Bastos, R., Pante, N., and Burke, B. (1995). Nuclear pore complex proteins. *Int. Rev. Cytol.* **162B**, 257–302.

Belmont, A. S., Zhai, Y., and Thilenius, A. (1993). Lamin B distribution and association with peripheral chromatin revealed by optical sectioning and electron microscopy tomography. *J. Cell Biol.* **123**, 1671–1685.

Benavente, R., and Krohne, G. (1985). Change of karyoskeleton during spermatogenesis of Xenopus: Expression of lamin L_{IV}, a nuclear lamina protein specific for the male germ line. *Proc. Natl. Acad. Sci. USA* **82**, 6176–6180.

Benavente, R., Krohne, G., and Franke, W. W. (1985). Cell type–specific expression of nuclear lamina proteins during development of *Xenopus laevis*. *Cell* **41**, 177–190.

Berrios, M., Fisher, P. A., and Matz, E. C. (1991). Localization of a myosin heavy chain-like polypeptide to *Drosophila* nuclear pore complexes. *Proc. Natl. Acad. Sci. USA* **88**, 219–223.

Berrios, M., Meller, V. H., McConnell, M., and Fisher, P. A. (1995). *Drosophila* gp210, an invertebrate nuclear pore complex glycoprotein. *Eur. J. Cell. Biol* **67**, 1–7.

Blow, J. J. (1993). Preventing re-replication of DNA in a single cycle: Evidence for a replication licensing factor. *J. Cell Biol* **122**, 993–1002.

Blow, J. J., and Laskey, R. A. (1986). Initiation of DNA replication in nuclei and purified DNA by a cell-free extract of *Xenopus* eggs. *Cell* **47**, 577–587.

Blow, J. J., and Laskey, R. A. (1988). A role for the nuclear envelope in controlling DNA replication within the cell cycle. *Nature* **332**, 546–548.

Blow, J. J., and Sleeman, A. M. (1990). Replication of purified DNA in *Xenopus* egg extract is dependent on nuclear assembly. *J. Cell Sci.* **95**, 383–391.

Bossie, C. A., and Sanders, M. M. (1993). A cDNA from *Drosophila melanogaster* encodes a lamin C–like intermediate filament protein. *J. Cell Sci.* **104**, 1263–1272.

Burke, B. (1990). The nuclear envelope and nuclear transport. *Curr. Opin. Cell Biol.* **2**, 514–520.

Burke, B., and Gerace, L. (1986). A cell-free system to study reassembly of the nuclear envelope at the end of mitosis. *Cell* **44**, 639–652.

Callaini, G., and Riparbelli, M. G. (1996). Fertilization in *Drosophila melanogaster*: Centrosome inheritance and organization of the first mitotic spindle. *Dev. Biol.* **176**, 199–208.

Cameron, L. A., and Poccia, D. L. (1994). *In vitro* development of the sea urchin male pronucleus. *Dev. Biol.* **162**, 568–578.

Chong, J. P. J., Mahbubani, H. M., Khoo, C.-Y., and Blow, J. J. (1995). Purification of an MCM-containing complex as a component of the DNA replication licensing system. *Nature* **375**, 418–421.

Collas, P., and Poccia, D. (1995). Formation of the sea urchin male pronucleus in vitro: Membrane-independent chromatin decondensation and nuclear envelope–dependent nuclear swelling. *Mol. Rep. Dev.* **42**, 106–113.

Collas, P., Pinto-Correia, C., and Poccia, D. L. (1995). Lamin dynamics during sea urchin male pronuclear formation *in vitro*. *Exp. Cell Res.* **219**, 687–698.

Comings, D. E. (1980). Arrangement of chromatin in the nucleus. *Hum. Genet.* **53**, 131–143.

Cothren, C. C., and Poccia, D. L. (1993). Two steps required for male pronucleus formation in the sea urchin egg. *Exp. Cell Res.* **205**, 126–133.

Coverley, D., Downes, C. S., Romanowski, P., and Laskey, R. A. (1993). Reversible effects of nuclear membrane permeabilization on DNA replication: Evidence for a positive licensing factor. *J. Cell Biol.* **122,** 985–992.

Cowin, P., and Burke, B. (1996). Cytoskeleton–membrane interactions. *Curr. Opin. Cell Biol.* **8,** 56–65.

Cremer, T., Cremer, C., Baumann, H., Luedtke, E.-K., Sperling, V., and Zorn, C. (1982). Rabl's model of the interphase chromosome arrangement tested in Chinese hamster cells by premature chromosome condensation and laser-UV-microbeam experiments. *Human Genet.* **60,** 46–56.

Crevel, G., and Cotterill, S. (1991). DNA replication in cell-free extracts from *Drosophila melanogaster.* *EMBO J.* **10,** 4361–4369.

Dabauvalle, M.-C., Loos, K., Merkert, H., and Scheer, U. (1991). Spontaneous assembly of pore complex–containing membranes ("annulate lamellae") in *Xenopus* egg extract in the absence of chromatin. *J. Cell Biol.* **112,** 1073–1082.

Davis, L. I., and Fink, G. R. (1990). The NUP1 gene encodes an essential component of the yeast nuclear pore complex. *Cell* **61,** 965–978.

Dessev, G. N. (1992). Nuclear envelope structure. *Curr. Opin. Cell Biol.* **4,** 430–435.

Dingwall, C., and Laskey, R. (1992). The nuclear membrane. *Science* **258,** 942–947.

Fabre, E., and Hurt, E. (1994). Nuclear transport. *Curr. Opin. Cell Biol.* **6,** 335–342.

Fawcett, D. W. (1966). On the occurrence of a fibrous lamina on the inner aspect of the nuclear envelope in certain cells of vertebrates. *Am. J. Anat.* **119,** 129–146.

Firmbach-Kraft, I., and Stick, R. (1993). The role of CaaX-dependent modifications in membrane association of *Xenopus* nuclear lamin B3 during meiosis and the fate of B3 in transfected mitotic cells. *J. Cell Biol.* **123,** 1661–1670.

Foe, V. E., Odell, G. M., and Edgar, B. A. (1993). Mitosis and morphogenesis in the *Drosophila* embryo: Point and counterpoint. In "The Development of *Drosophila melanogaster*" (M. Bate and A. Martinez Arias, eds.), vol. 1, pp. 149–300. Cold Spring Harbor Laboratory Press, Plainview, NY.

Foisner, R., and Gerace, L. (1993). Integral membrane proteins of the nuclear envelope interact with lamins and chromosomes, and binding is modulated by mitotic phosphorylation. *Cell* **73,** 1267–1279.

Forbes, D. (1992). Structure and function of the nuclear pore complex. *Annu. Rev. Cell Biol.* **8,** 495–527.

Fuller, M. T. (1993). Spermatogenesis. "The Development of *Drosophila melanogaster.*" (M. Bate and A. Martinez Arias, eds.), vol. 1, pp. 71–147. Cold Spring Harbor Laboratory Press, Plainview, New York.

Furukawa, K., and Hotta, Y. (1993). cDNA cloning of a germ cell–specific lamin B3 from mouse spermatocytes and analysis of its function by ectopic expression in somatic cells. *EMBO J.* **12,** 97–106.

Furukawa, K., Inagaki, H., and Hotta, Y. (1994). Identification and cloning of an mRNA coding for a germ cell–specific A-type lamin in mice. *Exp. Cell Res.* **212,** 426–430.

Furukawa, K., Pante, N., Aebi, U., and Gerace, L. (1995). cDNA cloning of lamina-associated polypeptide 2 (LAP2) and identification of regions that specify targeting to the nuclear envelope. *EMBO J.* **14,** 1626–1636.

Fussell, C. P. (1992). Rabl distribution of interphase and prophase telumeres in *Allium cepa* not altered by colchicine and/or ultracentrifugation. *Am. J. Botany* **79,** 771–777.

Georgatos, S. D., Meier, J., and Simos, G. (1994). Lamins and lamin-associated proteins. *Curr. Opin. Cell Biol.* **6,** 347–353.

Gerace, L. (1986). Nuclear lamina and organization of nuclear architecture. *TIBS* **11,** 443–446.

Gerace, L., and Burke, B. (1988). Functional organization of the nuclear envelope. *Ann. Rev. Cell Biol.* **4,** 335–374.

Gerace, L., and Foisner, R. (1994). Integral membrane proteins and dynamic organization of the nuclear envelope. *Trends Cell Biol.* **4,** 127–131.

Gerace, L., Ottaviano, Y., and Kondor-Koch, C. (1982). Identification of a major polypeptide of the nuclear pore complex. *J. Cell Biol.* **95,** 826–837.

Glass, C. A., Glass, J. R., Taniura, H., Hasel, K. W., Blevitt, J. M., and Gerace, L. (1993). The alpha-helical rod domain of human lamins A and C contains a chromatin binding site. *EMBO J.* **12,** 4413–4424.

Glass, J. R., and Gerace, L. (1990). Lamins A and C bind and assemble at the surface of mitotic chromosomes. *J. Cell. Biol.* **111,** 1047–1058.

Goldberg, M. W., and Allen, T. D. (1995). Structural and functional organization of the nuclear envelope. *Curr. Opin. Cell Biol.* **7,** 301–309.

Goldberg, M., Jenkins, H., Allen, T., Whitfield, W. G. F., and Hutchison, C. J. (1995). Xenopus lamin B3 has a direct role in the assembly of a replication competent nucleus: Evidence from cell-free egg extracts. *J. Cell Sci.* **108,** 3451–3461.

Gorlich, D., and Mattaj, I. W. (1996). Nucleocytoplasmic transport. *Science* **271,** 1513–1518.

Greber, U. F., Senior, A., and Gerace, L. (1990). A major glycoprotein of the nuclear pore complex is a membrane-spanning polypeptide with a large lumenal domain and a small cytoplasmic tail. *EMBO J.* **9,** 1495–1502.

Gruenbaum, Y., Landesman, Y., Drees, B., Bare, J. W., Saumweber, H., Paddy, M. R., Sedat, J. W., Smith, D. E., Benton, B. M., and Fisher, P. A. (1988). *Drosophila* nuclear lamin precursor Dm_0 is translated from either of two developmentally regulated mRNA species apparently encoded by a single gene. *J. Cell Biol.* **106,** 585–596.

Harel, A., Zlotkin, E., Nainudel-Epszteyn, S., Feinstein, N., Fisher, P., and Gruenbaum, Y. (1989). Persistence of major nuclear envelope antigens in an envelope-like structure during mitosis in *Drosophila melanogaster* embryos. *J. Cell Sci.* **94,** 463–470.

Hass, M., and Jost, E. (1993). Functional analysis of phosphorylation sites in human lamin A controlling lamin disassembly, nuclear transport and assembly. *Eur. J. Cell Biol.* **62,** 237–247.

Heald, R., and McKeon, F. (1990). Mutations of phosphorylation sites in lamin A that prevent nuclear lamina disassembly in mitosis. *Cell* **61,** 579–589.

Hennekes, H., and Nigg, E. A. (1994). The role of isoprenylation in membrane attachment of nuclear lamins. *J. Cell Sci.* **107,** 1019–1029.

Hiraoka, Y., Minden, J. S., Swedlow, J. R., Sedat, J. W., and Agard, D. A. (1989). Focal points for chromosome condensation and decondensation revealed by three-dimensional in vivo time-lapse microscopy. *Nature* **342,** 293–296.

Hochstrasser, M., and Sedat, J. W. (1987). Three-dimensional organization of *Drosophila melanogaster* interphase nuclei. II. Chromosome spatial organization and gene regulation. *J. Cell Biol.* **104,** 1471–1483.

Hochstrasser, M., Mathog, D., Gruenbaum, Y., Saumweber, H., and Sedat, J. W. (1986). Spatial organization of chromosomes in the salivary gland nuclei of *Drosophila melanogaster. J. Cell Biol.* **102,** 112–123.

Hoger, T. H., Krohne, G., and Kleinschmidt, J. A. (1991). Interaction of *Xenopus* lamins A and L-II with chromatin in vitro mediated by a sequence element in the carboxy-terminal domain. *Exp. Cell. Res.* **197,** 280–289.

Holtz, D., Tanaka, R. A., Hartwig, J., and McKeon, F. (1989). The CaaX motif of lamin A functions in conjunction with the nuclear localization signal to target assembly to the nuclear envelope. *Cell* **59,** 969–977.

Holy, J. M., Wessel, G., Berg, L., Gregg, R. G., and Schatten, G. (1995). Molecular characterization and expression patterns of a B-type nuclear lamin during sea urchin embryogenesis. *Dev. Biol.* **168,** 464–478.

Hutchison, C. J., Bridger, J. M., Cox, L. S., and Kill, I. R. (1994). Weaving a pattern from disparate threads: Lamin function in nuclear assembly and DNA replication. *J. Cell Sci.* **107,** 3259–3269.

Izaurralde, E., Mirkovitch, J., and Laemmli, U. K. (1988). Interaction of DNA with nuclear scaffolds in vitro. *J. Mol. Biol.* **200,** 111–125.

Jenkins, H., Holman, T., Lyon, C., Lane, B., Stick, R., and Hutchison, C. (1993). Nuclei that lack a lamina accumulate karyophilic proteins and assemble a nuclear matrix. *J. Cell Sci.* **106,** 275–285.

Jenkins, H., Whitfield, W. G. F., Goldberg, M. W., Allen, T. D., and Hutchison, C. J. (1995). Evidence for the direct involvement of lamins in the assembly of a replication competent nucleus. *Acta Biochimica Polonica* **42,** 133–144.

Jongens, T. A., Hay, B., Jan, L. Y., and Jan, Y. N. (1992). The germ cell–less gene product: A posteriorly localized component necessary for germ cell development in *Drosophila. Cell* **70,** 569–584.

Jongens, T. A., Ackerman, L. D., Swedlow, J. R., Jan, L. Y., and Jan, Y. N. (1994). Germ cell–less encodes a cell type–specific nuclear pore–associated protein and functions early in the germ-cell specification pathway of *Drosophila. Genes Dev.* **8,** 2123–2136.

Kitten, G. T., and Nigg, E. A. (1991). The CaaX motif is required for isoprenylation, carboxyl methylation, and nuclear membrane association of lamin B2. *J. Cell Biol.* **113,** 13–23.

Krohne, G., and Benavente, R. (1986). The nuclear lamins, a multigene family of proteins in evolution and differentiation. *Exp. Cell Res.* **162,** 1–10.

Krohne, G., Waizenegger, I., and Hoger, T. H. (1989). The conserved carboxy-terminal cysteine of nuclear lamins is essential for lamin association with the nuclear envelope. *J. Cell Biol.* **109,** 2003–2011.

Kubota, Y., Mimura, S., Nishimoto, S., Takisawa, H., and Nojima, H. (1995). Identification of the yeast MCM3–related protein as a component of the *Xenopus* DNA replication licensing factor. *Cell* **81,** 601–609.

Lehner, C. F., Stick, R., Eppenberger, H. M., and Nigg, E. A. (1987). Differential expression of nuclear lamin proteins during chicken development. *J. Cell Biol.* **105,** 577–587.

Leno, G. H., and Laskey, R. A. (1991). The nuclear membrane determines the timing of DNA replication in *Xenopus* egg extracts. *J. Cell Biol.* **112,** 557–566.

Lin, H., and Wolfner, M. F. (1991). The *Drosophila* maternal-effect gene *fs(1)Ya* encodes a cell cycle–dependent nuclear envelope component required for embryonic mitosis. *Cell* **64,** 49–62.

Lin, L., and Fisher, P. A. (1990). Immunoaffinity purification and functional characterization of interphase and meiotic *Drosophila* nuclear lamin isoforms. *J. Biol. Chem.* **265,** 12596–12601.

Lindsley, D. L., and Tokuyasu, K. T. (1980). Spermatogenesis. *In* "Genetics and Biology of *Drosophila*" (M. Ashburner and T. R. F. Wright, eds.), vol. 2d, pp. 225–294. Academic Press, New York.

Liu, J. (1996). Molecular and genetic analysis of YA, a developmentally regulated nuclear envelope protein of *Drosophila.* PhD dissertation. Cornell University, Ithaca, New York.

Liu, J., and Wolfner, M. F. Functional domains of YA, an essential, developmentally regulated nuclear lamina protein in *Drosophila.* Submitted.

Liu, J., Song, K., and Wolfner, M. F. (1995). Mutational analyses of *fs(1)Ya*, an essential, developmentally regulated, nuclear envelope protein in *Drosophila. Genetics* **141,** 1473–1481.

Liu, J., Lin, H., Lopez, J. M., and Wolfner, M. F. (1997). Formation of the male pronuclear lamina in *Drosophila melanogaster. Devel. Biol.* **184,** 187–196.

Lohka, M. J., and Masui, Y. (1984). Roles of cytosol and cytoplasmic particles in nuclear envelope assembly and sperm pronuclear formation in cell-free preparations from amphibian cells. *J. Cell Biol.* **98,** 1222–1230.

Longo, F. J. (1985). Pronuclear events during fertilization. *In* "Biology of Fertilization" (C. B. Metz and C. B. Monroy, eds.), pp. 251–293. Academic Press, New York.

Longo, F. J., Mathews, L., and Palazzo, R. E. (1994). Sperm nuclear transformation in cytoplasmic extracts from surf clam (*Spisula solidissima*) oocytes. *Dev. Biol.* **162,** 245–258.

Lopez, J. M. (1996). Functional studies of YA, a developmentally regulated nuclear envelope protein that is needed for *Drosophila* embryogenesis. PhD dissertation. Cornell University, Ithaca, New York.

Lopez, J. M., and Wolfner, M. F. (1997). The developmentally regulated *Drosophila* embryonic nuclear lamina protein "Young Arrest" (*fs(1)Ya*) is capable of associating with chromatin. *J. Cell Sci.* **110**, 643–651.

Lopez, J., Song, K., Hirshfeld, A., Lin, H., and Wolfner, M. F. (1994). The *Drosophila fs(1)Ya* protein, which is needed for the first mitotic division, is in the nuclear lamina and in the envelopes of cleavage nuclei, pronuclei and nonmitotic nuclei. *Devel. Biol.* **163**, 202–211.

Lourim, D., and Krohne, G. (1993). Membrane-associated lamins in *Xenopus* egg extracts: Identification of two vesicle populations. *J. Cell Biol.* **123**, 501–512.

Lourim, D., and Krohne, G. (1994). Lamin-dependent nuclear envelope reassembly following mitosis: An argument. *TIBS* **4**, 314–318.

Luderus, M. E. E., de Graaf, A., Mattia, E., den Blaauwen, J. L., Grande, M. A., de Jong, L., and van Driel, R. (1992). Binding of matrix attachment regions to lamin B_1. *Cell* **70**, 949–959.

Lutz, R. J., Trujillo, M. A., Denham, K. S., Wenger, L., and Sinensky, M. (1992). Nucleoplasmic localization of prelamin A: Implications for prenylation-dependent lamin A assembly into the nuclear lamina. *Proc. Natl. Acad. Sci. USA* **89**, 3000–3004.

Madine, M. A., Khoo, C.-Y., Mills, A. D., and Laskey, R. A. (1995). MCM3 complex required for cell cycle regulation of DNA replication in vertebrate cells. *Nature* **375**, 421–424.

Marshall, W. F., Dernburg, A. F., Harmon, B., Agard, D. A., and Sedat, J. W. (1996). Specific interactions of chromatin with the nuclear envelope: Positional determination within the nucleus in *Drosophila melanogaster*. *Mol. Biol. Cell* **7**, 825–842.

Mathog, D., Hochstrasser, M., Gruenbaum, Y., Saumweber, H., and Sedat, J. (1984). Characteristic folding pattern of polytene chromosomes in *Drosophila* salivary gland nuclei. *Nature* **308**, 414–421.

Maul, G. G., French, B. T., and Bechtol, K. B. (1986). Identification and redistribution of lamins during nuclear differentiation in mouse spermatogenesis. *Dev. Biol.* **115**, 68–77.

Maul, G. G., Schatten, G., Jimenez, S. A., and Carrera, A. E. (1987). Detection of nuclear lamin B epitopes in oocyte nuclei from mice, sea urchins, and clams using a human autoimmune serum. *Dev. Biol.* **121**, 368–375.

McKeon, F. (1991). Nuclear lamin proteins: Domains required for nuclear targeting, assembly, and cell-cycle-regulated dynamics. *Curr. Opin. Cell Biol.* **3**, 82–86.

Meier, J., Campbell, K. H. S., Ford, C. C., Stick, R., and Hutchison, C. J. (1991). The role of lamin L_{III} in nuclear assembly and DNA replication, in cell-free extracts of *Xenopus* eggs. *J. Cell Sci.* **98**, 271–279.

Moir, R. D., and Goldman, R. D. (1993). Lamin dynamics. *Curr. Opin. Cell Biol.* **5**, 408–411.

Moir, R. D., Montag-Lowy, M., and Goldman, R. D. (1994). Dynamic properties of nuclear lamins: Lamin B is associated with sites of DNA replication. *J. Cell Biol.* **125**, 1201–1212.

Moir, R. D., Spann, T. P., and Goldman, R. D. (1995). The dynamic properties and possible functions of nuclear lamins. *Int. Rev. Cytol.* **162B**, 141–182.

Morisato, D., and Anderson, K. V. (1995). Signaling pathways that establish the dorsal–ventral pattern of the *Drosophila* embryo. *Ann. Revs. Genet.* **29**, 371–399.

Moss, S. B., Donovan, M. J., and Bellve, A. R. (1987). The occurrence and distribution of lamin proteins during mammalian spermatogenesis and early embryonic development. *In* "Cell Biology of the Testis and Epididymis" (M.-C. Orgebin Crist and B. J. Danzo, eds.), pp. 75–89. Academy of Sciences, New York.

Moss, S. B., Burnham, B. L., and Bellve, A. R. (1993). The differential expression of lamin epitopes during mouse spermatogenesis. *Mol. Rep. Dev.* **34**, 164–174.

Nehrbass, U., Kern, H., Mutvei, A., Horstmann, H., Marshallsay, B., and Hurt, E. C. (1990). NSP1: A yeast nuclear envelope protein localized at the nuclear pores exerts its essential function by its carboxy-terminal domain. *Cell* **61**, 979–989.

Nehrbass, U., Fabre, E., Dihlmann, S., Herth, W., and Hurt, E. C. (1993). Analysis of nucleocytoplasmic transport in a thermosensitive mutant of the nuclear pore protein NSP1. *Eur. J. Cell Biol.* **62**, 1–12.

Newport, J. (1987). Nuclear reconstitution in vitro: Stages of assembly around protein-free DNA. *Cell* **48**, 205–217.

Newport, J. W., Wilson, K. L., and Dunphy, W. G. (1990). A lamin-independent pathway for nuclear envelope assembly. *J. Cell Biol.* **111**, 2247–2259.

Nigg, E. A. (1989). The nuclear envelope. *Curr. Opin. Cell Biol.* **1**, 435–440.

Nigg, E. (1992). Assembly–disassembly of the nuclear lamina. *Curr. Opin. Cell Biol.* **4**, 105–109.

Orr-Weaver, T. L. (1994). Developmental modification of the *Drosophila* cell cycle. *Trends Genet.* **10**, 321–327.

Padan, R., Nainudel-Epszteyn, S., Goitein, R., Fainsod, A., and Gruenbaum, Y. (1990). Isolation and characterization of the *Drosophila* nuclear envelope otefin cDNA. *J. Biol. Chem.* **265**, 7808–7813.

Paddy, M. R., Belmont, A. S., Saumweber, H., Agard, D. A., and Sedat, J. W. (1990). Interphase nuclear envelope lamins form a discontinuous network that interacts with only a fraction of the chromatin in the nuclear periphery. *Cell* **62**, 89–106.

Paddy, M. R., Saumweber, H., Agard, D. A., and Sedat, J. W. (1996). Time-resolved, in vivo studies of mitotic spindle formation and nuclear lamina breakdown in *Drosophila* early embryos. *J. Cell Sci.* **109**, 591–607.

Pante, N., and Aebi, U. (1994). Toward understanding the three-dimensional structure of the nuclear pore complex at the molecular level. *Curr. Opin. Cell Biol.* **4**, 187–196.

Pante, N., and Aebi, U. (1995). Exploring nuclear pore complex structure and function in molecular detail. *J. Cell Sci. Suppl.* **19**, 1–11.

Paulin-Levasseur, M., Blake, D. L., Julien, M., and Rouleau, L. (1996). The MAN antigens are non-lamin constituents of the nuclear lamina in vertebrate cells. *Chromosoma* **104**, 367–379.

Peter, M., and Nigg, E. A. (1991). Ectopic expression of an A-type lamin does not interfere with differentiation of lamin A-negative embryonal carcinoma cells. *J. Cell Sci.* **100**, 589–598.

Peter, M., Nakagawa, J., Doree, M., Labbe, J. C., and Nigg, E. A. (1990). In vitro disassembly of the nuclear lamina and M phase-specific phosphorylation of lamins by cdc2 kinase. *Cell* **61**, 591–602.

Peter, M., Heitlinger, E., Haner, M., Aebi, U., and Nigg, E. A. (1991). Disassembly of in vitro formed lamin head-to-tail polymers by CDC2 kinase. *EMBO J.* **10**, 1535–1544.

Peter, M., Sanghera, J. S., Pelech, S. L., and Nigg, E. A. (1992). Mitogen-activated protein kinases phosphorylate nuclear lamins and display sequence specificity overlapping that of mitotic protein kinases p34cdc2. *Eur. J. Biochem.* **205**, 287–294.

Powell, L., and Burke, B. (1990). Internuclear exchange of an inner nuclear membrane protein (p55) in heterokaryons: In vivo evidence for the interaction of p55 with the nuclear lamina. *J. Cell Biol.* **111**, 2225–2234.

Prather, R. S., and Schatten, G. (1992). Construction of the nuclear matrix at the transition from maternal to zygotic control of development in the mouse: An immunocytochemical study. *Mol. Rep. Dev.* **32**, 203–208.

Pundey, J. (1995). Spermatogenesis in nonmammalian vertebrates. *Microscopy Res. Techniques* **32**, 459–497.

Rabl, C. (1885). Über Zelltheilung. *Morpholog. Jahrbuch* **10**, 214–230.

Riemer, D., and Weber, K. (1994). The organization of the gene for *Drosophila* lamin C: Limited homology with vertebrate lamin genes and lack of homology versus the *Drosophila* lamin Dm_o gene. *Euro. J. Cell. Biol.* **63**, 299–306.

Riemer, D., Stuurman, N., Berrios, M., Hunter, C., Fisher, P. A., and Weber, K. (1995). Expression of *Drosophila* lamin C is developmentally regulated: Analogies with vertebrate A-type lamins. *J. Cell Sci.* **108**, 3189–3198.

Riparbelli, M. G., and Callaini, G. (1992). Distribution of a nuclear envelope antigen during the syncytial mitosis of the early *Drosophila* embryo as revealed by laser scanning confocal microscopy. *J. Cell Sci.* **102**, 299–305.

Rober, R. A., Weber, K., and Osborn, M. (1989). Differential timing of nuclear lamin A/C expression in the various organs of the mouse embryo and the young animal: A developmental study. *Development* **105,** 365–378.

Roth, S. Y., and Allis, C. D. (1992). Chromatin condensation: Does histone H1 dephosphorylation play a role? *TIBS* **17,** 93–98.

Schatten, G., and Schatten, H. (1987). Cytoskeletal alterations and nuclear architectural changes during mammalian fertilization. *Curr. Topics Dev. Biol.* **23,** 23–54.

Schatten, G., Maul, G. G., Schatten, H., Chaaly, N., Simerly, C., Balczon, R., and Brown, D. L. (1985). Nuclear lamins and peripheral nuclear antigens during fertilization and embryogenesis in mice and sea urchin. *Proc. Natl. Acad. Sci. USA* **82,** 4727–4731.

Schlenstedt, G., Hurt, E. C., Doye, V., and Silver, P. (1993). Reconstitution of nuclear pore protein transport with semi-intact yeast cells. *J. Cell Biol.* **123,** 785–798.

Senior, A., and Gerace, L. (1988). Integral membrane proteins specific to the inner nuclear membrane and associated with the nuclear lamina. *J. Cell Biol.* **107,** 2029–2036.

Sheehan, M. A., Mills, A. D., Sleeman, A. M., Laskey, R. A., and Blow, J. J. (1988). Steps in the assembly of replication-competent nuclei in a cell-free system from *Xenopus* eggs. *J. Cell Biol.* **106,** 1–12.

Smith, A., and Benavente, R. (1992). Identification of a structural protein component of rat synaptonemal complexes. *Exp. Cell Res.* **198,** 291–297.

Smith, A. V., and Orr-Weaver, T. L. (1991). The regulation of the cell cycle during *Drosophila* embryogenesis: The transition of polyteny. *Development* **112,** 997–1008.

Smith, D. E., and Fisher, P. A. (1989). Interconversion of *Drosophila* nuclear lamin isoforms during oogenesis, early embryogenesis, and upon entry of cultured cells into mitosis. *J. Cell. Biol.* **108,** 225–265.

Smith, D. E., Gruenbaum, Y., Berrios, M., and Fisher, P. A. (1987). Biosynthesis and interconversion of *Drosophila* nuclear lamin isoforms during normal growth and in response to heat shock. *J. Cell Biol.* **105,** 771–790.

Song, K. (1994). Developmental, genetic, and biochemical studies of *fs(1)Ya*, a nuclear envelope protein required for embryonic mitosis in *Drosophila*. PhD dissertation. Cornell University, Ithaca, New York.

Stafstrom, J. P., and Staehelin, L. A. (1984). Dynamics of the nuclear envelope and of nuclear pore complexes during mitosis in the *Drosophila* embryo. *Euro. J. Cell Biol.* **34,** 179–189.

Stewart, C., and Burke, B. (1987). Teratocarcinoma stem cells and early mouse embryos contain only a single major lamin polypeptide closely resembling lamin B. *Cell* **51,** 383–392.

Stick, R. (1988). cDNA cloning of the developmentally regulated lamin L_{III} of *Xenopus laevis*. *EMBO J.* **7,** 3189–3197.

Stick, R., and Hausen, P. (1985). Changes in the nuclear lamina composition during early development of *Xenopus laevis*. *Cell* **41,** 191–200.

Stick, R., and Schwarz, H. (1982). The disappearance of the nuclear lamina during spermatogenesis: An electron microscopic and immunofluorescence study. *Cell Differentiation* **11,** 235–243.

Stick, R., and Schwarz, H. (1983). Disappearance and reformation of the nuclear lamina structure during specific stages of meiosis in oocytes. *Cell* **33,** 949–958.

Stricker, S., Prather, R., Simerly, C., Schatten, H., and Schatten, G. (1989). Nuclear architecture changes during fertilization and development. *In* "The Cell Biology of Fertilization" (H. Schatten and G. Schatten, eds.), pp. 225–250. Academic Press, New York.

Stuurman, N., Maus, N., and Fisher, P. A. (1995). Interphase phosphorylation of the *Drosophila* nuclear lamin: Site-mapping using a monoclonal antibody. *J. Cell Sci.* **108,** 3137–3144.

Sudhakar, L., and Rao, M. R. (1990). Stage-dependent changes in localization of a germ cell–specific lamin during mammalian spermatogenesis. *J. Biol. Chem.* **265,** 22526–22532.

Sudhakar, L., Sivakumar, N., Behal, A., and Rao, M. R. S. (1992). Evolutionary conservation of a germ cell–specific lamin persisting through mammalian spermiogenesis. *Exp. Cell Res.* **198,** 78–84.

Sukegawa, J., and Blobel, G. (1993). A nuclear pore complex protein that contains zinc finger motifs, binds DNA, and faces the nucleoplasm. *Cell* **72**, 29–38.

Sweet, D. J., and Gerace, L. (1995). Taking from the cytoplasm and giving to the pore: Soluble transport factors in nuclear protein import. *Trends Cell Biol.* **5**, 444–447.

Taniura, H., Glass, C., and Gerace, L. (1995). A chromatin binding site in the tail domain of nuclear lamins that interacts with core histones. *J. Cell Biol.* **131**, 33–44.

Tye, B.-K. (1994). The MCM2-3-5 proteins: Are they replication licensing factors? *Trends Cell Biol.* **4**, 160–166.

Ulitzur, N., and Gruenbaum, Y. (1989). Nuclear envelope assembly around sperm chromatin in cell-free preparations from *Drosophila* embryos. *FEBS Lett.* **259**, 113–116.

Ulitzur, N., Harel, A., Feinstein, N., and Gruenbaum, Y. (1992). Lamin activity is essential for nuclear envelope assembly in a Drosophila embryo cell-free extract. *J. Cell Biol.* **119**, 17–25.

Vester, B., Smith, A., Krohne, G., and Benavente, R. (1993). Presence of a nuclear lamina in pachytene spermatocytes of the rat. *J. Cell Sci.* **104**, 557–563.

Ward, G. E., and Kirschner, M. (1990). Identification of cell cycle–regulated phosphorylation sites on nuclear lamin C. *Cell* **61**, 561–578.

Wente, S. R., and Blobel, G. (1993). A temperature-sensitive NUP116 null mutant forms a nuclear envelope seal over the yeast nuclear pore complex thereby blocking nucleocytoplasmic traffic. *J. Cell Biol.* **123**, 275–284.

Wilson, K. L., and Newport, J. (1988). A trypsin-sensitive receptor on membrane vesicles is required for nuclear envelope formation in vitro. *J. Cell Biol.* **107**, 57–68.

Wimmer, C., Doye, V., Grandi, P., Nehrbass, U., and Hurt, E. (1992). A new subclass of nucleoporins that functionally interacts with nuclear pore protein NSP1. *EMBO J.* **11**, 5051–5061.

Winey, M., and Byers, B. (1992). Spindle pole body of *Sacchromyces cerevisiae*: A model for genetic analysis of the centrosome cycle. *In* "The Centrosome" (V. I. Kalnins, ed.) pp. 197–218. Academic Press, San Diego, CA.

Winey, M., Hoyt, M. A., Chan, C., Goetsch, L., Botstein, D., and Byers, B. (1993). NDC1, a nuclear periphery component required for yeast spindle pole body duplication. *J. Cell Biol.* **122**, 743–751.

Worman, H. J., Yuan, J., Blobel, G., and Georgatos, S. D. (1988). A lamin B receptor in the nuclear envelope. *Proc. Natl. Acad. Sci. USA* **85**, 8531–8534.

Yamashita, M., Onozato, H., Nakanishi, T., and Nagahama, Y. (1990). Breakdown of the sperm nuclear envelope is a prerequisite for male pronucleus formation: Direct evidence from the gynogenetic crucian carp *Carassius auratus laangsdorfii*. *Dev. Biol.* **137**, 155–160.

Ye, Q., and Worman, H. J. (1994). Primary structure analysis and lamin B and DNA binding of human LBR, an integral protein of the nuclear envelope inner membrane. *J. Biol. Chem.* **269**, 11306–11311.

Yokoyama, N., Hayashi, N., Seki, T., Pante, N., Ohba, T., Nishii, K., Kuma, K., Hayashida, T., Miyata, T., and Aebi, U. (1995). RanBP2, a giant nucleopore protein that binds Ran-TC4. *Nature* **376**, 184–188.

Yuan, J., Simos, G., Blobel, G., and Georgatos, S. D. (1991). Binding of lamin A to polynucleosomes. *J. Biol. Chem.* **266**, 9211–921⁵

Zhao, K., Harel, A., Stuurman, N., Guedalia, D., and Gruenbaum, Y. (1996). Binding of matrix attachment regions to nuclear lamin is mediated by the rod domain and depends on the lamin polymerization state. *FEBS Let.* **380**, 161–164.

3

The EGFR Gene Family in Embryonic Cell Activities

Eileen D. Adamson
The Burnham Institute
La Jolla Cancer Research Center
La Jolla, California 92037

Lynn M. Wiley
Department of Obstetrics and Gynecology
University of California
Davis, California 95616

Current Topics in Developmental Biology, Vol. 35
Copyright © 1997 by Academic Press. All rights of reproduction in any form reserved.
0070-2153/97 $25.00

I. Introduction

The history of the epidermal growth factor receptor (EGFR) and ligand gene families is long by modern scientific standards. Recent work has added a number of receptor family members, and new ligands are being isolated that have distinct activities. Progress in understanding the roles of this group of genes has been relatively slow, until recently. A plethora of gene ablations in mice (knockouts) and a growing list of transgenic mice, joined with invertebrate findings and biochemical characterizations, have made a strong impact on this field. It is our intention to bring together the major new findings that have recently enriched our knowledge of the developmental roles and activities of the EGFR, or ErbB, family. The aim is to assess the strengths and weaknesses of the new work and to try to formulate hypotheses for further testing and to identify gaps in our understanding that need attention.

The major developmental areas that have been impacted recently are mammalian pre- and periimplantation and organogenesis and the roles played by the ErbB family of genes and their ligands in these processes. We will refer to the first member as *EGFR* and to the other *ErbBs* as the -2, -3, and -4 genes. The aberration of gene expression within this family has led to alterations of growth regulation and loss of differentiation processes. Therefore, the discussion will include cancerous processes and how they may reveal mechanisms for growth control in normal development. A large part of recent advances has involved the area of receptor crosstalk both within the ErbB family and with other receptor proteins. The ability of the receptors to communicate with the cytoskeleton, with other membrane proteins, with the cytosol, and with the nucleus has been recognized. The ErbB receptors comprise a major window for the cell to "see" outside stimuli and to communicate, both inwardly and outwardly, the cellular responses.

The EGFR may be one of the most influential of all receptor proteins because it is at the crux of several signaling pathways. It is activated in response to its growth factor ligands in a spatially and temporally appropriate manner during development to stimulate either proliferation, cell death, cell survival, cell locomotion, cell polarization, or differentiation, and it can share some of these responsibilities, if necessary, with its family members. The EGFR is activated in response to environmental stresses such as irradiation and hydrogen peroxide and generates signals that can lead to protective responses. EGFR is involved in activities vital to life, such as stem cell renewal and absorption of nutriments. EGFR can interact with all its family members, giving graded responses to individual ligands and hence specificity of responses. This family of receptor gene products can also autostimulate each other's synthesis. This last property may be at the basis of the frequency of EGFR overexpression observed in tumors of many kinds. It is not an exaggeration to state that the EGFR stands at the gateway of the normal and abnormal growth properties of cells. Therefore, understanding its activities remains an important field of study.

II. The ErbB Family of Genes

A. The Epidermal Growth Factor Receptor (EGFR, ErbB1, HER-1)

The developmental aspects of EGFR were reviewed last in this series in 1990 (Adamson, 1990. See also Adamson *et al.*, 1991; Carpenter, 1993). EGFR was the first member of the family detected, purified, and characterized (Cohen *et al.*, 1982; Downward *et al.*, 1984). Some characteristics of the ErbB genes and their products are shown in Table 1. They are all transmembrane kinases highly related to EGFR and able to bind to a variety of ligands.

The mature protein is observed in immunoprecipitates as a glycopolypeptide of Mr 170 kDa, with a smaller precursor at Mr 160 kDa and a proteolytic fragment of Mr 150 kDa. The extracellular cysteine-rich domain is able to bind several different ligands that have the characteristic EGF-like triple loops formed by six highly conserved cysteines. The membrane-spanning domain of EGFR precedes the intracellular tyrosine kinase domain and the carboxy-terminal cytoplasmic domain. Five phosphorylated tyrosines in the latter domain provide docking stations for signaling components. Ligand binding to the receptor starts a process of dimerization and higher-order oligomerization, giving high-affinity ligand binding sites. Tyrosine kinase activation leads to signal transduction to the nucleus, with subsequent activation and repression of characteristic sets of genes. The signal is generated by the tyrosine-phosphorylated receptor protein, forming complexes containing Shc, Sos, Grb2, PLCγ, c-src, and other components. Each connects to other intermediaries to give the pleiotropic responses typical of growth factor stimulation of a cell. Signals from platelet-derived growth factor receptor (PDGFR) follow a similar pathway but the responses are different because they occur in different cells and have different target genes. Responses include cell surface ruffling, shape changes, migration of the cell, and these cytoplasmic effects are integrated by or with the actin cytoskeleton. The signal is transduced by c-ras, c-raf-1, and MAPK (Burgering and Bos, 1995; Schlessinger and Bar-Sagi, 1994) and reaches the nucleus to produce many different effects on gene promoters. One of the direct effectors that activates genes is serum response factor (SRF), which becomes phosphorylated and forms the ternary activation complex binding at and inducing the serum response elements (SRE) in serum reponsive genes (Shaw *et al.*, 1989). Another mechanism for activation by EGF involves an additional activator, SIF, that binds to the sis-inducible element, or SIE, located close to the SRE of the c-fos promoter (Sadowski and Gilman, 1993). The first effects are, therefore, on the "immediate early" genes, the *fos* family, the *jun* family, and the *Egr* family of genes and eventually about a hundred other later genes. Soon, newly synthesized AP-1 (formed from dimers of the fos and jun families) can stimulate the induction of the early and late early genes, such as c-*myc*, which regulates cell cycle events. The immediate early transcription factors regulate a vast number of genes that lead to cellular re-

Table 1 Characteristics of the ErbB Genes

Character gene	Chromosome		mRNA	Protein size (kDa)	Activity (binding)	Location of gene product	Gene regulation
	Human	Mouse					
EGFR	7p11-13	11	10.5 kb 5.8 2.8	170 Pre 160	Grb2,PLCγ	Neurons, glia, skin, gut, kidney, liver, most tissues	Up by estrogen, EGF via Sp1, Ap2 Down by RA
ErbB2	17p11-q21	?	4.6 kb	185	GAP, PLCγ, β-catenin, src	Neurons, glia, heart, pancreas, kidney, placenta, skin	Up by Sp1, Ap2 Down by RA, T3
ErbB3	5	?	6.2 kb	160	PI3K^{p85}, Grb2, GAP, CSK, src	Brain, glia, skin, bronchi, adrenal cortex, ovary	Shorter half-life than EGFR Up by Ap2
ErbB4	2q33.3-34	?	?	180	Grb2, PLCγ	Neurons and glia, heart	?

A hierarchy of ErbB dimerized proteins for induction of proliferation is 3 + 2 > 4 + 2 > 1 + 2 > 1 + 1 > 1 + 3. From Pinkas-Kramarski *et al.* (1996 a and b).

sponses that are graded to the signal type, signal intensity, and cell type. For a review of signal transduction from TK receptors, see Ullrich and Schlessinger (1990).

B. The ErbB Family Members

The ErbB2 (neu, HER-2) receptor (Hung *et al.*, 1986) is distinguished because it has no specific ligand that binds to activate the tyrosine kinase. Instead, ErbB2 kinase is activated by dimerization with other members of the receptor family. Neu was first isolated from a rat neuroblastoma, where it was constitutively activated by a single amino acid change in the transmembrane domain (Bargmann *et al.*, 1986; Schechter *et al.*, 1984). Constitutive kinase activity is the basis for the tumorigenic properties of cells that express this activated form of ErbB2. Since then, this oncogene has been shown to be overexpressed in many types of tumors, including mammary, ovary, and prostate. Both EGFR and ErbB2 are considered to be protooncogenes (see section VII).

The ErbB3 receptor gene was isolated and cloned recently (Lemoine *et al.*, 1992). It is distinguishable from the other members of the group by having little or no intrinsic kinase activity; but upon dimerization, it synergizes with its partner to generate a signal in ligand bound cells. The receptor is also unique in the family in being able to recruit the enzyme PI3K to an SH2 docking sequence. ErbB3 can bind ligands of the heregulin/neuregulin type but not the EGF analogs.

The ErbB4 receptor (Plowman *et al.*, 1993a and b) has an active kinase domain and can form homo- or heterodimers with the other members of the family. It has a wide distribution in tissues, with great importance in the heart and brain. The neuregulins are the ligands that bind this receptor to activate the kinase domain. Only betacellulin of the EGF analogs can bind to ErbB4.

C. The EGF-like Ligands

The EGF-like domain has a wide occurrence in growth factors and in nonligand molecules. The three-loop structure that is formed by six conserved cysteines is able to bind to the EGFR only if certain spatial rules are obeyed. For instance, a growth factor that is incorrectly grouped in this family, Cripto-1(Cr-1), cannot bind to the EGFR-related receptors because it lacks one loop and has a truncated second loop in the EGF-like domain. This growth factor may have a receptor that is in the FGFR family of genes (Kinoshita *et al.*, 1995). True members so far isolated and characterized as EGFR-binding ligands (Table 2) are EGF (Cohen and Cohen, 1983); transforming growth factor-alpha (TGFα) (Shin *et al.*, 1994), amphiregulin (Ar) (Plowman *et al.*, 1990; Shoyab *et al.*, 1988, 1989), heparin-binding EGF (HB-EGF) (Besner *et al.*, 1992; Higashiyama *et al.*, 1993), beta-cellulin (BTC) (Barnard *et al.*, 1993; Sasada *et al.*, 1993; Seno *et al.*, 1996), and epiregulin (Toyoda *et al.*, 1995a and b). Several ligands that bind to the ErbB family of receptors are now called the neuregulins (NRG) or heregulins (HRG) and are all homologs of a gene with alternate spliced products (Adelaide *et al.*, 1994; Pinkas-Kramarski *et al.*, 1994). The ligands were named according to their activities or sources and include neural differentiation factor (NDF), acetylcholine receptor-inducing activity (ARIA), glial growth factor (GDGF), and ciliary neurotrophic factor (CNTF).

All of these ligands are produced as a larger precursor molecule that has a transmembrane domain and an intracellular domain. In most cases, the mechanism of the release of the growth factor/EGF domain is not known, but in the case of TGFα the precursor has been shown to have mitogenic activities in the membrane-bound form (Brachman *et al.*, 1989; Wong *et al.*, 1989). Betacellulin is known to bind to the EGFR, but in addition can bind directly to and activate ErbB4, the only ligand so far to have this dual property (Riese *et al.*, 1996). All the ligands have strong mitogenic activity on a variety of cell types. Each has slightly different concentration maxima based on its affinity for the receptor and the number of the receptors. The cell response differs according to the number of receptors and the interacting and signaling components present in each specific cell type. For example, in mammary tumor cells, low levels of ligand can stimulate cell proliferation, whereas high levels stimulate their differentiation.

Several features of HB-EGF and amphiregulin (Ar) make them distinct from their family members. For instance, they are glycoproteins that can bind to heparin and heparin sulfate proteoglycans (HSPG) and become incorporated into matrix. Both require HSPG for maximal presentation to the EGFR and maximal effect (Aviezer and Yayon, 1994). Both are known to play important roles in angiogenesis, an important and critical feature of tumorigenesis as well as normal embryogenesis. Ar also has two nuclear localization signals, and has been detected in nuclei and nucleoli, as well as the cytoplasm (Ebert *et al.*, 1994; G. Johnson *et al.*, 1991). This domain may be important for eliciting a mitogenic signal (G. Johnson and Wong, 1994; Piepkorn *et al.*, 1994). Ar and several of the ErbB proteins have been detected in the nucleus, but a role there is unclear. The activity of amphiregulin is reduced by the presence of heparin, a diagnostic feature of its presence in culture media. It is possible that some reports of differential activity (usually resulting in reports of low activity of Ar compared to its relatives) of Ar in culture systems are misleading, in view of the necessity for the amino-terminal (NT) nuclear localization region, the precursor region, and the extension of the mature active polypeptide to at least 87 aa in order to achieve maximum activity (Adam *et al.*, 1995; S. Thompson *et al.*, 1996; Thorne and Plowman, 1994). The origin of the name *amphiregulin* is notable, since Ar appeared to stimulate the proliferation of some normal cells but inhibit the growth of other cells. This needs reinvestigation (Adam *et al.*, 1996), because the

Table 2 The ErbB Ligands

Character gene	Chromosome Human	Mouse	mRNA (kb)	Protein (kDa)	Location
EGF	4q25	3	5	165/154/130 kDa mature 53aa, 6kDa	Saliv. gl., kidney, teeth, lung, uterus
TGFα	2p13	6	4.5, 1.4	25 kDa (160aa) prepro form, 18 to 6 kDa secreted (mature 50aa)	Ant. pituitary, mammary, brain
Amphiregulin	4q13-21	5	1.4	243/87aa mature 20 kDa	Skin, placenta, mammary, ovary, uterus
HB-EGF	5q21	18	2.5	160/27/14–19 kDa	Macrophages, smooth muscle, uterus
Betacellulin	4q13-21	5	?	177/80aa 32/18 kDa	Kidney, liver, pancreas, gut, mammary CA, insulinoma
Epiregulin	?	?	4.8	162aa/46aa	Embryo fibroblasts, 7-d embryo
Heregulins/ neuregulins	8p12-22	?	10 isoforms, by alt. splicing	45 kDa/30 kDa	Neurons, muscle, heart

form first isolated and sequenced was that secreted after phorbol ester treatment of human mammary carcinoma cell line, MCF7, an 80 aa form that has lower activity than carboxy-extended forms.

The heregulins/neuregulins are ligands for the ErbB2, -3, and -4 polypeptides. They cannot bind to EGFR directly, but can activate it as a heterodimer with another ErbB receptor. It is largely accepted that the EGFR and its related receptors are active only as dimers and possibly as higher oligomers. It is notable that the 10 or so active mitogens that were isolated from tissues and cell lines and, that had activity in brain, glia, and muscle cells turned out to be related by differential splicing of one gene, the heregulin gene. The different gene products from this gene have different target cells, but they concentrate in the electrically excitable tissues: skeletal, cardiac and smooth muscle, brain, and peripheral nerves.

Many of the members of the ErbB family of genes and their ligands also are expressed in the placenta, in the uterus, and in the ovary, indicating a strong role in the processes of development and reproduction.

III. The Oocyte and Preimplantation Embryo

A. The Oocyte and Its Attendant Cumulus Cells Express Active EGFR

Successful development of the female gamete into a developmentally competent oocyte requires reciprocal interactions with (and the concomitant maturation of) its surrounding cumulus cells. EGFR plays a role in these interactions and is present on oocytes and cumulus cells in the pig oocyte (Singh *et al.*, 1995), human oocyte (Maruo *et al.*, 1993), and mouse oocyte (Wiley *et al.*, 1992), and added EGF stimulates meiotic maturation of the oocyte by porcine cumulus-enclosed oocytes (Coskun and Lin, 1995) and both oocyte maturation and cumulus cell expansion in human (Gomez *et al.*, 1993), mouse (Boland and Gosden, 1994; Eppig and O'Brien, 1996), and bovine (Lorenzo *et al.*, 1994) cultured cumulus–oocyte complexes. Whereas the oocyte promotes cumulus expansion by secreting an extracellular cumulus "expansion enabling factor" (Eppig *et al.*, 1993a, 1993b), meiotic arrest of the oocyte appears to result from purines passing from the cumulus cells through connecting gap junctions to the oocyte (reviewed by Eppig, 1991). EGF can overcome meiotic arrest of cumulus-enclosed rat (Dekel and Sherizly, 1985) and mouse oocytes (Downs, 1989). Because EGF can decrease gap junction communication between cells (Lau *et al.*, 1992; Xie and Hu, 1994), it is possible that EGF's capacity to overrule meiotic arrest results from its ability to reduce gap junction communication and, consequently, the associated passage of cumulus cell-derived purines into the oocyte (see also Endo *et al.*, 1992). Consistent with this idea is the observation that the mouse oocyte can spontaneously resume meiosis when removed from its cumulus cells and their supply of purines (Chesnel *et al.*, 1994).

Some intriguing studies illustrate the conservation of molecular function in studies utilizing the frog *Xenopus laevis* oocytes. EGFR (but not p185neu) mRNA microinjected into the oocytes promote maturation and germinal vesicle breakdown (GVBD) upon addition of EGF. The kinase domain of the receptor is required for this activity (Narasimhan *et al.*, 1992). Collectively, these observations agree with the hypothesis that EGFR in the oocyte is primarily responsible for the effects of added EGF on oocyte maturation and, perhaps, also on oocyte-dependent cumulus expansion. EGFR in the follicular cells, on the other hand, may be more involved in steroidogenesis and in the response of follicular cells to gonadotropins (Hubbard, 1994; Luciano *et al.*, 1994; Tekpetey *et al.*, 1995). It is not yet certain whether oocytes and follicular cells, including cumulus cells, from any species express only EGFR or whether other ErbB receptors are present. Furthermore, whether EGF and/or one of the EGF-like ligands for EGFR is/are the physiological ligand *in vivo* for successful oocyte development is not yet certain. Future studies should clarify these uncertainties.

B. EGFR and Its Ligands Are Active during Preimplantation Development

The newly created zygote has two critical missions to accomplish before implantation. These are to achieve the capacity to implant, and to achieve the capacity to form a fetus. These missions are performed by the two cell lineages that are established during preimplantation development, the trophectoderm, which initiates implantation and development of the fetus, and inner cell mass (ICM), which initiates embryogenesis and development of the fetus. The establishment of these two cell lineages requires cell proliferation and cell differentiation, two of the known consequences of EGFR activation (see the review Wiley et al., 1995). Both cell proliferation and cell differentiation processes are stimulated by EGFR ligands, including EGF, TGF-α, and amphiregulin (Paria and Dey, 1990; Dardik and Schultz, 1991; Paria et al., 1991: Brice et al., 1993; Rappolee et al., 1992; Tsark et al., 1997). Since the embryos express both EGFR and some ligands in mouse (Rappolee et al., 1988; Tsark et al., 1997) and human (Chia et al., 1995), it is highly probable that expression of both EGFR and ligand contribute to the notable autonomy of mammalian preimplantation embryos in culture.

As is the case for oocyte EGFR, it has not yet been determined whether one or multiple ErbB genes are expressed during preimplantation development and whether the different EGFR ligands play different or similar roles in vivo. Gene ablations suggest that the different ligands can offset each other's absence and that the other growth factor receptors expressed by preimplantation embryos can partially offset the absence of embryonic EGFR. In vitro culture of embryos is always retarded compared to the in vivo process, and may mimic some aspects of embryo development with ablated genes. Under these less favorable conditions, functional EGFR and the corresponding ligands are required for optimal embryonic growth and differentiation of the zygote into a blastocyst that is capable of initiating implantation and embryogenesis.

C. Possible Roles for the EGFR and Its Ligands in Implantation of the Blastocyst

The mammalian embryo arrives in the uterus at the early blastocyst stage. The uterine fluid contains maternal EGF to increase the survival of cells and to aid in expansion of the blastocoel cavity and hatching of the blastocyst out of the zona pellucida (Dardik and Schultz, 1991). At this stage, programmed cell death, or apoptosis, is rarely seen; but in the late blastocyst, it may be the principal method to adjust the number of inner cell mass (ICM) cells to the permitted optimal level. At around this stage, embryos that have been aggregated to form giant blastocysts, when transferred into a pseudopregnant uterus, adjust to normal size before further development. The mechanism for this is unknown, but the survival of cells is normally influenced by the expression of EGFR on the trophoblast and ICM cells; the autocrine production of TGFα and Ar could be acting on the receptors to evoke this response.

The EGFRs are observed to be divided between the outer trophectoderm cell membranes and the inner membranes facing the blastocoelic cavity (Dardik et al., 1992) and on the ICM cells. The outer-facing receptors are thought to be important in the process of positioning and adhesion of the blastocyst to the maternal uterine wall. The model presented by Dey and co-workers is that the EGFRs on the embryo are used for implantation of the blastocyst on the uterine epithelium via the pro-form of the ligand inserted into the apical membrane of the epithelium. This group of researchers has made a special focus on the uterus during the periimplantation period, in order to define the roles of EGFR and ligands in the implantation process (Das *et al.*, 1994a; Paria *et al.*, 1993). They have established that few if any full-length EGFRs are presented on the uterine epithelium, although the production of a truncated form of the EGFRs has been detected as a mRNA species (Tong *et al.*, 1996). If the truncated form is translated and secreted, it could serve to bind and remove soluble EGF ligands. The truncated form might also bind to membrane-inserted EGF-like ligands in the uterine epithelium and could explain why other studies reported finding [^{125}I]EGF binding sites on the pregnant uterine epithelium (Brown *et al.*, 1989).

Since there are EGFRs on the blastocyst and apparently none on the uterine wall, are there ligands on the uterine epithelium to bind and activate the embryo EGFR? In a series of elegant papers, the Dey group has used the delayed implantation model to determine the time of the appearance and disappearance of molecules that might regulate the process of implantation in response to estrogen and progesterone. Ovariectomized females cannot implant embryos, but delayed implantation maintained by progesterone (P4) can allow the blastocysts to remain viable and to grow in size in the uterus without attaching. The delayed blastocyst down-regulates its EGFRs, an important characteristic in establishing an EGFR-dependent mechanism for implantation. Restoration of estrogen (E2) together with P4 allows reappearance of the EGFR, implantation of the delayed blastocyst, and decidualization of the uterine stroma. Interestingly, there is some indication that EGF can replace E2 under certain conditions, a finding that indicates that EGF-like ligands act downstream of the signaling process. Embryo implantation in the hypophysectomized delayed-implantation pregnant rat can be stimulated merely by puncturing the uterus and injecting intrauterine EGF or even by IV injection of EGF (Johnson and Chatterjee, 1993). The mechanism is thought to be via production of prostaglandins in the uterus, but this would not account for the location of the implantation site, a process that appears to involve directly at least two EGF ligands, as described next.

During the process of implantation, Ar is found in progesterone-regulated sites on the uterine epithelium only on day 4 of pregnancy and only at implantation sites. The implication is that membrane-bound Ar molecules could be the adhesive sites for attachment of embryo EGFR (Das *et al.*, 1995; Das *et al.*, 1994a). A similar window of production of membrane-bound HB-EGF occurs at 6–7 h before attachment at 22.00 to 23.00 h on day 4 of pregnancy in mouse (Das *et al.*, 1994b; Wang *et al.*, 1994). From the carefully measured times of expression, the case is strong that both these ligands could play a role as adhesion receptors for the blastocyst. In particular, the tight localization of the expression of HB-EGF on the uterine wall precisely defines the site of implantation (Raab *et al.*, 1996). What decides the placement of the implantation site is still not accounted for. Is it regulated by anatomical features such as the presence of an arteriole? The activities and adhesivity of both ligands may be increased by the heparin sulfate proteoglycans (HSPG) that is also present on the trophectoderm surface. The angiogenic properties of HB-EGF and HSPG could also play a role in building up blood flow to the area. An *in vitro* system verified that HB-EGF expressed on cells could act as a substrate for blastocyst attachment (Raab *et al.*, 1996). Adhesion of blastocysts to cells that overexpress transmembrane HB-EGF is inhibited by TGFα, while delayed blastocysts on which EGFRs are down-regulated do not attach. Treatment of blastocysts with heparinase or a peptide that corresponds to the heparin-binding domain also inhibits attachment and indicates that more than one mechanism for attachment of the blastocyst to the uterine wall is likely to be in operation.

The close apposition of the ligands in the uterus and the receptors on the blastocyst may signal to increase metabolic activity in both directions. The uterus initiates the rapidly progressive decidualization process, the response of the uterine epithelium and stroma to the attachment of the embryo to create a vascular site for nutrient supplies. The ICM proliferates into the egg cylinder, and the endoreduplication of trophoblast DNA produces giant cells that secrete gonadotropins and tissue-type plasminogen activator (tPA) (Medcalf and Schleuning, 1991; Sappino *et al.*, 1989), urokinase plasminogen activator (Harvey *et al.*, 1995), and other proteinases that contribute to successful implantation. This process is stimulated by EGF-like ligands that are produced in increasing abundance in the uterine stroma and decidua (Han *et al.*, 1987; Lysiak *et al.*, 1995). The proteinases act to increase the invasive activity of embryonic cells into the uterine stroma and so form the future placenta. Decidualization is also the signal for maternal production of EGF ligands, which can act in placental development or may diffuse to increase the survival of the developing embryo. The arguments just outlined indicate the involvement of the EGFR at several steps in the process of mammalian embryo implantation, and they are summarized in Fig. 1.

The implantation process is one of the most vulnerable in mammalian fertility and reproduction. At this stage most reproductive losses occur without the knowledge of the mother. The process is well protected to an exact degree so that the adaptive and plastic embryo will survive to develop if at all possible. The Dey group has shown that the expression of TGFα (Paria *et al.*, 1994) as well as HB-EGF and Ar on the apical surface of the luminal epithelial cells is also regulated by the steroid hormones in an implantation-dependent manner (Tamada *et al.*, 1991). This abundance (redundancy) of the EGF ligands in the pregnant uterus is surely an indication of the importance of the interactions between the EGFR and its ligands at this stage in gestation. The expression of a number of genes that have similar functions is suggestive that a vital program must be sustained.

The outcome of experiments to abrogate the expression of the EGFR gene

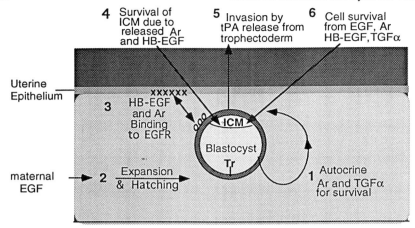

Fig. 1 EGFR and implantation. (1) Ar and TGFα are produced by the embryo and act to increase rate of development and survival (Tsark *et al.*, 1997); (2) EGF/HB-EGF stimulate blastocyst expansion and hatching *in vitro* (Dardik and Schultz, 1993; Das *et al.*, 1994); (3) ADHESION mode of action: EGFR on trophectoderm surface bind to HB-EGF and Ar at precise sites on day 4 uterine epithelium to initiate implantation (Das *et al.*, 1994, 1995); (4) Only blastocysts at the precise sites are close enough to receive EGF, Ar, and HB-EGF from the uterus, needed for ICM survival; (5) Trophectoderm cells bound via EGFR to uterine epithelium are stimulated to secrete proteinases to allow invasion into the uterine stroma (Bass *et al.*, 1994; Cross *et al.*, 1994); (6) Invasion of the trophectoderm cells allows them to reach the decidual cells that produce TGFα. This stimulates the growth of primitive ectoderm cells.

provides evidence that a partnership between EGFR and its ligands is required at the precisely correct level. In certain strains of mouse, such as the outbred CF1 mouse, the process of implantation has a small window of success. And in the mouse model of loss of EGFR by gene targeting in this strain, the majority of embryos die at this stage (Threadgill *et al.*, 1995). Other strains presumably have sufficient levels of other compensatory molecules that allow survival until the next crisis point. In CF-1 EGFR($-/-$) mice, the uterus undergoes decidualization but the ICM cells do not survive, while the trophoblast cells do not get further than the formation of giant cells and also show abnormal adhesivity. In this mouse strain, presumably the embryo without EGFRs is able to make alternate attachment to the uterus, such as the increased production of HSPG deposited on the surface of the blastocyst that will still be able to attach to HB-EGF and Ar anchored on the uterine wall. Decidualization can occur from nonspecific stimulation of foreign material in the uterus and does not depend on the presence of an embryo. Here, the embryo itself is unable to receive proliferation signals in the absence of the EGFR; the ICM cells degrade, perhaps because of the lack of survival signals from incomplete ligand binding.

In these studies, some strains of mice survive the implantation stage. Using semiquantitative RT-PCR, Threadgill and co-workers find no increased maternal endowment of EGFR transcripts in the oocytes of the surviving strains. They also show, by transferring them to uteri of other genetic backgrounds, that the effect is embryo specific. Therefore, to explain the rescue of certain genetic strains, it is necessary to invoke the assistance of other, related genes [neither the ErbBs nor HRG genes have so far been reported as expressing in the blastocyst, although ErbB2 is expressed in uterine epithelium, in response to E2 (Lim *et al.*, 1997)] or even unrelated genes that may have similar activities. One possibility is that PDGF-α-R, which is also expressed in the ICM and trophectoderm cells, together with PDGF-A, can at least maintain survival of the cells in the blastocyst by autocrine activity. The role of this receptor in attraction of the blastocyst to the uterine epithelium (Bidwell *et al.*, 1995) might be explained in terms of the chemotactic effects of either PDGF-A or -B from the uterine cells (PDGF-α-R can bind both forms of PDGF), or vice versa, because the PDGF receptors are also expressed in the mouse uterus under the control of estrogen (Bidwell *et al.*, 1995). Some assistance may also derive from the leukemia inhibitory factor (LIF)/LIF receptor system, which is active at this stage. Ablation of both LIF genes in the female mouse affects implantation negatively through an essential role in the uterus (Bhatt *et al.*, 1991; Stewart *et al.*, 1992) that has not yet been defined (Cullinan *et al.*, 1996).

IV. Roles for ErbB Genes in Fetal and Postnatal Development

A. EGFR Gene Ablation Effects in "Knockout" Mice

In spite of the accepted limitations of the gene targeting approach, a wealth of information has been produced from the ablation of the EGFR gene in three laboratories (Miettinen *et al.*, 1995; Sibilia and Wagner, 1995; Threadgill *et al.*, 1995). For the first time, a clear indication of the different effects of the loss of gene expression in different mouse strains has been demonstrated. The interpretation is that each species and strain exerts different levels of gene activities that combine to produce quite a different menu in the same-cell type. Where the level of replacement gene products is sufficient, there may be no defect in action until the next crisis point. The result of the abrogation of the EGFR gene is that three different phenotypes are produced. The longest lived managed to get to several weeks of age, the shortest lived did not get through the blastocyst/implantation stage. The tissues most affected by the loss of EGFR expression were documented by all three laboratories. For mouse strains that die at the midgestation stage (129Sv background), the placenta is severely affected, although there is some difference of opinion about the main target: the spongiotrophoblast (the outermost region of the placenta) (Sibilia and Wagner, 1995) or both the spongiotrophoblast and the labyrinthine (inner) region (Threadgill *et al.*, 1995). This might also depend on genetic background. As a result of inadequate placental development, smaller and less well-developed embryos are produced; organogenesis is incomplete, leading to death. The homozygous EGFR($-/-$) offspring that survive to postnatal life are more instructive of the nature of the roles of EGFR. The skin, hair and whiskers, eyelids, and eyes are severely retarded in development and altered in arrangement. Mice are born without eyelids; and later these appear and close over the eyes, the reverse of normal events. The disorientation of the hair follicles leads to mice with a hairless or scruffy appearance. The skin in the third postnatal week is thinner, with flattened basal cells and altered zone proportions. One natural mutation in mouse, called flaky skin, *fsn/fsn*, has elevated levels of EGFR and increased thickness and scaliness of the skin, but whether this is caused directly by the overexpression of the *EGFR* gene is not yet known (Nanney *et al.*, 1996).

The overall picture of the defects in postnatal animals can be summarized as ectodermal and endodermal epithelial disruption. The liver epithelial cords are thickened and sinusoids narrow, the kidney collecting tubules are dilated and the cells flattened. The colonic epithelium is distorted. The lungs are immature, with thicker alveolar septae and less saccule dilation. These are the epithelial tissues that express the highest levels of EGFR and are known to be EGFR-regulated from earlier studies. The central nervous system (CNS) is also affected developmentally by the loss of EGFR. Changes in distributions of cell layers in the anterior cortex, atrophy resulting in loss of the entire cortex, reduction in the size of the cerebellum, rotation of the hippocampus, and specific loss of mitral and tufted cells in the olfactory bulb all occur. The results indicate that because the thickness of the ventricular zone and germinative zones is increased, disturbance of migration and loss of neurons both occur.

On the other hand, no effect on the heart, skeleton, muscle, teeth, pancreas, or gonads was reported for the homozygous EGFR($-/-$) mice. We expected no effect on muscle, because the expression of a dominant negative EGFR construct in embryonic stem (ES) cells also predicts this result, as judged from the plentiful muscle formed in differentiated mutant ES cell cultures and teratocarcinomas (Wu and Adamson, 1996).

B. Effects of ErbB Gene Knockouts and Their Ligands in Development

In contrast to EGFR, lack of the other ErbB receptors and their major ligand neuregulin (NRG) leads to major heart defects and embryonic death at E10-11. All three phenotypes resulting from the loss of ErbB2 (K. Lee *et al.*, 1995), ErbB4 (Gassmann *et al.*, 1995), and NRG (Meyer and Birchmeier, 1995) are similar. The heart fails to form muscle trabeculae in the ventricles, and embryonic blood flow is decreased. Normally, the endocardium is the source of neuregulin and stimulates the development of the musculature of the heart.

Neural crest cells (NCC) express both NRG and ErbB2, which may contribute to development, differentiation, and migration of NCC. Meyer and Birchmeier (1995) disrupted the neuregulin gene with a lacZ-neo combination targeting vector and were also able to map the sites of activity of this growth factor gene by staining for β-galactosidase activity. The cranial nerves in NRG($-/-$) embryos have a severely altered morphology, while peripheral projections are unaffected. They also looked at the sites of mRNA production for ErbB3 and ErbB4, which occur in a nonoverlapping pattern. Cranial NCC express neuregulin and ErbB2 and -3, but not ErbB4. In trunk NCC, ErbB3 is expressed, but not ErbB4 and NRG. The loss of NRG in knockout animals suggests that it is a survival factor for neurogenic cells of the neural crest, and that in its absence cells that normally come to express ErbB3 in cranial structures are much reduced in number.

In addition to the heart effects, sensory and motor nerves are affected in null mutation animals, pointing to predominant roles for the ErbB2, -3, -4, and NRG genes in neural and cardiac development. This phenotype is quite different from that of the EGFR mutation, particularly in respect of the development of heart muscle, where, in EGFR mutants, there is no heart defect. Whether there is overlap in their activities in the brain is not clear, because all these genes are expressed there and all ErbB gene-targeted animals have CNS defects (see section D6 for further discussion). Because the ErbB2, -4, and NRG null embryos only live to 11 days before resorption, the later stages of brain development cannot be observed. It is possible that there would be redundancy in this gene family, if the heart lethality were surmounted, for instance, by the production of knockout animals in a cre-lox system that allows expression in the heart but inducing gene inactivation in the brain.

C. Natural Mutations in Mice and Studies Using Transgenic Mice

Mice with natural mutations that affect the skin, eyes, whiskers, and fur have been known for some time. Two were similar and called *wa*-1 and *wa*-2 (waved) and another called *marcel* has died out. All have the same skin and fur characteristics as the EGFR knockout mice, but their defects are not lethal and development is normal. The abrogation of the transforming growth factor-alpha (*TGFα*) gene in mice (Luetteke *et al.*, 1993; Mann *et al.*, 1993) produced an exact fit to the *wa*-1 mouse, and both genes map to chromosome 6. Mutation of the *TGFα* gene was confirmed in the mutant mouse, which produces only 10–20% levels of normal TGFα transcripts in the brains of homozygous animals. The pups have wavy fur, curly whiskers, and a low frequency of eye defects, including open eyelids at birth, with defects of the cornea, retina, and lens. The mice are fertile and able to suckle their young. Although TGFα is absent from milk in these animals, there are sufficient amounts of other growth factors, such as EGF, Ar, and Cripto-1, to replace it. The *TGFα* gene in man appears to play a specific role in craniofacial development, at least under conditions where its activity is affected by genetic mutation. Two studies of restriction fragment length polymorphisms (RFLPs) in families with risk of cleft lip with or without cleft palate have shown a statistical probability of connection with the *TGFα* gene (Chenevix-Trench *et al.*, 1992; Holder *et al.*, 1992). These cases could account for one-third of all cases of cleft palate. However, a different conclusion was reached by another group of workers, and so the question is still open (Vintiner *et al.*, 1992).

The *wa*-2 mouse was shown to map to chromosome 11, where *EGFR* is located. Investigation by two laboratories showed the mutation is indeed in the *EGFR* gene, with a single amino acid change close to the ATP binding site, which produces a kinase activity only 10% of normal (Fowler *et al.*, 1995; Luetteke *et al.*, 1994). As predicted from the name and appearance of the mutant mouse, the phenotype is very similar to null TGFα mice. *In situ* hybridization confirmed that the mRNA for EGFR is found on the outer root sheath of active hair follicles, and for TGFα is on the inner hair shaft, thus providing strong circumstantial evidence for a close paracrine effect on the development of the hair follicle (Luetteke *et al.*, 1994). The *in vitro* effect of the *wa*-2–defective EGFR is greater than the effect *in vivo*, suggesting that compensatory effects from interacting genes are playing a role. *In vitro*, while the receptors are expressed at normal levels, the tyrosine phosphorylation of the mutant form, EGFR-V743G, is reduced to low levels. The receptors can bind EGF, but the affinities are low and the rate of internalization after binding to ligand is reduced. The defect likely is caused by the alteration of the ATP binding pocket by the change of the single amino acid. The *in vivo* effects of both the TGFα and EGFR mutant mice suggests that the hair follicle and the eyes are the most sensitive to reduced EGFR activation.

One transgenic mouse model causes a truncated EGFR vector to be expressed specifically in the skin (Murillas *et al.*, 1995). This kinase-deleted protein acts to decrease the activity of the EGFR by forming unproductive dimers. The hair follicles are disrupted, and the skin becomes hairless, with disrupted hair growth cycles. This sort of study with tissue-specific deletion of EGFR expression is greatly needed to understand the role of the gene in a single tissue where the complication of malnutrition does not mask the specific effects. Other transgenic mice that over- or underexpress EGFR or other ErbB-related genes have been largely restricted to those directed to mammary gland expression and will be discussed under that heading.

D. The Role of ErbB Genes in Tissues and Organs

1. The Placenta

The placenta is a most essential structure for the correct development of the fetus. It provides the nutrients and oxygen and removes the waste. The EGF family of genes has a large role in the successful development of the intimate contacts and interactions between maternal and fetal cells that are needed for development of the placenta and the mammalian embryo. EGFRs expressed on the trophectoderm cells in the blastocyst continue to be expressed on the giant cells that they form in the ectoplacental cone. The maternal component of the developing placenta, the decidualized stroma of the uterus, forms the outer portion of the placenta and gradually thins during gestation. The giant cells are seen at the inner border of the deciduum in the mouse. The spongiotrophoblast is the middle region of the placenta, where the maternal blood sinuses widen and break up into the tributaries that ensure nutrient exchange in the labyrinthine inner portion of the placenta. All of the cells in these tissue compartments express EGFR. In the human placenta, the nutrient interchange occurs through the fingerlike projections of fetal villi that are bathed in maternal blood. The villi are lined with two layers of cells, the inner cytotrophoblast and the outer syncytial trophoblast, both of which express EGFR. In the basal epithelial cells, the EGFR are basolaterally placed, but the syncytial cells to which the basal cells give rise by their differentiation have receptors also on the apical surface facing the maternal blood spaces. The differentiation of cytotrophoblast cells is also stimulated by EGF (Bass *et al.*, 1994). The absorption of nutrients occurs from the syncytium to the basal layer, and it has long been a controversial subject, still not resolved, as to whether maternal EGF can pass to the fetus through the placenta. At the preplacenta stages, the embryo can bind EGF ligands from the uterine fluid, and small molecules pass through the parietal and visceral yolk sacs. However, the embryo is also well provided with autocrine/paracrine EGF ligands, starting with Ar and TGFα at the blastocyst stage. The invasion of trophoblast cells into the maternal uterine stroma is an essential event in the establishment of a placenta. EGF has been shown to be a specific and potent stimulator of the invasion of first-trimester human cytotrophoblast cells *in vitro*, but these cells do not themselves produce EGF or TGFα (Bass *et al.*, 1994). The mechanism may be via the stimulation of proteases that are induced by maternal EGF and other growth factors (Matrisian and Hogan, 1990). The placenta is a rich source of ligands: EGF has been reported as an early product of the human pregnant endometrium stromal cells during decidualization (Sakakibara *et al.*, 1994) and provides the first stimulus. In addition, the placenta becomes invaded by macrophage-like bone-marrow-derived cells that can synthesize a host of growth factors, such as CSF-1. This may be why there is disagreement about the ligands synthesized by the placental cells proper and the use of antibodies that could be detecting Fc receptors rather than specific antigens. The cells of the uterus, luminal epithelium, glandular epithelium, stroma, and decidua all appear to produce EGF and TGFα (Haining *et al.*, 1991). In addition the later stages, the placenta expresses ErbB2 (Muhlhauser *et al.*, 1993), and this may assist in similar ways to EGFR. The reproductive tissues are very sensitive to steroid-hormone-regulated induction of the EGFR and its ligands. In the normal cycling human uterus, the expression of EGF, TGFα, and EGFR rise and become highest in the late follicular and early luteal stages (Imai *et al.*, 1995). Retinoic acid (RA) has the opposite effect, down regulating the synthesis of EGFR in cultured human trophoblast cells (Roulier *et al.*, 1994). The total effect of RA is uncertain, because both increased (K. Thompson and Rosner, 1989) and decreased transcription (Hudson *et al.*, 1990b) of the EGFR and TGFα (Grandis *et al.*, 1996) genes have been reported for different cell types.

In summary, EGFR and its ligands stimulate invasion and proliferation in the early stages (Mochizuki and Maruo, 1992) and differentiation of trophoblast cells later (Amemiya *et al.*, 1994). The results of the EGFR knockout studies show that midgestation death appears to be attributable to the defective placenta, supporting the conclusion that EGFR is one of the two most important growth factor systems for development of the placenta (the other is IGF-R, which may interact with EGFR; see later, Section VIB.3).

2. Liver, Kidney, and Gastrointestinal Tract

The mouse liver has the highest expression levels of EGFR of all mouse tissue. The male liver expresses a higher level than the female, and these levels can be manipulated using testosterone (Kashimata *et al.*, 1988). Interestingly, the liver is not greatly affected in the knockout mouse model, indicating a role of not much importance in development. The fetal liver from the thirteenth to the eighteenth day of gestation in mouse is the main hematopoietic center, and appears to be normal in EGFR-null mice. The sinusoid structure is altered somewhat in knock-out mice, but functions up to 3 weeks of age without difficulty. However, mal-functioning of the liver may contribute to the general ill health that leads to early death in these animals. A likely role for the adult liver is in absorption and removal of EGF-like peptides from the blood, thereby ensuring the very short half-life of these ligands in the serum (approximately 1 min for EGF; Tyson *et al.*, 1989). Liver is one of the few identified tissues that produces the secreted form of the EGFR by alternate splicing of the gene (Petch *et al.*, 1990) which argues for a unique requirement for neutralization of the proliferative signal that would be given to a tissue with such high expression of EGFR. The secreted EGFR would act toward this end by binding EGF ligands and removing them from the responsive cell. This hypothesis has some merit because the non-proliferative adult rat brain also makes this form of truncated EGFR (Nieto-Sampedro, 1988). Injury reduces truncated EGFR expression, and astrocytes that overexpress full-length surface EGFR now appear in response to injury.

In transgenic mice with reduced EGFR activity experimentally effected by dominant–negative EGFR expression, adults live to at least 4 months, and nearly all mice with small stature that fail to thrive have livers with vacuolated hepato-cytes, with small nuclei occurring as clusters of several hundred cells in a repeated pattern among more normal cells. The conclusion is that these cells are malfunctioning and contribute to ill health and eventually liver failure before 4 months of age (J. X. Wu and E. D. Adamson, unpublished observations). The liver is also the site of ErbB3 expression, regulated by insulin, with a role thought to be nutritional in nature (Carver *et al.*, 1996).

The kidney expresses EGFR in most cell types; but in its absence, the EGFR-null animal develops normally except for widened collecting ducts. The function of the kidney is disturbed in an inherited polycystic kidney disease (PKD) that appears to be caused by the loss of EGF expression in a mouse model of PKD. A few days into postnatal development, enlarged cysts arise through the increased expansion of the epithelial layers, with decreased excretory function (Gattone *et al.*, 1990; Orellana *et al.*, 1995). Although this disease in mouse is ameliorated by EGF (Gattone *et al.*, 1995), the introduction of excess TGFα increases the kidney enlargement (Gattone *et al.*, 1996). The basis for this is still uncertain but may indicate another example of the different effects of specific EGF-like ligands.

The maximal expression and activation of the EGFR in the intestine occurs during prenatal development (J. Thompson *et al.*, 1994). mRNA for EGF and TGFα (at earlier stages than EGF) are present and presumably form an autocrine loop (Dvorak *et al.*, 1994). Moreover, each of the EGF-like ligands can increase the expression of the other ligands in an *in vitro* study using an intestinal cell line (Barnard *et al.*, 1995). The thickness of the intestine lining is reduced in knock-out mice, but all the cell types appear to be represented. The epithelial cells that line the intestine are continually sloughed off and have to be renewed. The cells of the small intestinal crypts, for example, are "born" at the base; and as the cells mature, they move up the villi and come to a position where they are lost into the contents of the gut. These epithelial cells express basolateral EGFRs that are able to respond to EGF or TGFα in the gut lumen (especially in suckling animals) or in the blood, to stimulate proliferation, wound repair, nutrient transport, differentiated function, and regeneration (Saxena *et al.*, 1992). Tumors of the intestinal tract often overexpress EGFR, ErbB2, and ligands. An autocrine loop can support excessive cell proliferation and locomotion as cells become metastatic (Issing *et al.*, 1996).

3. Reproductive Tract

EGF is known to be present in the developing fetal reproductive tract in both males and females, and EGFR levels are higher in the fetal male reproductive system compared to the female (Gupta, 1996). If EGF or EGFR levels are induced in pregnant mice or rats, sexual development is altered, and the Wolffian duct persists in females. EGF and androgens both stimulate and stabilize the Wolffian duct in female fetal mice and stimulate the development of the male reproductive tract, while antiandrogens inhibit, indicating that EGF acts via androgen receptors (Gupta *et al.*, 1996). In cases of undescended testis in rats, treatment with EGF relieves the defect (Cain *et al.*, 1994). Male mice have large stores of EGF in the salivary glands that are regulated by testosterone levels. After sialoadenectomy, the number of spermatids and mature sperm decline 50%, and this can be restored by EGF, indicating a role for EGF in meiotic maturation (Tsutsumi *et al.*, 1986). EGF and EGFRs are all present in the human ovary, in the preantral and antral follicle, and in granulosa cells during the 1st and 2nd trimesters of gestation (Yeh *et al.*, 1993). TGFα is expressed in the preantral follicle. This declines later, but the granulosa cells become rich in this growth factor, and it may play a role in steroid hormone production (S. Li *et al.*, 1994; Maruo *et al.*, 1993). Preimplantation development may be assisted by the growth factors present in the oviduct fluid (see Fig. 1). The mouse oviduct synthesizes TGFα and HB-EGF on days 1–4 of pregnancy at a useful time to support morula-stage development, but synthesis is not hormone induced in this tissue (Dalton *et al.*, 1994).

TGFα and EGFR are present in early prepubertal testis development and decrease during puberty, suggesting they may have a role in early development. Although EGFR is expressed on Sertoli cells, the receptors appear to be inactive: EGFR is found on spermatogenic cells and can respond to TGFα, but whether there is a physiological role is unclear (Mullaney and Skinner, 1992).

4. Heart and Muscle

[See also the earlier section on ErbB gene ablation studies and reviews by Mohun (1992) and by Shastry *et al.*, (1994).] As indicated, the heart is an organ that requires for its development the expression and interaction of all of the ErbB receptors except perhaps EGFR. Loss of any of ErbB2, ErbB3, or NRG inhibits the formation of muscle structures in the ventricle. Although the heart muscular wall is formed, the muscular trabeculae are not developed, the heart beat is weak and slow, and edema soon develops. The animals die, due to the inability of the heart to pump blood effectively. NRG is normally produced in the endocardial cushions, and its loss is similar to ErbB2 and ErbB4 knockout hearts, suggesting that the ligand acts locally in heart development.

Of the three types of muscle, skeletal muscle appears to be the least affected by disruption of any of the ErbB receptors or ligands during development. If the ErbB2 and ErbB4 knockout mice were able to survive long enough, problems would probably arise because of the loss of neural stimulation and interaction that maintains muscular tone. It is known that ErbB2 and ErbB3, but probably not EGFR or ErbB4, are expressed in skeletal muscle. ErbB2 is not specifically located like ErB3, which is found at postsynaptic membranes (Ho *et al.*, 1995; Jo *et al.*, 1995). Heregulin, or at least one isoform of heregulin (acetylcholine receptor-inducing protein, or ARIA) is made by muscle cells and is also concentrated at synaptic sites in adult muscle fibers. In animals that lack the AChR clustering, the cytoskeletal molecule rapsyn, the AChRs, and ErB3 fail to cluster at synapses (Moscoso *et al.*, 1995). This model supports the hypothesis that ErbB2 and HRG provide the proliferative stimulus for muscle cell production and the tyrosine kinase activity that is needed for this process, while ARIA and ErbB3 (together with ErbB2; Altiok *et al.*, 1995) are concerned with synapse location and activity.

Fetal mouse myoblast cell lines such as C2C12 and MM14 in culture also express EGFR and respond to EGF by proliferating. Other growth factors will sustain this phase, such as bFGF. When the conditions for fusion are present, the resulting myotubes express little or no EGFR or FGFR (Olwin and Hauschka, 1988). It is probable that satellite cells, the stem cells of muscle regeneration, do express EGFR, as a means of mitogenic regenerative response to the release of EGF-type growth factors from injured muscle. The idea that EGFRs are not required for muscle development is also supported by the results of the differentiation in culture of ES cells that express dominant–negative EGFR. In these cultures, and also in the teratocarcinomas formed from them, more muscle, both skeletal and cardiac, are produced than in parental or control cultures (Wu and Adamson, 1996). This can be interpreted as meaning that other growth factors were able to sustain myoblast growth and commitment. In these cells, differentiation to other tissues that are EGFR-dependent is underrepresented, because EGFR activity is needed either for differentiation or for survival. The latter is part of the explanation because increased apoptosis is seen in mutant ES cultures compared to controls.

Heart development and cardiac muscle differentiation are most severely affected by the disruption of the *erbB2*, *erbB4*, and *NRG* genes, as stated earlier. Muscle is not totally absent, but the trabeculae-thickened walls are required for heart function. Precise roles for each of these genes is a subject of much attention, but autocrine effects seem likely.

Smooth muscle is much more clearly a target of EGFR activity (Yang *et al.*, 1994) and its ligands. EGF, TGFα, and HB-EGF (as well as bFGF and PDGF) all stimulate smooth muscle cell cultures to proliferate in culture (Grainger *et al.*, 1994). Smooth muscle cells changing from contractile to prolerative express 10 times more high-affinity EGFR (Saltis *et al.*, 1995); therefore, EGF ligand activity is much more likely to stimulate cell mitogenesis than differentiation.

5. Lungs

It has been known for 17 years that the final stages of lung maturation in rabbit fetuses requires the presence of EGF (Catterton *et al.*, 1979). EGF protects fetal rat lungs from hyperoxic toxemia by stimulating surfactant synthesis (Price *et al.*, 1993). The expression of the EGFR in fetal rat lungs is negatively regulated by androgens. Male rats and humans have fewer lung EGFRs than females (about twofold), a finding that relates to the higher incidence of surfactant lung disease in male babies born prematurely (Klein and Nielsen, 1993). The developmental process of lung morphogenesis is stimulated by interaction of the surrounding mesoderm and is regulated by matrix depositions and by metalloproteinases that focally degrade matrix to initiate branching (Ganser et al., 1991). The expression of TGFα, EGF, and the EGFR occurs in late-gestation fetal rat and mouse (Snead et al., 1989), in human lung, and in cultured fetal rat lung cells. TGFα, EGF, and EGFR are colocalized in epithelial and smooth muscle cells of bronchioles and bronchi and in epithelial cells of saccules. Epithelial cells cultured from late-gestation fetal rat lung transcribe TGFα and EGFR mRNA and produce TGFα and EGFR proteins. Cultured fibroblasts contain EGFR mRNA but no detectable TGFα mRNA. These results demonstrate that fetal lung epithelial cells produce TGFα that might act through an autocrine or paracrine mechanism between epithelial and mesenchymal cells. The colocalization of TGFα and EGF suggests that these growth factors might act in parallel during lung development (Strandjord et al., 1993, 1994).

Lung development can be studied *in vitro* by organ culture in which morphological changes such as branching of ductules and lobular-alveolar formation in response to experimental manipulation can be observed. Experimental adjustment of growth factors in the medium has shown that EGF increases lung development (Strum et al., 1993) and antiserum to EGF inhibits lung development (Yasui et al., 1993). EGF enhances DNA synthesis in the tracheal and bronchial tree of chick embryo lung cultures and induces lung buds when applied to the bronchus as an implant (Goldin and Opperman, 1980), as does fetal mesenchyme. PreproEGF mRNA is detected in E11–E17-day mouse embryo cultures and is specifically localized around primitive airways. Immunocytochemical assays show the colocalization of EGF and EGFR. Addition of EGF dose dependently enhances the morphogenesis of lung explants, a process that is inhibited by tyrphostin, an inhibitor of tyrosine phosphokinases (Warburton et al., 1992). Amphiregulin also stimulates lung development (Schuger et al., 1996), and TGFβ inhibits lung development by reducing the production of amphiregulin (Bennett et al., 1992). Another putative morphogen that may act via up-regulation of EGFR in lung morphogenesis is retinoic acid (RA), with a maximal stimulation at a dose of 1 μM (Schuger et al., 1993). This dose is nonphysiological unless produced locally, but it could be synergized by unidentified factors. The effects of RA are complex and conflicting since part of the effect is on the increase of EGFR protein in cells such as fibroblasts caused by an unknown mechanism not involving synthesis (Jetten, 1980), while the direct effect on the gene appears to be inhibitory (but see section V).

6. Brain and Nervous System

The central nervous system (CNS) is a rich source of all four ErbB gene products and most ligands. CNS development is affected in EGFR, ErbB2, ErbB4, and NRG knockout mice, as indicated earlier. EGF, TGFα, and all the neuregulin variants are expressed here, both in adult and in fetal stages. In the developing chick, regional distributions of ErbB products have been mapped by immunoblotting. ErbB2 is detected at all stages of embryonic brain development and in the spinal cord; ErbB3 is expressed strongly at very early stages and transiently and weakly later in the eye, spinal cord, and sciatic nerve. On the other hand, ErbB4 appears in chick embryo brain only on day 6, continuing at high levels thereafter, and is strongest in the cerebellum, the hemispheres, and the spinal cord but not in the eye or sciatic nerve at embryonic day 14. However, retinal cultures from earlier stages do express ErbB4. In these studies, neurons are most usually predominant in ErbB4 expression (Francoeur et al., 1995). Studies of the localization of ErbB expression in the mammalian brain show that the hippocampus, cortex, telencephalon, and cerebellum express immunoreactive EGFR. In the hippocampus, the CA2 field in postnatal day 7 and adult rats has intensive-staining EGFR in the cell bodies of the neurons. In the adult rat, the pyramidal cell body of the hippocampus is a major site of expression (Tucker et al., 1993). While the acute administration of EGF to mice does not affect memory processes tested by passive avoidance or habituation learning in mice (Ukai et al., 1994), the study should also be repeated using NRG, since this seems to be the ligand favored in the CNS. In mice at day 9.5 in gestation, all parts of the brain express ErbB4. In the hindbrain, expression is localized to rhombomeres 3 and 5 and the dorsal third of the neural tube, as well as in cells associated with the forming ventral motor neurons. There are extensive hindbrain alterations in innervation in ErbB4-null mice that are not seen in ErbB2- and NRG-null mice, suggesting a role in creating a barrier between the rhombomeres in the developing hindbrain. Several defects in this region are similar to those seen in Krox-20 knockout mice, suggesting that this transcription factor may regulate ErbB4 (see section V).

Although proliferation ceases in the brain around birth, adult brain tissue is capable of proliferating in primary cultures, and cell division is stimulated by EGF and bFGF (Gage et al., 1995) but not by PDGF, or NGF (Reynolds and Weiss, 1992). In fetal rat brain cultures, EGF stimulates glial cells to proliferate in culture, and cholinergic neuronal cells are indirectly affected (Kenigsberg et al., 1992). Different isoforms of HRG directs neural crest cells to differentiate into glial cells (Shah et al., 1994), and ARIA causes bipotential glial progenitor cells to develop into oligodendrocytes. The NDF form of NRG induces the maturation of astrocytes and has a survival role for cultured cells (Pinkas-Kramarski et al., 1994). Since both the receptors and their ligands are found together in the brain, it is supposed that interaction occurs and plays a role in the developmental process. In the mutant wa-1 mouse, the striatum of the brain has much reduced levels of glial fibrillary acidic protein (GFAP) mRNA and protein, suggesting that glial development is impaired in the absence of TGFα (Weickert and Blum, 1995). Not only the ligand type alters the differentiation fate of retinal neurons, but also the level of expression of the receptor. Overexpression of EGFR in these cells causes differentiation to proceed toward Müller glial fate rather than rod photoreceptors (Lillien, 1995). ErbB4 may be important in axonal signaling leading to myelination in the peripheral nervous system (PNS). In the Schwann cells derived from the sciatic nerves, myelination appears to occur via activation of the ErbB4 receptor by NDF. The latter is a product of the neural tube and activates the transcription factor Krox-20 (Egr-2). Contact of the Schwann cells with neurons leads to myelination activity that can be mimicked by the application of NDFβ. In animals that lack Krox-20, myelination does not occur (Murphy *et al.*, 1996).

Since the EGFR and ligand genes are steroid hormone responsive and involved in many stages of development, they may play roles in feedback regulation processes. The hypothalamus and anterior pituitary are a neuroendocrine system involved in the regulation of steroid hormone production. The female hypo-thalamus responds to EGF and TGFα by releasing luteinizing hormone releasing factor (LHRH), a process that is blocked by RG50864, an inhibitor of the EGFR (Ojeda *et al.*, 1990). Both the EGFR and TGFα are expressed in anterior pituitary tissue, and EGF is known to increase the differentiated expression of these cells by stimulating the production of prolactin and growth hormone in pituitary cell cultures. In cells derived from bovine pituitary gland, proteinkinase C (PKC) is an intermediary in the production of EGFR and TGFα, and hormones that acti-vate the phosphatidyl inositol pathway will also likely stimulate the activity of the EGFR system (Mueller *et al.*, 1989). Another link in the chain is illustrated by lesions of the female hypothalamus that cause precocious sexual maturation via the increased production of LHRH. This process appears to be mediated by TGFα (Junier *et al.*, 1993). Such lesions are always located in the region contain-ing LHRH-producing cells, and activated cells in this region are known to ex-press increased levels of the full-length EGFR mRNA and protein. Because the cells with the highest kinase activity are glial cells, the idea is that an autocrine effect of increased EGFR and TGFα in glial cells activates the adjacent LHRH-producing neurons by paracrine stimulation. In addition, the pituitary is EGF/TGFα/EGFR autoregulatory (Fan and Childs, 1995), independent of steroid hormones (Felix *et al.*, 1995; Mouihate and Lestage, 1995).

A number of *in vitro* model systems have advanced our understanding of the activities of EGF ligands in neuronal development. TGFα or EGF induce neuro-nal differentiation of PC12 pheochromocytoma cells (derived from the rat adre-nal cortex) (Nakafuku and Kaziro, 1993) under certain conditions, such as low serum and culture on a highly adhesive plastic surface. In normal culture condi-tions, EGF induces proliferation, not differentiation. More recently, a subclone of PC12 cells also shows EGF-induced neuronal differentiation and sustained sig-naling via mitogen-activated phosphokinase (MAPK) (Yamada *et al.*, 1996). This appears to be caused by a higher level of signaling of EGFR than the parental cells and indicates that the fate of cells can be altered by the level of activity of its ErbB receptors. This was also the case in another culture system, P19 embryonal carcinoma (EC) cells that differentiate to neurons when induced by high concentrations of retinoic acid. P19 EC cells do not express EGFR until differentiation occurs. An exogenous EGFR vector expressed in these cells induced differentiation into neuronal cells without the need for RA (den Hertog *et al.*, 1991). In contrast, when EGFR expression and activity is decreased in these cells by the expression of a dominant–negative mouse EGFR construct or by the expression of an antisense EGFR RNA, differentiation into neurons is much reduced (Wu and Adamson, 1993). These studies strongly support the idea that to achieve differentiation to neuronal cells, activation of the EGFR is a requirement. RA-treated P19 cells can also differentiate into oligodendrocytes that are capable of myelination. An immortal cell line derived from RA-treated P19 cultures in serum-free conditions is dependent on EGF for proliferation. In this case, the removal of EGF is the trigger for differentiation of these cells into astrocytes and oligodendrocytes (Staines *et al.*, 1996), and EGF is entirely restricted to a mitogenic mode of action. Other examples indicate that the number of high-affinity receptors expressed changes the cellular response, and this is supported by the variable penetration of the mutant phenotype in EGFR-null mice and in *wa-2* mice.

7. Mammary Glands

The EGFR-null mouse strain that lives to postnatal stages has various epidermal tissue defects, but as yet there is no information concerning the development of the mammary glands [see the review by Medina (1996)]. It is likely that this tissue will have disturbed development, because multiple experiments have indicated the importance of the EGFR in this tissue (Coleman *et al.*, 1988). The presence of TGFα (Moorby *et al.*, 1995), Ar (S. Li *et al.*, 1992), and Cripto-1 (Kenney *et al.*, 1995) in postnatal glands is important for autocrine and paracrine stimulation during the proliferative stages. Combinations of EGF ligands with other growth factors, such as TGFβ, which inhibits ductal growth (Silberstein and Daniel, 1987), and keratinocyte growth factor [KGF, or fibroblast growth factor-7 (FGF-7)], which stimulates it, all contribute to branching morphogenesis to produce a mammary tree that fills the fat pad (Cunha and Hom, 1996). The extracellular matrix plays an important part in mammary gland morphogenesis, in part because of the ability to store growth factors (TGFβ and FGFs) and to present growth factors (Ar and FGFs) during mesenchymal– or stromal-epithelial interactions that occur in gland development.

There are five stages of mammary gland development to consider: prenatal, postnatal, pregnant, lactating, and regressing. The prenatal stage of a simple ductule arising from the nipple and with five branches arrests further development at the 17th day of gestation, and development is continued only later, with the onset of puberty, under the influence of the steroid hormones estrogen, progesterone, and hydrocortisone. The ductal system develops into a branching tree and comes to the edge of the fat pad and stops. End buds disappear in the virgin mouse at about 10 weeks of age. Further development requires prolactin to stimulate the proliferation of lobuloalveolar structures to prepare for pregnancy and lactation. The EGF-like ligands, EGF, TGFα, and Ar are present in virgin mammary gland, are inducible by estrogen administration (Liu *et al.*, 1987; McAndrew *et al.*, 1994), and are active at all stages of development of the mammary gland (Plaut, 1993; Darcy *et al.*, 1995; Kenney *et al.*, 1995). The distribution of each ligand is distinct; for example, TGFα is found in the cap cells of the terminal end-buds and stroma, while EGF is found in luminal epithelium of the ducts and end buds but not in periductal stroma or ductal myoepithelial cells (Snedeker *et al.*, 1991). EGFRs are present on all types of mammary epithelium (ductal, myoepithelium, cap, and end-bud cells) as well as the stromal cells that are in contact with the epithelium. EGFRs are also induced by estrogen in the mature mammary gland (Haslam *et al.*, 1992). Autocrine / paracrine EGF ligands in conjunction with other growth factors stimulate the epithelial–mesenchymal interactive process (Coleman and Daniel, 1990; Coleman *et al.*, 1988; Venkateswaran *et al.*, 1993; Vonderhaar, 1987; Cunha and Hom, 1996). TGFβ is present before puberty, where it functions to prevent ductal growth and branching (Silberstein *et al.*, 1992; Soriano *et al.*, 1996), and is produced during regression after lactation, to cause apoptosis and a return to the resting state (Coleman and Daniel, 1990; Daniel and Robinson, 1992; Jhappan *et al.*, 1993; Kordon *et al.*, 1995). Retinoic acid also influences the differentiated expression of mammary cells and can replace EGF in stimulating casein expression in a differentiating mammary cell culture model (P. Lee *et al.*, 1995), suggesting that their signaling pathways converge.

The transgenic mouse model to study the effect of growth factors on mammary cell hyperplasia, dysplasias, and tumors has long been popular because two useful promoters (although somewhat promiscuous) can be used to overexpress gene products specifically in the mammary gland. As a result of the overexpression of specific genes, it is possible to evaluate their potential to affect normal development as well as progression to tumor formation, a process that requires secondary mutation events (Cardiff, 1996). The mouse mammary tumor virus (MMTV) promoter is used to study expression during the developmental stages in mammary and salivary glands as well as in skin (Halter *et al.*, 1992), while the whey acidic protein (WAP) promoter is useful for expression during pregnancy and lactation. Transgenic mice that express TGFα from the MMTV promoter develop a range of abnormalities, from hyperplastic nodules at 4 weeks of age to adenocarcinoma in 3-month-old females. The up-regulation of the EGFR accompanies this progression, suggesting that excess TGFα production, together with hormonal augmentation, can predispose mice to adenocarcinoma (Matsui *et al.*, 1990). When TGFα is expressed from the metallothionein promoter in transgenic mice, the liver and pancreas are predominantly affected, giving dysplasia of the mammary gland at much later times (Jhappan *et al.*, 1990). Overexpression of TGFα is also accompanied by increased cyclin D1 expression (Sandgren *et al.*,

1995). With the use of *in situ* hybridization with labeled antisense RNA probes, the level of expression and the cell type can be identified. For TGFα, the cystic structures of the nonneoplastic mammary gland express the highest levels (Halter *et al.*, 1992), and tumors arise stochastically and are slow to grow. However, synergy with other genes, such as c-*src*, may occur either spontaneously or experimentally to induce more rapid tumor formation (Guy *et al.*, 1994).

In transgenic mice that overexpress HRG, also a normal growth factor expressed in the mammary gland (Ethier *et al.*, 1996), the end buds persist, when normally they regress in virgin mice older than 8 weeks (Krane and Leder, 1996). When the mice are bred, the levels of HRG increases, and adenocarcinomas develop at about 1 year of age, accompanied by ErbB3 phosphorylation, suggesting that this pair of receptor/ligands operate together in mammary tumors. All four c-*erbB* genes are expressed in the mammary gland, and all ligands have been detected as products either in the normal gland or in breast cancer. Multifocal synchronous appearance of the cancerous state is observable in mammary glands of the ErbB2/neu oncogene when the activated gene is overexpressed (Guy *et al.*, 1992). Activation consists of a single amino acid change in the transmembrane domain of the rat protein. In one MMTV-human c-ErbB2 transgenic mouse line, lactation is defective, and adenocarcinomas arise in all mammary epithelia at 3 months. In another strain, expression in kidney, lung, mammary, and salivary glands gives rise to preneoplastic lesions followed by kidney and lung failure and death at 4 months of age (Stocklin *et al.*, 1993).

8. Skeleton, Bone, and Tooth

The original assay for EGF was the injection of newborn mice with the sample and observation of the precocious eruption of the incisor teeth and opening of the eyelids; therefore, overexpression of the normal ligand can exaggerate the functions. EGF, TGFα, and EGFR are all expressed in the embryonic first branchial arch during mammalian development and are important to craniofacial development (Slavkin, 1993). An autosomal dominant disorder in man called the Rieger syndrome is tightly linked to the EGF gene on chromosome 4q. The syndromes linked to this disorder are absent maxillary incisor teeth and malformations of the eye. The former defect can be mimicked using tooth germ cultures. EGF is normally produced in the tooth cusp and is detected in serum-free, defined cultures on day E16. EGFR is also expressed, and when its activity is suppressed by the addition of tyrphostin, subsequent inhibition of DNA synthesis and tooth growth occurs. Antisense oligonucleotides to EGFR can also effectively reduce EGFR, with the same result, while addition of EGF stimulates DNA and RNA synthesis (Hu *et al.*, 1993). Cartilage and bone, tissues of mesodermal origin, appear to be unaffected by the loss of EGFR in null mice, but tooth development requires enamel organ epithelia as well as dental ectomesenchyme. The data suggest that the differentiation and regulation of the functions of mesenchymal cells is by multiple overlapping growth factor systems, whereas epithelial development is much more critically dependent on EGFR activities.

V. The Regulation of the EGFR Gene Family

The EGFR gene has no TATA or CAAT box, but has about six transcriptional start sites in a GC-rich domain upstream of translation start in exon 1, which codes for a signal peptide. A large first intron contains several regulatory elements, as does the promoter region. Many DNaseI-sensitive sites in the promoter have been described, of which the most prominent are multiple Sp1 sites (Ishii *et al.*, 1985; Kageyama *et al.*, 1988), with a basal CCCGCC sequence. Sp1 is ubiquitously expressed and is probably responsible for the basal activity of the promoter through interaction with TFIID (Smale *et al.*, 1990). The basal activity is down-regulated by repressor factors, such as the transcription factor that is mutated in Wilm's tumor, WT-1 (Madden *et al.*, 1991). WT-1 codes for a tumor suppressor transcription factor that can bind in several positions in the EGFR promoter, and might be an important negative influence in the regulation of EGFR during gestation in tissues such as kidney and the reproductive organs. WT-1 is active in the urinary and reproductive tract at precise developmental stages and is known to inhibit the activity of the EGFR promoter (Englert *et al.*, 1995) as well PDGFs and insulin growth factor receptors. However, WT-1 activity is limited and cannot explain the universally low transcriptional state of the EGFR gene. Two other WT-1-related transcription factors, Egr-1 (Krox-24) and Egr-2 (Krox-20), may also regulate EGFR expression by virtue of their identical DNA-binding sites (Christy and Nathans, 1989; LeMaire *et al.*, 1988), but this has not yet been tested. A new finding, that the transactivating factor, Ap-2, binds to a region (-165 to -105) and activates the EGFR promoter, is promising, since it may interact with other factors in the regulatory region, although it is not clear if a tumor promoter stimulus is needed to induce it (A. Johnson, 1996). Other transactivating factors that affect EGFR transcription are the GC factor (Kitadai *et al.*, 1993), TCF, and ETF1 and p53 (Deb *et al.*, 1994; Merlo *et al.*, 1994). Interestingly, both wild-type p53 and mutant types can transcriptionally activate the EGFR gene, through different binding sites. Wild-type p53 binds to a consensus site at -239 to -265, while the mutant p53 does not but has a different target site at -20 to -104 (Ludes-Meyers *et al.*, 1996). This means that at least some mutations of the *p53* gene do not alter its ability to stimulate *EGFR* expression, a finding with implications in how cells convert to cancer cells with growth advantages. The very large first intron of the gene contains several regulatory elements that might also interact with the promoter (Chrysogelos *et al.*, 1994), and these still have to be defined.

Although the regulation of the *EGFR* gene has been studied for 12 years since its cDNA was cloned (Merlino *et al.*, 1984; Simmen *et al.*, 1984; Ullrich *et al.*,

1984) from A431 human vulval epithelial carcinoma cells, its regulation is complex and still imprecisely understood. Transcriptional up-regulation of the *EGFR* gene by EGF and TGFα occurs in breast cancer and most other cell types. Transcriptional and posttranscriptional regulation by thyroid hormone and retinoic acid is reported, both up and down (Hudson *et al.*, 1990b; Roulier *et al.*, 1994; K. Thompson *et al.*, 1992; Xu *et al.*, 1993), depending on the cell type. In most cases, it appears that up-regulation of EGFR activity by thyroid hormone or RA is caused by an increase in the level of high-affinity EGF-receptor protein, rather than a direct effect on the promoter (Jacobberger *et al.*, 1995; Mantzouris *et al.*, 1993). In breast cancer, both estrogen-receptor-negative and -positive cells produce increased levels of EGFR message and protein in reponse to EGF. Surprisingly, in breast cells, TGFβ1 also up-regulates EGFR expression while decreasing growth for cells that have TGFβ1 receptors. The tumor promoter 12-*o*-tetradecanoylphorbol-13-acetate (TPA) can also up-regulate EGFR expression (Bjorge *et al.*, 1989; Hudson *et al.*, 1990c) or down-regulate by phosphorylation of threonine 653 (Lund *et al.*, 1990). In fact, a repressor element in the EGFR receptor promoter (-919 to -870) containing a TTCGAGGG motif is required for binding and repression of the receptor gene by an as-yet-uncharacterized 128-kDa factor present in extracts from HeLa and other cells (Hou *et al.*, 1994a and b).

The most obvious lack in understanding of the EGFR gene regulation is the location of estrogen-responsive elements, because direct transcriptional activation does occur in the mouse and rat uterus (Mukku and Stancel, 1985). This question has been investigated in human breast cancer cells in which the estrogen receptor (ER) expression was constitutively overexpressed, resulting in increased EGFR and TGFα mRNA (Sheikh *et al.*, 1994). Later work indicates that the effect of estrogen is biphasic, first increasing EGFR mRNA and protein levels independent of protein synthesis, and then decreasing them. Apparently there are three imperfect estrogen receptor element (ERE) palidromic half-sites in the human EGFR promoter that are capable of binding ER at low affinity (Yarden *et al.*, 1996). Whether the promoter will stimulate reporter gene activity in response to ER binding and whether these sites are conserved in the mouse EGFR is not known. In the uterus and in some cell lines, EGF can replace estrogen in growth promotion, suggesting that paracrine or autocrine loops can bypass estrogen dependency. In fact, Curtis *et al.*, (1996) have shown that in ER KO mice there is no estrogen-like response to EGF (DNA proliferation and progesterone receptor expression), although normal *c-fos* expression is elicited. This means that somewhere in these two signaling pathways, there is crosstalk from EGFR to activate the ER. The human *TGFα* gene promoter also contains two imperfect 13 bp palindromic EREs that are nearly homologous to other imperfect EREs. The promoter responds to estrogen presumably by receptor binding at these sites (Saeki *et al.*, 1991).

Thus a complex picture emerges for the regulation of the expression of the

EGFR. Since EGFR plays a signaling role for virtually all cells and especially during development, this is not surprising. A multiple set of conditions modulate it, and it has proved difficult to pinpoint a single important regulator other than Sp1. The transcriptional up-regulation of the *EGFR* gene by its ligands is likely to be indirect, because there are no obvious serum response elements (SREs) in the known sequences. The most likely mechanism is through the induction of *c-fos*, *c-jun* (AP-1), and *Egr-1*, which respond through their SREs and perhaps at the SIE (sis-inducible element). The immediate early gene products then activate or inhibit the *EGFR* promoter, depending on the levels and types of transactivators present. In studies using the endogenous *EGFR* gene, various results have been obtained, with posttranscriptional up-regulation suggested as a major regulator of the receptor by its ligands. The other area that is still not understood is the requirement that EGF-like ligands must be in contact with the receptor for several hours in order to achieve the mitogenic signal for cell proliferation. This implies that either the strength of the signal has to build up or only the later pathways generate the highly modulated signal pathways that require a continuous signal in order to act as a double-check for continuation of this important step. Another interpretation is that moving through a late G1 checkpoint is needed to commit to autonomous cell cycle progression. Clearly the *EGFR* gene merits a complex interacting regulatory mechanism that largely keeps it repressed in adult cells but that is acutely responsive to multiple stimuli.

The *erbB2* gene promoter is also regulated by Sp1 (Chen and Gill, 1994), by OB2-1 [better known as Ap-2 transcription factors (Bosher *et al.*, 1995)], by EGF, and by tumor promoters (Hudson *et al.*, 1990a). There is an indication that Ap-2 may be important in *erbB* gene regulation because abrogation of the *Ap-2* gene in mice produces a phenotype with some similarities to *erbB2* and *EGFR* knockouts, although much more severe (Zhang *et al.*, 1996). ErbB2/neu is also negatively auto-regulated (Zhao and Hung, 1992).

VI. Interactions of the ErbB Family

A. Crosstalk Between EGFR and ErbBs

The EGFR can be modulated in activity by homodimerization or with a member of the ErbB family, triggered by ligand or antibody binding. Conformational changes occur to assume the active form of the receptor, to activate the kinase domain, and hence to phosphorylate its partner and send signals along several pathways. The docking of proteins onto the activated EGFR is specific for certain intermediaries that have appropriate SH2-binding sites, and differs between the ErbB proteins. This accounts for some of the specificity of responses. During development, EGFR may be the only member of the family expressed until about day 8 of gestation, when the other ErbB proteins start to be expressed in the heart and neural tissues. All members of the family are expressed henceforth in these excitable tissues and also in mammary glands and other epithelial tissues. The opportunity to form heterodimers occurs in these tissues when the heregulin/ neuregulin ligands are also present. Some major differences in how they may present different signals to the cell have been indicated (earlier) in their affinities for each ligand, their innate kinase activity, and their ability to attract different signaling intermediaries.

The complexities of cell responses to ErbB ligands is increased by the presence in the cell of more than one ErbB gene product. The possibilities are that homodimers and heterodimers might form. If EGFR is present, the EGF ligands bind and activate homodimers best. If ErbB2 is also present, the heterodimer is activated by EGF ligands and gives a stronger signal. NDF signaling in cells with EGFR and other ErbB proteins is mediated by heterodimers even when there is low kinase activity, as in ErbB3. The presence of EGFR gives weaker signaling responses to EGF compared to NDF ligands because EGFR is rapidly endocytosed while the other ErbBs are not, and this extends the duration of activation (Baulida et al., 1996). Some of the hierarchy that occurs between the heterodimers of this family has been worked out by transfecting expression vectors for the ErbB genes into CHO, or into NIH3T3 cells that have no EGFR (NR6 cells), or into HC11 mouse normal mammary cells. Heterodimers containing ErbB2 are the most active in generating signals, the kinase activity is the highest, and complexes are more stable and therefore more potent (Tzahar et al., 1996). The hierarchy of crosstalk between the receptors in inducing proliferation is ErbB3 + 2, > 4 + 2, > 1 + 2, > 1 + 1, > 1 + 3 (higher with NDF than EGF) (Pinkas-Kramarski et al., 1996a and b). EGFR can bind to ErbB4, but only the dimer formed in response to NDF is stable enough to detect the signal (Tzahar et al., 1996). In a different kind of hierarchical study, T47D breast cancer cells that contain all four of the ErbB receptors were treated with all the known ligands. The growth-promoting responses were evaluated and compared. Under these conditions most receptors are activated to some degree, thus illustrating the remarkable range of signaling that emanates from ligand-stimulated cells (Beerli and Hynes, 1996). Each ErbB receptor binds to different subsets of SH2-containing proteins in order to generate the signal to the nucleus. However, coupling of Grb2 to ErbB2 does not seem to occur, and other components appear to be involved.

B. Crosstalk Between EGFR and Other Signaling Systems

1. Stress Signals

Signals that can be received by EGFRs are not restricted to EGF-like ligands. These "nonligand" signals include cold stress, ionizing and nonionizing radiation, and chemical toxicants that are tumorigenic and teratogenic. Another type of stress has been shown to stimulate EGF ligand and EGFR expression in the pituitary within 30 min in the whole animal (rat) exposed to cold stress and in pituitary cell cultures (Fan *et al.*, 1995). Nonionizing radiation such as ultraviolet irradiation can cause cells to undergo rapid EGFR activation, tyrosine auto-phosphorylation, and the generation of several signal pathways, leading to imme-diate early gene stimulation (Huang *et al.*, 1996). Both MAPK and SAPK/JNK pathways are likely involved, thus mobilizing a large array of regulatory genes to respond to the insult. Reactive oxidant species (ROS) in the cell are upstream of EGFR activation in this process, and reagents such as hydrogen peroxide can activate EGFR directly within 2 minutes. How UV generates oxidants (from membrane lipids?) is not precisely known, but oxidants likely work by inducing dimerization or clustering of receptors (Rosette and Karin, 1996), possibly by forming intermolecular cysteine dimers. Activation of the transcription of large numbers of genes provides an important clue to how irradiation or toxins in the environment may contribute to carcinogenic initiation or cancer progression.

Ionizing radiation can also affect EGFR function (Kwok and Sutherland, 1991, 1992; Schmidt-Ullrich *et al.*, 1994). Sensitivity to ionizing radiation is reduced by the preexposure of MCF7 breast cancer cells in culture to EGF. Since EGF is a mitogen for these cells, it cannot be acting to increase G1 phase of the cell cycle, thus aiding in DNA repair. Instead, it may be increasing the levels of EGFR-signaling pathways so that stress responses can allow the cell to survive (Woll-man *et al.*, 1994). Some cells release growth factors, including TGFα after irradiation, that may play a similar role. Increased radiosensitivity to EGF occurs when the cells (A431) have a high density of EGFR and are growth inhibited by EGF, so signaling events may be reduced to explain this example based on the same principles. During early embryonic development, an acute exposure of 1.0 Gy or less of γ rays produces a delay in blastocyst development without affecting cell proliferation rate. This delay is prevented by adding TGFα (Peters *et al.*, 1996).

EGFR function is also sensitive to several chemical tumor promoters, includ-ing dioxin, the most poisonous man-made molecule known (dioxin, 2,3,7,8-tetrachlorodibenzo-*p*-dioxin; TCDD). The binding of EGF by EGFR is substan-tially reduced in the liver by TCDD within 48 h after a single dose in rats (Madhukar *et al.*, 1984). Interestingly, this reduction in ligand binding is accom-panied by an increase in protein kinases associated with EGFR in hepatic plasma membranes (Madhukar *et al.*, 1988). This might explain TCDD's EGF-like stim-ulatory effect on keratinocytes *in vitro* (Kawamoto *et al.*, 1989). Whether these agents can disrupt the developmental roles of the EGFR remains unknown. TCDD is a powerful teratogen in mammals (Couture *et al.*, 1990), and it alters EGFR function and epithelial differentiation in developing embryonic palatal shelves (Abbott and Birnbaum, 1990). TCDD also affects mouse blastocyst formation *in vitro*, presumably by affecting the rate at which the trophectoderm acquires its epithelial phenotype and function (Blankenship *et al.*, 1993). An important role of EGFR is in trophectoderm formation (Brice *et al.*, 1993; Dardik *et al.*, 1992; Wiley *et al.*, 1992); therefore it is possible that the effect of TCDD is via alteration of EGFR function in the development of epithelia.

2. Cell Morphology and Adhesion

It has long been observed that a rapid response of the cell to EGF is the "ruffling" of the membrane, alterations in the cytoskeleton that allow the leading-edge plasma membrane to expand during cell migration [reviewed by Dedhar and Hannigan (1996)]. The cell-membrane-inserted integrins can detach and reattach to move the cell along a substrate. EGFR signals to the cytoskeleton in a number of ways. First, the receptor protein is able to bind to actin (den Hartigh *et al.*, 1992; van Bergen en Henegouwen *et al.*, 1992), and this leads to actin polymerization (Rijken *et al.*, 1995). The interaction with EGFR causes the phosphorylation of actin (van Delft *et al.*, 1995), and these components appear to locate in the membrane ruffles together with phospholipaseCγ (PLCγ) (Diakonova *et al.*, 1995). The interaction with actin might be two-way under certain circumstances; for example, the cytoskeleton may respond physically to the stress of zero gravity and down-regulate the effect of EGF by as-yet-unknown mechanisms (Rijken *et al.*, 1994). Down-regulated responses of the EGF signaling related to stress effects on the cytoskeleton might also occur in the embryo when cells become overcrowded and physically deformed, and could be a way of reducing c-fos, c-jun, and Egr-1 activities when the proliferation phase has filled the space available. Regulation of the cytoskeleton, lamellipodia, and filopodia in EGF-stimulated cells occurs through specific GTPases, the Rho proteins, via p120 Ras-GAP (Nobes and Hall, 1995). Interaction of different signals at the c-ras or Rho stage is an obvious possible intersection point.

It would be interesting if the interaction of EGFR with actin plays a role in inducing cell locomotion during morphogenesis. Although the EGFR is closely associated with actin, it is not clear if the receptor can be found in adhesion plaques, as is c-src protein (Tsukita *et al.*, 1991). Since both integrins and the ErbB proteins cluster after ligand binding, a motive force would be generated for cell locomotion. Isolated focal adhesion complexes contain intermediates of the signaling pathway for both adhesion and growth factors. The presence of c-src, FAK, Grb-2, Shc, and PLCγ as well as the high-affinity receptor for FGF, flg, sets the scene for integration of adhesion and growth-factor signaling on the cytoskeleton (Plopper *et al.*, 1995). One possible mechanism specific to brain, heart, lung, and colon involves activated EGFR tyrosine kinase binding to the SH3 protein, Wiskott–Aldrich syndrome protein (WASP), that regulates the depolymerization of actin, a process that is dependent on PIP2 (Miki *et al.*, 1996). Upon cell binding to a substrate, the integrins become phosphorylated on tyrosine and then bind to Shc and Grb2, thus linking the cytoskeleton to the c-ras pathway (Mainiero *et al.*, 1996). Addition of EGF causes tyrosine phosphorylation of the β4-integrin in epithelial cells, suppresses the recruitment of Shc, and causes the deterioration of hemidesmosomes. In addition, the cells now move faster toward a laminin substrate, suggesting that EGF can promote processes such as wound healing and metastasis of tumor cells. Laminin contains an EGF-like motif and can stimulate cell proliferation by contact with that fragment (M. Lin and Bertics, 1995). A unique cell system for examining the interaction of EGFR with laminin receptor (with integrin α6) was isolated by selection of cells that overexpress EGFR and can adhere to laminin with increased spreading and motility (M. Lin and Bertics, 1995). This model may reveal useful mechanisms for understanding receptor interactions, signal transduction pathways, and cell responses to ligands.

3. Interaction with Other Growth Factor Receptors

For cells in culture, it is known that the addition of platelet-derived growth factor (PDGF), bombesin, phorbol esters, bFGF, and aFGF can lead to the down-regulation of the EGFR affinity, binding, and hence signaling. Some of these effects act via protein kinase C or protein kinase A and include Thr^{734} phosphorylation that reduces EGFR affinity. This is probably a competitive effect when the signal intermediaries are already activated and more signals are prevented by as-yet-unknown mechanisms.

An intriguing interaction is that between insulinlike growth factor–I receptor (IGF-IR) and EGFR (Burgaud and Baserga, 1996; Coppola et al., 1994). Embryo fibroblasts devoid of IGF-IR due to targeted abrogation of that gene cannot grow in serum-free medium, even if overexpressing EGFR. Wild-type cells with IGF-IR and overexpressed EGFR can grow in serum-free medium in the presence of EGF and can grow in soft agar. Only the reintroduction of IGF-IR with an intact ATP-binding site (other mutations were irrelevant) into the IGF-IR-null cells is able to stimulate anchorage-independent growth. Even mutations of IGF-IR that are incapable of giving a mitogenic signal are able to stimulate anchorage-independent-growth phenotype. IGF-IR mutants that are mitogenic but not transforming allow growth in EGF but do not allow transformed growth. In summary, growth in soft agar requires an activated EGFR and an IGF-IR capable of transmitting specifically the transforming signal. Although there is no detectable physical interaction of the two receptors, functional interaction does occur, and EGF can stimulate the phosphorylation of the insulin receptor substrate-1 (IRS-1), when EGFRs are overexpressed. In cells with mutant IGF-IR and over-expressed EGFR, EGF can still stimulate cell proliferation even though IGF-I cannot. Yet EGF cannot produce a signal if there are no IGF-IR present. Although these results are obtained in extremely artificial conditions, they tell us that different receptors can crosstalk and also that IGF-IR is important for promotion of the transformed phenotype. This situation could be important to the onset of uncontrolled growth leading to tumorigenesis.

Crosstalk between these receptors could occur through Shc adapter protein, because there are two tyrosine-binding sites, one at the NT is a phosphotyrosine-binding domain (PTB), while the carboxyterminal (CT) has an src-homology domain (SH2) with different binding targets. The PTB is necessary for IGF-IR signaling, and the SH2 is specific for the EGFR (Ricketts *et al.*, 1996). Both use their specific coupling domain to signal through the ras/mitogen activated phosphokinase (MAPK) pathway: If it is possible for both receptors to bind to the same Shc molecule, then they could be linked together in this manner. If the Shc becomes activated, it could convey the signals from either or both sources. A nice example of the convergence of signaling pathways in the responses of a cell is that of PC12 cells stimulated with EGF that normally gives a proliferative response. But in cells with overexpressed EGFR (or insulinR), the addition of EGF (or insulin) leads to differentiation as efficiently as the addition of nerve growth factor (NGF). The sustained elevation of MAPK is necessary for this response, and its subsequent phosphorylation and migration to the nucleus (Traverse *et al.*, 1994).

The preimplantation mouse embryo is in a position to make similar interactions, since EGFR and IGF-IR are expressed together with endogenous TGFα, Ar, IGF-II, and maternal insulin. If one of these systems is compromised, then the other components may be able to compensate. This flexibility may include the PDGF-αR, which is also expressed together with PDGF-A. It might also explain the response of preimplantation embryos to ionizing radiation (Wiley *et al.*, 1994). An acute exposure of 1 Gy has no effect on the proliferation rate of conventionally cultured embryos, although there is an negative proliferative effect when the irradiated embryos are challenged by direct cell–cell contact with unirradiated embryos in chimeras. The reduced cell-proliferation rate of irradiated blastomeres is rescued by the addition of IGF but not by the addition of TGFα (Peters *et al.*, 1996), suggesting that the IGF-IR system is important in situations where proliferation is the monitored endpoint, as in embryonic development and in transformation studies. Other studies indicate that in preimplantation development, EGFR is more important for differentiation (Brice *et al.*, 1993). In this light, it is notable that irradiated preimplantation embryos develop into blastocysts more slowly than unirradiated and that added TGFα can restore the normal development rate (Peters *et al.*, 1996).

4. Interaction with Signaling Components

After EGF or HRG stimulation, a number of proteins that contain SH2 or SH3 domains become associated with the phosphorylated receptor protein (reviewed by Carraway and Cantley, 1994). Shc, Sos, Grb-2, and PLCγ can bind as a large complex to activated EGFR, and signaling is almost entirely through the MAPK pathway rather than via STAT proteins and JNK (Pinkas-Kramarski *et al.*, 1996b). c-Src appears to be necessary to give the signal for DNA synthesis after

EGF stimulation through the Shc/Grb-2 intermediaries, and extended interaction of 8 h is needed to complete the pathway that enables the G1 restriction point to be traversed. At least three DNA-binding complexes signal to genes in the nucleus. One is the formation of the ternary complex of TCF (Elk1, a transcription factor in the Ets family) and SRF on the serum response element in promoters of responsive genes such as c-fos (Hipskind *et al.*, 1994). Another mechanism involves the sis-inducible element (SIE) and the proteins that bind to it, SIF (Ruff-Jamison *et al.*, 1994). Both elements appear to be necessary for full activation of the *c-fos* gene promoter. Some conditions stimulate the STAT factors, for example, the administration of EGF or interferon-γ (INF-γ) to mice or to certain cells results in the appearance of STAT1α and STAT1β that are phosphorylated on tyrosine in the cytoplasm and then migrate to the nucleus to bind to specific DNA elements. An element in the *β-casein* promoter binds STAT1 and STAT5 in mammary glands stimulated with EGF, and accounts for the stimulation of differentiated expression (Ruff-Jamison *et al.*, 1995).

Several proteins that bind to the EGFR and that have regulatory functions have been cloned recently. A binding component that appears to be specific to EGFR is c-Cbl, a 120-kDa protooncogene first discovered in immune cell receptor signaling. c-Cbl becomes attached to the EGFR complex, probably through binding to the SH3 domain of Grb-2, and endows an inhibitory activity on the EGFR (Fukazawa *et al.*, 1996). c-Cbl becomes tyrosine phosphorylated and associates with several other signaling components, including PI3kinase and Crk. This connection gives another dimension to possible signaling choices in the cell. c-Cbl also associates with v-src in *v-src*-transformed cells (Odai *et al.*, 1995) and is one of many proteins that are tyrosine phosphorylated in transformed cells. It remains to be proved that a zinc-finger protein, ZPR1, that binds to the non-activated EGFR protein plays a role both in basal activity of the receptor and in signaling to the nucleus (Galcheva-Gargova *et al.*, 1996). The 51-kDa protein is released from its binding site in the kinase domain of EGFR (and PDGFR) after ligand binding to the receptor and, after its activation by phosphorylation, migrates from the cytoplasm to the nucleus. This promises to be an important new mode of signaling. The Eps8 protein associates with and is a substrate for EGFR and other RTKs. This 97-kDa protein has a nuclear targeting domain and an SH3 domain, although it appears to bind to the EGFR at a juxta-membrane site not involving an SH2 domain (Castagnino *et al.*, 1995). After activation of the EGFR, the internalized protein is removed by degradation in the lysosomes. Modulation of this process occurs by the covalent linkage of EGFR to ubiquitin and the degradation pathway (Galcheva-Gargova *et al.*, 1995).

ErbB-2 also strongly couples to the MAPK pathway, *via* Shc, but Grb-2 and c-Cbl are absent from these complexes. Of significance to the tumorigenic process, overexpressed c-ErbB2 interacts directly with and activates c-src in mammary cells (Muthuswamy and Muller, 1995). This interaction of the SH2 domain of src with neu is cell-type-specific and does not occur in fibroblasts. In addition, although EGFR may not interact directly with c-src, when EGF is added to mammary cells, c-src is also activated, via heterodimerization with ErbB2/neu (Muthuswamy and Muller, 1995). Activated ErbB3 heterodimers are coupled to Shc and PI3Kinase but not to PI3K and Cbl, thus providing diverse signaling in responsive cells.

VII. The Occurrence and Importance of the ErbB Genes in Cancer

The EGFR gene was discovered by its overexpression in erythroblastic leukemia in chickens and named ErbB for this reason. The accompanying overexpressed ErbA was later characterized as a member of the steroid receptor family of genes. ErbB protein is a truncated form of EGFR, with a short extracellular domain that is unable to bind EGF, but the kinase domain is constitutively active. Although the EGFR is not expressed in any adult hematopoietic cells, an early precursor is thought to have expressed it, and an invading retrovirus selected these two genes for its own survival to create an infective virus that would multiply and eventually kill the host. Hence, normal genes that are subjugated in a variety of ways and establish themselves permanently in the retroviral genome are called viral oncogenes. The host cell with the original gene has a version of that gene called the cellular proto-oncogene. Both EGFR and ErbB2 are in this category, as are many of the signal transduction intermediaries that were also established by other retroviruses.

ErbB2 was first discovered as the *neu* oncogene in a rat neuroblastoma. Its normal counterpart is named ErbB2 (HER2 in human) and is highly conserved evolutionarily. Both of these genes are frequently found to be overexpressed in carcinomas of mammary, ovary, prostate, colon, etc. In some cases, the genes are amplified. In others, transcription is induced and the protein overexpressed. The appearance of these two markers is significant for poor prognosis and appears to occur late in the progression of tumor aggressiveness. Interestingly, some of the same carcinomas also overexpress ErbB-3 and -4, suggesting that they have a common mode of induction, perhaps based on steroid hormone effects.

The first studies of tumors that overexpress the EGFR gene showed that, in many cases, the gene is amplified, and the prominent example is A431 vulval carcinoma cell line that has multiple copies of the gene. Amplification is not necessary for overexpression, because, instead, both mRNA and protein may be elevated in tumor cells. This event alone is not enough to cause transformation, and at least the overexpression of a ligand is required to provide the next step in tumor progression. Similarly, the overexpression of one of the ErbB proteins alone has little effect. But if two are overexpressed in mammary cells, they may synergize to give neoplastic transformation and mammary carcinomas. For example, ErbB-2 and -3 cooperate and heterodimerization is necessary to activate the kinase domain of ErbB3. This combination often occurs in mammary tumor cell lines. The heterodimer can be chronically activated by heregulin production in the same cell (Alimandi *et al.*, 1995). The auto- and cross-induction within the EGF family of ligands has been studied in a nontransformed intestinal cell line. The results suggest that EGF, HB-EGF, Ar, TGFα, and BTC all induce the expression of the others. For Ar and HB-EGF the induction is in part direct (Barnard *et al.*, 1994). The interaction of ErbB2 and c-myc or c-Hras also leads to increased tumorigenicity (W. Li *et al.*, 1996).

There are suggestions of a role for overexpression of EGFR in the progression to hormone independence in more aggressive breast cancers (Chrysogelos *et al.*, 1994). Overexpression of one gene in this family escalates to more and may be part of progression, or the increased susceptibility, to carcinogens. For instance, male TGFα transgenic mice succumb rapidly to hepatocarcinogenesis after exposure to diethylnitrosamine or phenobarbital (Takagi *et al.*, 1993). In addition, the stomach of the TGFα transgenic mouse frequently develops dysplasia and hyperplasia, symptoms similar to a human disease, Menetrier's disease (Takagi *et al.*, 1992). Overexpression of the EGFR has major implications for the overphosphorylation of cellular proteins that modulate adhesion, such as the inhibition of the activities of integrins. Epithelial cells then show reduced cell spreading, and loss of epithelial phenotype in response to EGF. During this process, β-catenin/E-cadherin complexes are tyrosine phosphorylated and become more soluble in detergent (Fujii, 1996). In these cells the cytoskeleton is less stable and the cells more motile, characteristics that enhance metastatic behavior.

Overexpression of ErbB2 leads to decreased α2β1 integrin expression, a process that involves Spl activity inhibition, and that could lead to disruption of tissue architecture in breast cancer (Ye *et al.*, 1996). Increased ErbB2 expression also is correlated with decreased E-cadherin expression, a result that increases the instability of epithelial structures. The CT domain of E-cadherin normally interacts with β-catenin and helps in maintaining epithelial integrity. This interaction is weakened by the interaction of overexpressed ErbB2 with β-catenin and its phosphorylation, a common situation in adenocarcinoma cell lines. A dominant negative version of β-catenin prevents this interaction and returns the cells to a more normal epithelial morphology and noninvasive behavior (Shibata *et al.*, 1996).

VIII. EGFR in Nonmammalian Development

A. EGFR and Ligands in *Caenorhabditis elegans*

The nematode worm *C. elegans* also expresses a highly homologous set of genes to the EGFR, with developmental functions that have been best worked out for vulval development (reviewed by Kayne and Sternberg, 1995). The TGFα equivalent is a membrane-bound protein encoded by Lin-3, which binds to and activates, by phosphorylation, the Let-23 gene product, or the EGFR. The equivalent of Grb2 is Sem-5: Ras, Let-60; Raf-1, Lin-45; MAPK, Sur-1/MPK-1. Thus, signaling travels a highly analogous pathway to reach the nucleus, with subsequent transcription factor induction. The developmental consequences of null mutations in the pathway are larval lethality, sterility, no vulval development, abnormal male spicule development, and abnormal posterior ectodermal fates. The level of Ras expression acts as a switch to fate, with high levels necessary for vulva development. Some negative influences are common to the mammalian signal pathway, such as c-Cbl or its *C. elegans* equivalent, Sli-1; UNC-101, which prevents correct EGFR interactions by acting in an adaptinlike protein for internalization.

B. EGFR and Ligands in *Drosophila*

Both *C. elegans* and *Drosophila* genetics and the unique ability to study molecular structure and function in these species have combined to make some interesting discoveries on the conservation of the developmentally important pathways (reviewed by Duffy and Perrimon, 1996; Wassarman *et al.*, 1995). For instance, for at least four of the receptor kinase adaptor proteins, Grb2, Raf, MEK, and MAPK, chimeric molecules containing mammalian sequences in place of the normal protein functions correctly. The EGFR is conserved as DER, or the torpedo gene; the DER ligands Spitz (Spi, equivalent to TGFα) and Gurken function in axis determination in the oocyte. If DER is activated by Gurken, a dorsal fate is set and the dorsal/ventral axis established. The activity of DER, Gurken, and Cornichon (both ligands for DER) is needed to retain a posterior fate. The signal pathway components so far recognized are EGFR, DER; Grb2, Drk; Shc, Dshc; Sos, RasGNRF; Ras, Ras1; GAP, Gap1, Raf, Draf; MAPKK, Dsor1; MAPK, Rolled. The most studied system that this signaling pathway regulates is the development of the ommatidia of the eye (Chang *et al.*, 1994). A number of intriguing regulatory molecules akin to TGFα, such as Rhomboid and Star, are transmembrane ligands of DER that can be secreted in a soluble form with marked alteration in cell fates. This type of effect may have its equivalent in mammalian tissues, such as in the uterus, to guide the process of implantation, by either soluble forms of ligands or soluble forms of the EGFR (Tong *et al.*, 1996). The influences of mesenchyme on various adjacent epithelia are examples of inductive interactions that are based on a combination of secreted and transmembrane components with positive and negative effectors integrated by unknown mechanisms that may be based on the levels and duration of signal pathway activation.

IX. Conclusions

The aim here was to highlight the recent findings on the common threads connecting the activities of the EGFR and its ligands in embryonic development, in tissue function, and in malignant transformation. We also indicated the gaps and uncertainties, especially in understanding mechanisms of action. The intention was to build up the concept of multilevel integration of the receptor, with its homologs and multiple ligands, to give subtle and specific cellular responses. Importantly, the EGFR system does not signal along one pathway. In fact, it is one of the most interactive systems, with far-reaching effects on the cytoskeleton, locomotion, differentiated function, and cell polarity, with the extracellular matrix and secreted factors that further increase its long reach. Finally, evidence was presented to highlight the reciprocity between the EGFR signaling and the signaling of other growth factor receptors. The result is a complex physiology that is normally, and of necessity, well regulated. The activities of the ErbB family that pertain in development are vital to cell shape, movement, and morphological processes such as organ and tissue shaping, as well as regulating function. In adults, ErbB roles appear to be focused on stem cell renewal, detoxification, and nutrition. Recent advances in mapping the activities of the components of signaling pathways are helping to explain both redundancy in genes and their specificity of actions that characterize the interplay between EGFR and its fellow growth factor receptors in mammals.

Abbreviations

Ar, amphiregulin; aFGF, acidic fibroblast growth factor; bFGF, basic fibroblast growth factor; BTC, betacellulin; CNS, central nervous system; ES, embryonic stem; EC, embryonal carcinoma; EGFR, epidermal growth factor receptor (also called ErbB1, HER-1); HB-EGF, heparin-binding epidermal growth factor; HRG, heregulin; ICM, inner cell mass; IGF-I, IGF-II, insulinlike growth factors; IV, intravenous; JNK, c-jun kinase; KO, knockout; LIF, leukemia inhibitory factor; LHRH, luteinizing hormone releasing factor; MAPK, mitogen-activated phosphokinase; NCC, neural crest cells; Neu, neuroblastoma gene product (ErbB2 or HER-2); NDF, neural differentiation factor; NRG, neuregulin; PDGF, platelet-derived growth factor; PI3K, phosphoinositol-3 kinase; PKC, phosphokinase C; PLCγ, phospholipase Cγ; PNS, peripheral nervous system; SC, spinal cord; SIE, sis-inducible element; SRE, serum response element; SRF, serum response factor; TGFα, transforming growth factor-α.

Acknowledgments

We offer many apologies to those whose work could not be mentioned because of the restrictions in space available. We thank Drs. Ian DeBelle and Dan Mercola for critical reading of the manuscript.

This work was supported by a grant from the DOD, DAMD17-94-J-4286, and by grants from the USPHS, CA28427 (EDA) and ES05409 (LMW).

References

Abbott, B. D., and Birnbaum, L. S. (1990). *Toxicol. Appl. Pharmacol.* **106**, 418–432.

Adam, R., Drummond, D. R., Solic, N., Holt, S. J., Sharma, R. P., Chamberlin, S. G., and Davies, D. E. (1995). *Biochim. Biophys. Acta* **1266**, 83–90.

Adam, R. M., Chamberlin, S. G., and Davies, D. E. (1996). *Growth Factors* **13**, 193–203.

Adamson, E. D. (1990). *Current Topics in Developmental Biology* **24**, pp. 1–29.

Adelaide, J., Penault-Llorca, F., Dib, A., Yarden, Y., Jacquemier, J., and Birnbaum, D. (1994). *Genes Chromosomes Cancer* **11**, 66–69.

Alimandi, M., Romano, A., Curia, M. C., Muraro, R., Fedi, P., Aaronson, S. A., Di Fiore, P. P., and Kraus, M. H. (1995). *Oncogene* **10**, 1813–1821.

Altiok, N., Bessereau, J. L., and Changeux, J. P. (1995). *Embo. J.* **14**, 4258–4266.

Amemiya, K., Kurachi, H., Adachi, H., Morishige, K. I., Adachi, K., Imai, T., and Miyake, A. (1994). *J. Endocrinol.* **143**, 291–301.

Aviezer, D., and Yayon, A. (1994). *Proc. Natl. Acad. Sci. USA* **91**, 12173–12177.

Bargmann, C. I., Hung, M. C., and Weinberg, R. A. (1986). *Nature* **319**, 226–230.

Barnard, J., Shing, Y., Christofori, G., Hanahan, D., Ono, Y., Sasada, R., Igarashi, K., and Folkman, J. (1993). *J. Pediatr. Gastroenterol. Nutr.* **17**, 343–344.

Barnard, J. A., Graves-Deal, R., Pittelkow, M. R., DuBois, R., Cook, P., Ramsey, G. W., Bishop, P. R., Damstrup, L., and Coffey, R. J. (1994). *J. Biol. Chem.* **269**, 22817–22822.

Barnard, J. A., Beauchamp, R. D., Russell, W. E., Dubois, R. N., and Coffey, R. J. (1995). *Gastroenterology* **108**, 564–580.

Bass, K. E., Morrish, D., Roth, I., Bhardwaj, D., Taylor, R., Zhou, Y., and Fisher, S. J. (1994). *Devel. Biol.* **164**, 550–561.

Baulida, J., Kraus, M. H., Alimandi, M., Di Fiore, P. P., and Carpenter, G. (1996). *J. Biol. Chem.* **271**, 5251–5257.

Beerli, R. R., and Hynes, N. E. (1996). *J. Biol. Chem.* **271**, 6071–6076.

Bennett, K. L., Plowman, G. D., Buckley, S. D., Skonier, J., and Purchio, A. F. (1992). *Growth Factors* **7**, 207–213.

Besner, G. E., Whelton, D., Crissman-Combs, M. A., Steffen, C. L., Kim, G. Y., and Brigstock, D. R. (1992). *Growth Factors* **7**, 289–296.

Bhatt, H., Brunet, L. J., and Stewart, C. L. (1991). *Proc. Natl. Acad. Sci. USA* **88**, 11408–11412.

Bidwell, M. C., Eitzman, B. A., Walmer, D. K., McLachlan, J. A., and Gray, K. D. (1995). *Endocrinology* **136**, 5189–5201.

Bjorge, J. S., Paterson, A. J., and Kudlow, J. E. (1989). *J. Biol. Chem.* **264**, 4021–4027.

Blankenship, A. L., Suffia, M. C., Matsumura, F., Walsh, K. J., and Wiley, L. M. (1993). *Reprod. Toxicol.* **7**, 255–261.

Boland, N. I., and Gosden, R. G. (1994). *J. Reprod. Fertil.* **101**, 369–374.

Bosher, J. M., Williams, T., and Hurst, H. C. (1995). *Proc. Natl. Acad. Sci. USA* **92**, 744–747.

Brachman, R., Lindquist, P. B., Nagashima, M., Kohr, W., Lipari, T., Napier, M., and Derynck, R. (1989). *Cell* **56**, 691–700.

Brice, E. C., Wu, J.-X., Muraro, R., Adamson, E. D., and Wiley, L. M. (1993). *Dev. Genet.* **14**, 174–184.

Brown, M. J., Zogg, J. L., Schultz, G. S., and Hilton, F. K. (1989). *Endocrinology* **124**, 2882–2887.

Burgaud, J. L., and Baserga, R. (1996). *Exp. Cell Res.* **223**, 412–419.

Cain, M. P., Kramer, S. A., Tindall, D. J., and Husmann, D. A. (1994). *Urology* **43**, 375–378.

Cardiff, R. D. (1996). *J. Mammary Gland Biol. Neoplasia* **1**, 61–73.
Carpenter, G. (1993). *Curr. Opin. Cell Biol.* **5**, 261–264.
Carraway, K. L. R., and Cantley, L. C. (1994). *Cell* **78**, 5–8.
Carver, R. S., Sliwkowski, M. X., Sitaric, S., and Russell, W. E. (1996). *J. Biol. Chem.* **271**, 13491–13496.
Castagnino, P., Biesova, Z., Wong, W. T., Fazioli, F., Gill, G. N., and Di Fiore, P. P. (1995). *Oncogene* **10**, 723–729.
Catterton, W. Z., Escobedo, M. B., and Sexson, W. R. (1979). *Ped. Res.* **13**, 104–108.
Chang, H. C., Karim, F. D., O'Neill, E. M., Rebay, I., Solomon, N. M., Therrien, M., Wassarman, D. A., Wolff, T., and Rubin, G. M. (1994). *Cold Spring Harb. Symp. Quant. Biol.* **59**, 147–153.
Chen, Y., and Gill, G. N. (1994). *Oncogene* **9**, 2269–2276.
Chenevix-Trench, G., Jones, K., Green, A. C., Duffy, D. L., and Martin, N. G. (1992). *Am. J. Hum. Genet.* **51**, 1377–1385.
Chesnel, F., Wigglesworth, K., and Eppig, J. J. (1994). *Develop. Biol.* **161**, 285–294.
Chia, C. M., Winston, R. M. L., and Handyside, A. H. (1995). *Development* **121**, 299–307.
Chrysogelos, S. A., Yarden, R. I., Lauber, A. H., and Murphy, J. M. (1994). *Breast Cancer Res. Treat.* **31**, 227–236.
Cohen, S. (1983). *Cancer* **51**, 1787–1791.
Cohen, S., Fava, R. A., and Sawyer, S. T. (1982). *Proc. Natl. Acad. Sci. USA* **79**, 6237–6241.
Coleman, S., and Daniel, C. W. (1990). *Dev. Biol.* **137**, 425–433.
Coleman, S., Silberstein, G. B., and Daniel, C. W. (1988). *Dev. Biol.* **127**, 304–315.
Coppola, D., Ferber, A., Miura, M., Sell, C., D'Ambrosio, C., Rubin, R., and Baserga, R. (1994). *Mol. Cell. Biol.* **14**, 4588–4595.
Coskun, S., and Lin, Y. C. (1995). *Molec. Reprod. Develop.* **42**, 311–317.
Couture, L. A., Abbott, B. D., and Birnbaum, L. S. (1990). *Teratology* **42**, 619–627.
Cullinan, E. B., Abbondanzo, S. J., Anderson, P. S., Pollard, J. W., Lessey, B. A., and Stewart, C. L. (1996). *Proc. Natl. Acad. Sci. USA* **93**, 3115–3120.
Cunha, G. R., and Hom, Y. K. (1996). *J. Mamm. Gland Biol. Neoplasia* **1**, 21–35.
Curtis, S. W., Washburn, T., Sewall, C., DiAugustine, R., Lindzey, J., Couse, J. F., and Korach, K. S. (1996). *Proc. Natl. Acad. Sci. USA* **93**, 12626–12630.
Dalton, T., Kover, K., Dey, S. K., and Andrews, G. K. (1994). *Biol. Reprod.* **51**, 597–606.
Daniel, C. W., and Robinson, S. D. (1992). *Mol. Reprod. Dev.* **32**, 145–151.
Dardik, A., and Schultz, R. M. (1991). *Development* **113**, 919–930.
Dardik, A., Smith, R., and Schultz, R. M. (1992). *Dev. Biol.* **154**, 396–409.
Dardik, A., Doherty, A. S., and Schultz, R. M. (1993). *Molec. Reprod. Develop.* **34**, 396–401.
Das, S. K., Tsukamura, H., Paria, B. C., Andrews, G. K., and Dey, S. K. (1994a). *Endocrinology* **134**, 971–981.
Das, S. K., Wang, X. N., Paria, B. C., Damm, D., Abraham, J. A., Klagsbrun, M., Andrews, G. K., and Dey, S. K. (1994b). *Development* **120**, 1071–1083.
Das, S. K., Chakraborty, I., Paria, B. C., Wang, X. N., Plowman, G., and Dey, S. K. (1995). *Mol. Endocrinol.* **9**, 691–705.
Deb, S. P., Munoz, R. M., Brown, D. R., Subler, M. A., and Deb, S. (1994). *Oncogene* **9**, 1341–1349.
Dedhar, S., and Hannigan, G. E. (1996). *Curr. Opin. Cell Biol.* **8**, 657–669.
Dekel, N., and Sherizly, I. (1985). *Endocrinology* **116**, 406–409.
den Hartigh, J. C., van Bergen en Henegouwen, P. M., Verkleij, A. J., and Boonstra, J. (1992). *J. Cell. Biol.* **119**, 349–355.
den Hertog, J., De Laat, A. W., Schlessinger, J., and Kruijer, W. (1991). *Cell Growth Differ.* **2**, 155–164.
Diakonova, M., Payrastre, B., van Velzen, A. G., Hage, W. J., van Bergen en Henegouwen, P. M., Boonstra, J., Cremers, F. F., and Humbel, B. M. (1995). *J. Cell Sci.* **108**, 2499–2509.

Downs, S. M. (1989). *Biol. Reprod.* **41**, 371–379.

Downward, J., Yarden, Y., Mayes, E., Scrace, G., Totty, N., Stockwell, P., Ullrich, A., Schlessinger, J., and Waterfield, M. D. (1984). *Nature* **307**, 521–527.

Duffy, J. B., and Perrimon, N. (1996). *Curr. Opin. Cell Biol.* **8**, 231–238.

Dvorak, B., Holubec, H., LeBouton, A. V., Wilson, J. M., and Koldovsky, O. (1994). *FEBS Lett.* **352**, 291–295.

Ebert, M., Yokoyama, M., Kobrin, M. S., Friess, H., Lopez, M. E., Buchler, M. W., Johnson, G. R., and Korc, M. (1994). *Cancer Res.* **54**, 3959–3962.

Endo, K., Atlas, S. J., Rone, J. D., Zanagnolo, V. L., Kuo, T-C., Dharmarajan, A. M., and Wallach, E. E. (1992). *Endocrinology* **130**, 186–192.

Englert, C., Hou, X., Maheswaran, S., Bennett, P., Ngwu, C., Re, G. G., Garvin, A. J., Rosner, M. R., and Haber, D. A. (1995). *Embo. J.* **14**, 4662–4675.

Eppig, J. J. (1991). *BioEssays* **13**, 569–547.

Eppig, J. J., and O'Brien, M. J. (1996). *Biol. Reprod.* **54**, 197–207.

Eppig, J. J., Peters, A. H. F. M., Telfer, E. E. and Wigglesworth, K. (1993a). *Molec. Reprod. Develop.* **34**, 450–456.

Eppig, J. J., Wigglesworth, K., and Chesnel, K. (1993b). *Develop. Biol.* **158**, 400–409.

Ethier, S. P., Langton, B. C., and Dilts, C. A. (1996). *Mol. Carcinog.* **15**, 134–143.

Fan, X., and Childs, G. V. (1995). *Endocrinology* **136**, 2284–2293.

Fan, X., Nagle, G. T., Collins, T. J., and Childs, G. V. (1995). *Endocrinology* **136**, 873–880.

Felix, R., Meza, U., and Cota, G. (1995). *Endocrinology* **136**, 939–946.

Fowler, K. J., Walker, F., Alexander, W., Hibbs, M. L., Nice, E. C., Bohmer, R. M., Mann, G. B., Thumwood, C., Maglitto, R., Danks, J. A., Chetty, R., Burgess, A. W., and Dunn, A. R. (1995). *Proc. Natl. Acad. Sci. USA* **92**, 1465–1469.

Francoeur, J. R., Richardson, P. M., Dunn, R. J., and Carbonatto, S. (1995). *J. Neurosci. Res.* **41**, 836–845.

Fujii, K. (1996). *J. Invest. Dermatol.* **107**, 195–202.

Fukazawa, T., Miyake, S., Band, V., and Band, H. (1996). *J. Biol. Chem.* **271**, 14554–14559.

Galcheva-Gargova, Z., Theroux, S. J., and Davis, R. J. (1995). *Oncogene* **11**, 2649–2655.

Galcheva-Gargova, Z., Konstantinov, K. N., Wu, I. H., Klier, F. G., Barrett, T., and Davis, R. J. (1996). *Science* **272**, 1797–1802.

Ganser, G. L., Stricklin, G. P., and Matrisian, L. M. (1991). *Int. J. Dev. Biol.* **35**, 453–461.

Gassmann, M., Casagranda, F., Orioli, D., Simon, H., Lai, C., Klein, R., and Lemke, G. (1995). *Nature* **378**, 390–394.

Gattone, V. H., 2nd., Andrews, G. K., Niu, F. W., Chadwick, L. J., Klein, R. M., and Calvet, J. P. (1990). *Dev. Biol.* **138**, 225–230.

Gattone, V. H., 2nd., Lowden, D. A., and Cowley, B. D., Jr. (1995). *Dev. Biol.* **169**, 504–510.

Gattone, V. H., 2nd., Kuenstler, K. A., Lindemann, G. W., Lu, X., Cowley, B. D., Jr., Rankin, C. A., and Calvet, J. P. (1996). *J. Lab. Clin. Med.* **127**, 214–222.

Goldin, G. V., and Opperman, L. A. (1980). *J. Embryol. Exp. Morphol.* **60**, 235–243.

Gomez, E., Tarin, J. J., and Pellicer, A. (1993). *Fert. Steril.* **60**, 40–46.

Grainger, D. J., Witchell, C. M., Weissberg, P. L., and Metcalfe, J. C. (1994). *Cardiovasc. Res.* **28**, 1238–1242.

Grandis, J. R., Zeng, Q., and Tweardy, D. J. (1996). *Nature Medicine* **2**, 237–240.

Gupta, C. (1996). *Endocrinology* **137**, 905–910.

Gupta, C., Chandorkar, A., and Nguyen, A. P. (1996). *Mol. Cell. Endocrinol.* **123**, 89–95.

Guy, C. T., Webster, M. A., Schaller, M., Parsons, T. J., Cardiff, R. D., and Muller, W. J. (1992). *Proc. Natl. Acad. Sci. USA* **89**, 10578–10582.

Guy, C. T., Muthuswamy, S. K., Cardiff, R. D., Soriano, P., and Muller, W. J. (1994). *Genes Dev.* **8**, 23–32.

Haining, R. E., Schofield, J. P., Jones, D. S., Rajput-Williams, J., and Smith, S. K. (1991). *J. Mol. Endocrinol.* **6,** 207–214.

Halter, S. A., Dempsey, P., Matsui, Y., Stokes, M. K., Graves-Deal, R., Hogan, B. L., and Coffey, R. J. (1992). *Am. J. Pathol.* **140,** 1131–1146.

Han, V. K. M., Hunter, E. S., Pratt, R. M., Zendegui, J., and Lee, D. C. (1987). *Mol. Cell. Biol.* **7,** 2335–2343.

Harvey, M. B., Leco, K. J., Arcellana-Panlilio, M. Y., Zhang, X., Edwards, D. R., and Schultz, G. A. (1995). *Development* **121,** 1005–1014.

Haslam, S. Z., Counterman, L. J., and Nummy, K. A. (1992). *J. Cell. Physiol.* **152,** 553–557.

Higashiyama, S., Abraham, J. A., and Klagsbrun, M. (1993). *J. Cell. Biol.* **122,** 933–940.

Hipskind, R. A., Baccarini, M., and Nordheim, A. (1994). *Mol. Cell. Biol.* **14,** 6219–6231.

Ho, W. H., Armanini, M. P., Nuijens, A., Phillips, H. S., Osheroff, P. L. (1995). *J. Biol. Chem.* **270,** 14523–14532.

Holder, S. E., Vintiner, G. M., Farren, B., Malcolm, S., and Winter, R. M. (1992). *J. Med. Genet.* **29,** 390–392.

Hou, X., Johnson, A. C., and Rosner, M. R. (1994a). *J. Biol. Chem.* **269,** 4307–4312.

Hou, X., Johnson, A. C., and Rosner, M. R. (1994b). *Cell Growth Differ.* **5,** 801–809.

Hu, C. C., Sakakura, Y., Sasano, Y., Shum, L., Bringas, P. J., Werb, Z., and Slavkin, H. C. (1993). *Int. J. Dev. Biol.* **36,** 505–516.

Huang, R.-P., Wu, J.-X., Fan, Y., and Adamson, E. D. (1996). *J. Cell. Biol.* **133,** 211–220.

Hubbard, C. J. (1994). *J. Cell. Physiol.* **160,** 227–232.

Hudson, L. G., Ertl, A. P., and Gill, G. N. (1990a). *J. Biol. Chem.* **265,** 4389–4393.

Hudson, L. G., Santon, J. B., Glass, C. K., and Gill, G. N. (1990b). *Cell* **62,** 1165–1175.

Hudson, L. G., Thompson, K. L., Xu, J., and Gill, G. N. (1990c). *Proc. Natl. Acad. Sci. USA* **87,** 7536–7540.

Hung, M. C., Schechter, A. L., Chevray, P. Y., Stern, D. F., and Weinberg, R. A. (1986). *Proc. Natl. Acad. Sci. USA* **83,** 261–264.

Imai, T., Kurachi, H., Adachi, K., Adachi, H., Yoshimoto, Y., Homma, H., Tadokoro, C., Takeda, S., Yamaguchi, M., Sakata, M., et al. (1995). *Biol. Reprod.* **52,** 928–938.

Ishii, S., Xu, Y.-H., Stratton, R. H., Roe, B. A., Merlino, G. T., and Pastan, I. (1985). *Proc. Natl. Acad. Sci. USA* **82,** 4920–4924.

Issing, W. J., Liebich, C., Wustrow, T. P., and Ullrich, A. (1996). *Anticancer Res.* **16,** 283–288.

Jacobberger, J. W., Sizemore, N., Gorodeski, G., and Rorke, E. A. (1995). *Exp. Cell. Res.* **220,** 390–396.

Jetten, A. M. (1980). *Nature* **284,** 626–629.

Jhappan, C., Stahle, C., Harkins, R. N., Fausto, N., Smith, G. H., and Merlino, G. T. (1990). *Cell* **61,** 1137–1146.

Jhappan, C., Geiser, A. G., Kordon, E. C., Bagheri, D., Hennighausen, L., Roberts, A. B., Smith, G. H., and Merlino, G. (1993). *Embo. J.* **12,** 1835–1845.

Jo, S. A., Zhu, X., Marchionni, M. A., and Burden, S. J. (1995). *Nature* **373,** 158–161.

Johnson, A. C. (1996). *J. Biol. Chem.* **271,** 3033–3038.

Johnson, D. C., and Chatterjee, S. (1993). *J. Reprod. Fertil.* **99,** 557–559.

Johnson, G. R., and Wong, L. (1994). *J. Biol. Chem.* **269,** 27149–27154.

Johnson, G. R., Saeki, T., Auersperg, N., Gordon, A. W., Shoyab, M., Salomon, D. S., and Stromberg, K. (1991). *Biochem. Biophys. Res. Commun.* **180,** 481–488.

Junier, M. P., Hill, D. F., Costa, M. E., Felder, S., and Ojeda, S. R. (1993). *J. Neurosci.* **13,** 703–713.

Kageyama, R., Merlino, G. T., and Pastan, I. (1988). *Proc. Natl. Acad. Sci. USA* **85,** 5016–5020.

Kashimata, M., Hiramatsu, M., and Minami, N. (1988). *Endocrinology* **122,** 1707–1714.

Kawamoto, T., Matsumura, F., Madhukar, B. V., and Bombick, D. W. (1989). *J. Biochem. Toxicol.* **4,** 173–182.

Kayne, P. S., and Sternberg, P. W. (1995). *Curr. Opin. Genet. Dev.* **5,** 38–43.

Kenigsberg, R. L., Mazzoni, I. E., Collier, B., and Cuello, A. C. (1992). *Neuroscience* **50,** 85–97.

Kenney, N. J., Huang, R. P., Johnson, G. R., Wu, J. X., Okamura, D., Matheny, W., Kordon, E., Gullick, W. J., Plowman, G., Smith, G. H., et al. (1995). *Mol. Reprod. Dev.* **41,** 277–286.

Kinoshita, N., Minshull, J., and Kirschner, M. W. (1995). *Cell* **83,** 621–630.

Kitadai, Y., Yamazaki, H., Yasui, W., Kyo, E., Yokozaki, H., Kajiyama, G., Johnson, A. C., Pastan, I., and Tahara, E. (1993). *Cell. Growth Differ.* **4,** 291–296.

Klein, J. M., and Nielsen, H. C. (1993). *J. Clin. Invest.* **91,** 425–431.

Kordon, E. C., McKnight, R. A., Jhappan, C., Hennighausen, L., Merlino, G., and Smith, G. H. (1995). *Dev. Biol.* **168,** 47–61.

Krane, I. M., and Leder, P. (1996). *Oncogene* **12,** 1781–1788.

Kwok, T. T., and Sutherland, R. M. (1991). *Br. J. Cancer* **64,** 251–254.

Kwok, T. T., and Sutherland, R. M. (1992). *Int. J. Radiat. Oncol. Biol. Phys.* **22,** 525–527.

Lau, A. F., Kanemitsu, M. Y., Kurata, W. E., Danesh, S., and Boynton, A. L. (1992). *Molec. Biol. Cell* **3,** 865–874.

Lee, K. F., Simon, H., Chen, H., Bates, B., Hung, M. C., and Hauser, C. (1995). *Nature* **378,** 394–398.

Lee, P. P., Darcy, K. M., Shudo, K., and Ip, M. M. (1995). *Endocrinology* **136,** 1718–1730.

Lemoine, N. R., Barnes, D. M., Hollywood, D. P., Hughes, C. M., Smith, P., Dublin, E., Prigent, S. A., Gullick, W. J., and Hurst, H. C. (1992). *Br. J. Cancer* **66,** 1116–1121.

Li, S., Plowman, G. D., Buckley, S. D., and Shipley, G. D. (1992). *J. Cell. Physiol.* **153,** 103–111.

Li, S., Maruo, T., Ladines-Llave, C. A., Samoto, T., Kondo, H., and Mochizuki, M. (1994). *Endocr. J.* **41,** 693–701.

Li, W., Park, J. W., Nuijens, A., Sliwkowski, M. X., and Keller, G. A. (1996). *Oncogene* **12,** 2473–2477.

Lillien, L. (1995). *Nature* **377,** 158–162.

Lim, H., Dey, S. K., and Das, S. K. (1997). *Endocrinology* (in press).

Lin, M. L., and Bertics, P. J. (1995). *J. Cell. Physiol.* **164,** 593–604.

Liu, S. C., Sanfilippo, B., Perroteau, I., Derynck, R., Salomon, D. S., and Kidwell, W. R. (1987). *Mol. Endocrinol.* **1,** 683–692.

Lorenzo, P. L., Illera, M. J., Illera, J. C., and Illera, M. (1994). *J. Reprod. Fertil.* **101,** 697–701.

Luciano, A. M., Pappalardo, A., Ray, C., and Pelluso, J. J. (1994). *Biol. Reprod.* **51,** 646–654.

Ludes-Meyers, J. H., Subler, M. A., Shivakumar, C. V., Munoz, R. M., Jiang, P., Bigger, J. E., Brown, D. R., Deb, S. P., and Deb, S. (1996). *Mol. Cell. Biol.* **16,** 6009–6019.

Luetteke, N. C., Qiu, T.-H., Peiffer, R. L., Oliver, P., Smithies, O., and Lee, D. C. (1993). *Cell* **73,** 263–278.

Luetteke, N. C., Phillips, H. K., Qiu, T. H., Copeland, N. G., Earp, H. S., Jenkins, N. A., and Lee, D. C. (1994). *J. Cell. Physiol.* **8,** 399–413.

Lund, K. A., Lazar, C. S., Chen, W. S., Walsh, B. J., Welsh, J. B., Herbst, J. J., Walton, G. M., Rosenfeld, M. G., Gill, G. N., and Wiley, H. S. (1990). *J. Biol. Chem.* **265,** 20517–20523.

Lysiak, J. J., Johnson, G. R., and Lala, P. K. (1995). *Placenta* **16,** 359–366.

Madden, S. L., Cook, D. M., Morris, J. F., Gashler, A., Sukhatme, V. P., and Rauscher, F. J. I. (1991). *Science* **253,** 1550–1553.

Madhukar, B. V., Brewster, D. W., and Matsumura, F. (1984). *Proc. Natl. Acad. Sci. USA* **81,** 7407–7411.

Madhukar, B. V., Ebner, K., Matsumura, F., Bombick, D. W., Brewster, D. W., and Kawamoto, T. (1988). *J. Biochem. Toxicol.* **3,** 261–277.

Mainiero, F., Pepe, A., Yeon, M., Ren, Y., and Giancotti, F. G. (1996). *J. Cell. Biol.* **134,** 241–253.

Mann, G. B., Fowler, K. J., Gabriel, A., Nice, E. C., Williams, L., and Dunn, A. R. (1993). *Cell* **73,** 249–261.

Mantzouris, N. M., Kaplan, J., Clark, G., and Hise, M. K. (1993). *Am. J. Kidney Dis.* **22,** 858–864.

Maruo, T., Ladines-Llave, C. A., Samoto, T., Matsuo, H., Manalo, A. S., Ito, H., and Mochizuki, M. (1993). *Endocrinology* **132,** 924–931.

Matrisian, L. M., and Hogan, B. L. (1990). *Curr. Top. Dev. Biol.* **24,** 219–259.

Matsui, Y., Halter, S. A., Holt, J. T., Hogan, B. L., and Coffey, R. J. (1990). *Cell* **61,** 1147–1155.

Medcalf, R. L., and Schleuning, W.-D. (1991). *Mol. Endocrinol.* **5,** 1773–1779.

Medina, D. (1996). *J. Mamm. Gland Biol. Neoplasia* **1,** 5–19.

Merlino, G. T., Xu, Y., Ishii, S., Clark, A. J. L., Semba, K., Toyoshima, K., Yamamoto, T., and Pastan, I. (1984). *Science* **224,** 417–419.

Merlo, G. R., Venesio, T., Taverna, D., Marte, B. M., Callahan, R., and Hynes, N. E. (1994). *Oncogene* **9,** 443–453.

Meyer, D., and Birchmeier, C. (1995). *Nature* **378,** 386–390.

Miettinen, P. J., Berger, J. E., Meneses, J., Phung, Y., Pedersen, R. A., Werb, Z., and Derynck, R. (1995). *Nature* **376,** 337–341.

Miki, H., Miura, K., and Takenawa, T. (1996). *Embo. J.* **15,** 5326–5335.

Mochizuki, M., and Maruo, T. (1992). *Nippon Naibunpi Gakkai Zasshi* **68,** 724–735.

Mohun, T. (1992). *Curr. Opin. Cell. Biol.* **4,** 923–928.

Moorby, C. D., Taylor, J. A., and Forsyth, I. A. (1995). *J. Endocrinol.* **144,** 165–171.

Moscoso, L. M., Chu, G. C., Gautam, M., Noakes, P. G., Merlie, J. P., and Sanes, J. R. (1995). *Dev. Biol.* **172,** 158–169.

Mouihate, A., and Lestage, J. (1995). *Neuroreport* **6,** 1401–1404.

Mueller, S. G., Kobrin, M. S., Paterson, A. J., and Kudlow, J. E. (1989). *Mol. Endocrinol.* **3,** 976–983.

Muhlhauser, D. W., Crescimanno, C., Kaufmann, P., Hofler, H., Zaccheo, D., and Castellucci, M. (1993). *J. Histochem. Cytochem.* **41,** 165–173.

Mukku, V. R., and Stancel, G. M. (1985). *J. Biol. Chem.* **260,** 9820–9826.

Mullaney, B. P., and Skinner, M. K. (1992). *Mol. Endocrinol.* **6,** 2103–2113.

Murillas, R., Larcher, F., Conti, C. J., Santos, M., Ullrich, A., and Jorcano, J. L. (1995). *Embo. J.* **14,** 5216–5223.

Murphy, P., Topilko, P., Schneider-Maunoury, S., Seitanidou, T., Baron-Van Evercooren, A., and Charnay, P. (1996). *Development* **122,** 2847–2857.

Muthuswamy, S. K., and Muller, W. J. (1995). *Oncogene* **11,** 271–279.

Nakafuku, M., and Kaziro, Y. (1993) *FEBS Lett.* **315,** 227–232.

Nanney, L. B., Sundberg, J. P., and King, L. E. (1996). *J. Invest. Dermatol.* **106,** 1169–1174.

Narasimhan, V., Hamill, O., and Cerione, R. A. (1992). *FEBS Lett.* **303,** 164–168.

Nieto-Sampedro, M. (1988). *Science* **240,** 1784–1786.

Nobes, C. D., and Hall, A. (1995). *Cell* **81,** 53–62.

Odai, H., Sasaki, K., Hanazono, Y., Ueno, H., Tanaka, T., Miyagawa, K., Mitani, K., Yazaki, Y., and Hirai, H. (1995). *Jpn. J. Cancer Res.* **86,** 1119–1126.

Ojeda, S. R., Urbanski, H. F., M. E., C., Hill, D. F., and Moholt-Siebert, M. (1990). *Proc. Natl. Acad. Sci. USA* **87,** 9698–9702.

Olwin, B. B., and Hauschka, S. D. (1988). *J. Cell. Biol.* **107,** 761–769.

Orellana, S. A., Sweeney, W. E., Neff, C. D., and Avner, E. D. (1995). *Kidney Internat.* **47,** 490–499.

Paria, B. C., and Dey, S. K. (1990). *Proc. Natl. Acad. Sci. USA* **87,** 4756–4760.

Paria, B. C., Tsukamura, H., and Dey, S. K. (1991). *Biol. Reprod.* **45,** 711–718.

Paria, B. C., Das, S. K., Andrews, G. K., and Dey, S. K. (1993). *Proc. Natl. Acad. Sci. USA* **90,** 55–59.

Paria, B. C., Das, S. K., Huet-Hudson, Y. M., and Dey, S. K. (1994). *Biol. Reprod.* **50,** 481–491.

Petch, L. A., Harris, J., Raymond, V. W., Blasband, A., Lee, D. C., and Earp, H. S. (1990). *Mol. Cell. Biol.* **10,** 2973–2982.

Peters, J. M., Tsark, E. C., and Wiley, L. M. (1996). *Radiat. Res.* **145,** 722–729.

Piepkorn, M., Lo, C., and Plowman, G. (1994). *J. Cell. Physiol.* **159,** 114–120.

Pinkas-Kramarski, R., Eilam, R., Spiegler, O., Lavi, S., Liu, N., Chang, D., Wen, D., Schwartz, M., and Yarden, Y. (1994). *Proc. Natl. Acad. Sci. USA* **91,** 9387–9391.

Pinkas-Kramarski, R., Shelly, M., Glathe, S., Ratzkin, B. J., and Yarden, Y. (1996a). *J. Biol. Chem.* **271,** 19029–19032.

Pinkas-Kramarski, R., Soussan, L., Waterman, H., Levkowitz, G., Alroy, I., Klapper, L., Lavi, S., Seger, R., Ratzkin, B. J., Sela, M., and Yarden, Y. (1996b). *Embo. J.* **15,** 2452–2467.

Plaut, K. (1993). *J. Dairy Sci.* **76,** 1526–1538.

Plopper, G. E., McNamee, H. P., Dike, L. E., Bojanowski, K., and Ingber, D. E. (1995). *Mol. Biol. Cell* **6,** 1349–1365.

Plowman, G. D., Green, J. M., McDonald, V. L., Neubauer, M. G., Disteche, C. M., Todaro, G. J., and Shoyab, M. (1990). *Mol. Cell. Biol.* **10,** 1969–1981.

Plowman, G. D., Culouscou, J. M., Whitney, G. S., Green, J. M., Carlton, G. W., Foy, L., Neubauer, M. G., and Shoyab, M. (1993a). *Proc. Natl. Acad. Sci. USA* **90,** 1746–1750.

Plowman, G. D., Green, J. M., Culouscou, J. M., Carlton, G. W., Rothwell, V. M., and Buckley, S. (1993b). *Nature* **366,** 473–475.

Price, L. T., Chen, Y., and Frank, L. (1993). *Ped. Res.* **34,** 577–585.

Raab, G., Kover, K., Paria, B. C., Dey, S. K., Ezzell, R. M., and Klagsbrun, M. (1996). *Development* **122,** 637–645.

Rappolee, D. A., Brenner, C. A., Schultz, R., Mark, D., and Werb, Z. (1988). *Science* **421,** 1823–1825.

Rappolee, D. A., Sturm, K. S., Behrendtsen O., Schultz G. A., Pedersen R. A., and Werb, Z. (1992). *Genes Develop.* **6,** 939–952.

Reynolds, B. A., and Weiss, S. (1992). *Science* **255,** 1707–1710.

Ricketts, W. A., Rose, D. W., Shoelson, S., and Olefsky, J. M. (1996). *J. Biol. Chem.* **271,** 26165–16169.

Riese, D. J. N., Bermingham, Y., van Raaij, T. M., Buckley, S., Plowman, G. D., and Stern, D. F. (1996). *Oncogene* **12,** 345–353.

Rijken, P. J., Boonstra, J., Verkleij, A. J., and de Laat, S. W. (1994). *Adv. Space Biol. Med.* **4,** 159–188.

Rijken, P. J., Post, S. M., Hage, W. J., Van Bergen En Henegouwen, P. M. P., Verkleij, A. J., and Boonstra, J. (1995). *Exp. Cell Res.* **218,** 223–232.

Rosette, C., and Karin, M. (1996). *Science* **274,** 1194–1196.

Roulier, S., Rochette-Egly, C., Rebut-Bonneton, C., Porquet, D., and Evain-Brion, D. (1994). *Mol. Cell. Endocrinol.* **105,** 165–173.

Ruff-Jamison, S., Zhong, Z., Wen, Z., Chen, K., Darnell, J. E., Jr., and Cohen, S. (1994). *J. Biol. Chem.* **269,** 21933–21935.

Ruff-Jamison, S., Chen, K., and Cohen, S. (1995). *Proc. Natl. Acad. Sci. USA* **92,** 4215–4218.

Sadowski, H. B., and Gilman, M. Z. (1993). *Nature* **362,** 79–83.

Saeki, T., Cristiano, A., Lynch, M. J., Brattain, M., Kim, N., Normanno, N., Kenney, N., Ciardiello, F., and

Sakakibara, H., Taga, M., Saji, M., Kida, H., and Minaguchi, H. (1994). *J. Clin. Endocrinol. Metab.* **79,** 223–226.

Saltis, J., Thomas, A. C., Agrotis, A., Campbell, J. H., Campbell, G. R., and Bobik, A. (1995). *Atherosclerosis* **118,** 77–87.

Sandgren, E. P., Schroeder, J. A., Qui, T. H., Palmiter, R. D., Brinster, R. L., and Lee, D. C. (1995). *Cancer Res.* **55,** 3915–3927.

Sappino, A. P., Huarte, J., Belin, D., and Vassalli, J. D. (1989). *J. Cell Biol.* **109**, 2471–2479.

Sasada, R., Ono, Y., Taniyama, Y., Shing, Y., Folkman, J., and Igarashi, K. (1993). *Biochem. Biophys. Res. Commun.* **190**, 1173–1179.

Saxena, S. K., Thompson, J. S., and Sharp, J. G. (1992). *Surgery* **111**, 318–325.

Schechter, A. L., Stern, D. F., Vaidyanathan, L., Decker, S. J., Drebin, J. A., Greene, M. I., and Weinberg, R. A. (1984). *Nature* **312**, 513–516.

Schmidt-Ullrich, R. K., Valerie, K. C., Chan, W., and McWilliams, D. (1994). *Int. J. Radiat. Oncol. Biol. Phys.* **29**, 813–819.

Schuger, L., Varani, J., Mitra, R. J., and Gilbride, K. (1993). *Dev. Biol.* **159**, 462–473.

Schuger, L., Johnson, G. R., Gilbride, K., Plowman, G. D., and Mandel, R. (1996). *Development* **122**, 1759–1767.

Seno, M., Tada, H., Kosaka, M., Sasada, R., Shing, Y., Folkman, J., Ueda, M., and Yamada, H. (1996). *Growth Factors* **13**, 181–191.

Shah, N. M., Marchionni, M. A., Isaacs, I., Stroobant, P., and Anderson, D. J. (1994). *Cell* **77**, 349–360.

Shastry, B. S., Dias, P., Dilling, M., and Houghton, P. (1994). *Mol. Cell. Biochem.* **136**, 171–182.

Shaw, P. E., Schröter, H., and Nordheim, A. (1989). *Cell* **56**, 563–572.

Sheikh, M. S., Shao, Z. M., Chen, J. C., Li, X. S., Hussain, A., and Fontana, J. A. (1994). *J. Cell. Biochem.* **54**, 289–298.

Shibata, T., Ochiai, A., Kanai, Y., Akimoto, S., Gotoh, M., Yasui, N., Machinami, R., and Hirohashi, S. (1996). *Oncogene* **13**, 883–889.

Shin, S. Y., Takenouchi, T., Yokoyama, T., Ohtaki, T., and Munekata, E. (1994). *Int. J. Pept. Protein Res.* **44**, 485–490.

Shoyab, M., McDonald, V. L., Bradley, J. G., and Todaro, G. J. (1988). *Proc. Natl. Acad. Sci. USA* **85**, 6528–6532.

Shoyab, M., Plowman, G. D., McDonald, U. L., Bradley, J. G., and Todaro, G. J. (1989). *Science* **243**, 1074–1076.

Sibilia, M., and Wagner, E. F. (1995). *Science* **269**, 234–238.

Silberstein, G. B., and Daniel, C. W. (1987). *Science* **237**, 291–293.

Silberstein, G. B., Flanders, K. C., Roberts, A. B., and Daniel, C. W. (1992). *Dev. Biol.* **152**, 354–362.

Simmen, F. A., Gope, M. L., Schulz, T. Z., Wright, D. A., Carpenter, G., and O'Malley, B. W. (1984). *Biochem. Biophys. Res. Comm.* **124**, 125–132.

Singh, B., Kennedy, T. G., Tekpetey, F. R., and Armstrong, D. T. (1995). *Mol. Cell. Endocrinol.* **113**, 137–143.

Slavkin, H. C. (1993). *Amer. J. Med. Gen.* **47**, 689–697.

Smale, S. T., Schmidt, M. C., Berk, A. J., and Baltimore, D. (1990). *Proc. Natl. Acad. Sci. USA* **87**, 4509–4513.

Snead, M. L., Luo, W., Oliver, P., Nakamura, M., Don-Wheeler, G., Bessen, C., Bell, G. I., Rall, L. B., and Slavkin, H. C. (1989). *Dev. Biol.* **134**, 420–429.

Snedeker, S. M., Brown, C. F., and DiAugustine, R. P. (1991). *Proc. Natl. Acad. Sci. USA* **88**, 276–280.

Soriano, J. V., Orci, L., and Montesano, R. (1996). *Biochem. Biophys. Res. Commun.* **220**, 879–885.

Staines, W. A., Craig, J., Reuhl, K., and McBurney, M. W. (1996). *Neuroscience* **71**, 845–853.

Stewart, C. L., Kaspar, P., Brunet, L. J., Bhatt, H., Gadi, I., Kontgen, F., and Abbondanzo, S. J. (1992). *Nature* **359**, 76–79.

Stocklin, E., Botteri, F., and Groner, B. (1993). *J. Cell Biol.* **122**, 199–208.

Strandjord, T. P., Clark, J. G., Hodson, W. A., Schmidt, R. A., and Madtes, D. K. (1993). *Am. J. Respir. Cell. Mol. Biol.* **8**, 266–272.

Strandjord, T. P., Clark, J. G., and Madtes, D. K. (1994). *Am. J. Physiol.* **267**, L384–L389.

Strum, J. M., DeSanti, A. M., and McDowell, E. M. (1993). *Tissue Cell* **25**, 645–655.

Takagi, H., Jhappan, C., Sharp, R., and Merlino, G. (1992). *J. Clin. Invest.* **90**, 1161–1167.

Takagi, H., Sharp, R., Takayama, H., Anver, M. R., Ward, J. M., and Merlino, G. (1993). *Cancer Res.* **53**, 4329–4336.

Tamada, H., Das, S. K., Andrews, G. K., and Dey, S. K. (1991). *Biol. Reprod.* **45**, 365–372.

Tekpetey F. R., Singh, B., Barbe, G., and Armstrong, D. T. (1995). *Molec. Cell. Endocrinol.* **110**, 95–102.

Thompson, J. F., van den Berg, M., and Stokkers, P. C. (1994). *Gastroenterology* **107**, 1278–1287.

Thompson, K. L., and Rosner, M. R. (1989). *J. Biol. Chem.* **264**, 3230–3234.

Thompson, K. L., Santon, J. B., Shephard, L. B., Walton, G. M., and Gill, G. N. (1992). *Mol. Endocrinol.* **6**, 627–635.

Thompson, S. A., Harris, A., Hoang, D., Ferrer, M., and Johnson, G. R. (1996). *J. Biol. Chem.* **271**, 17927–17931.

Thorne, B. A., and Plowman, G. D. (1994). *Mol. Cell. Biol.* **14**, 1635–1646.

Threadgill, D. W., Dlugosz, A. A., Hansen, L., Tennenbaum, T., Lichti, U., Yee, D., LeMantia, C., Mourton, T., Herrup, K., Harris, R. C., Barnard, J. A., Yuspa, S. H., Coffey, R. J., and Magnuson, T. (1995). *Science* **269**, 230–234.

Tong, B. J., Das, S. K., Threadgill, D., Magnuson, T., and Dey, S. K. (1996). *Endocrinology* **137**, 1492–1496.

Toyoda, H., Komurasaki, T., Ikeda, Y., Yoshimoto, M., and Morimoto, S. (1995a). *FEBS Lett.* **377**, 403–407.

Toyoda, H., Komurasaki, T., Uchida, D., Takayama, Y., Isobe, T., Okuyama, T., and Hanada, K. (1995b). *J. Biol. Chem.* **270**, 7495–7500.

Traverse, S., Seedorf, K., Paterson, H., Marshall, C. J., Cohen, P., and Ullrich, A. (1994). *Curr. Biol.* **4**, 694–701.

Tsark, E. C., Adamson, E. D., Withers, G. E., III, and Wiley, L. M. (1997). *Molec. Reprod. Devel.* (in press).

Tsukita, S. A., Oishi, K., Akiyama, T., Yamanashi, Y., Yamamoto, T., and Tsukita, S. H. (1991). *J. Cell. Biol.* **113**, 867–879.

Tsutsumi, O., Kurachi, H., and Oka, T. (1986). *Science* **233**, 975–977.

Tucker, M. S., Khan, I., Fuchs-Young, R., Price, S., Steininger, T., Greene, G., Wainer, B. H., and Rosner, M. R. (1993). *Brain Res.* **631**, 65–71.

Tzahar, E., Waterman, H., Chen, X., Levkowitz, G., Karungaran, D., Lavi, S., Ratzkin, B. J., and Yarden, Y. (1996). *Mol. Cell. Biol.* **16**, 5276–5287.

Ukai, M., Miura, M., and Kameyama, T. (1994). *Gen. Pharmacol.* **25**, 1157–1162.

Ullrich, A., and Schlessinger, J. (1990). *Cell* **61**, 203–212.

Ullrich, A., Coussens, L., Hayflick, J. S., Dull, T. J., Grey, A., Tam, A. W., Lee, J., Yarden, Y., Liberman, T. A., Schlessinger, J., Downward, J., Mayes, E. L. V., Whittle, N., Waterfield, M. D., and Seeburg, P. H. (1984). *Nature* **309**, 418–425.

van Bergen en Henegouwen, P. M., den Hartigh, J. C., Romeyn, P., Verkleij, A. J., and Boonstra, J. (1992). *Exp. Cell Res.* **199**, 90–97.

van Delft, S., Verkleij, A. J., Boonstra, J., and van Bergen en Henegouwen, P. M. (1995). *FEBS Lett.* **357**, 251–254.

Venkateswaran, V., Oliver, S. A., Ram, T. G., and Hosick, H. L. (1993). *Growth Regulation* **3**, 138–145.

Vintiner, G. M., Holder, S. E., Winter, R. M., and Malcolm, S. (1992). *J. Med. Genet.* **29**, 393–397.

Vonderhaar, B. K. (1987). *J. Cell. Physiol.* **132**, 581–584.

Wang, X. N., Das, S. K., Damm, D., Klagsbrun, M., Abraham, J. A., and Dey, S. K. (1994). *Endocrinology* **135**, 1264–1271.

Warburton, D., Seth, R., Shum, L., Horcher, P. G., Hall, F. L., Werb, Z., and Slavkin, H. C. (1992). *Dev. Biol.* **149,** 123–133.

Wassarman, D. A., Therrien, M., and Rubin, G. M. (1995). *Curr. Opin. Genet. Dev.* **5,** 44–50.

Weickert, C. S., and Blum, M. (1995). *Brain Res. Dev.* **86,** 203–216.

Wiley, L. M., Wu, J.-X., Harari, I., and Adamson, E. D. (1992). *Dev. Biol.* **149,** 247–260.

Wiley, L. M., Raabe, O. G., Khan, R., and Straume, T. (1994). *Mutat. Res.* **309,** 83–92.

Wiley, L. M., Adamson, E. D., and Tsark, E. C. (1995). *BioEssays* **17,** 839–846.

Wollman, R., Yahalom, J., Maxy, R., Pinto, J., and Fuks, Z. (1994). *Int. J. Radiat. Oncol. Biol. Phys.* **30,** 91–98.

Wong, S. T., Winchell, L. F., McCune, B. K., Earp, H. S., Teixido, J., Massague, J., Herman, B., and Lee, D. C. (1989). *Cell* **56,** 495–506.

Wu, J.-X., and Adamson, E. D. (1993). *Dev. Biol.* **159,** 208–222.

Wu, J.-X., and Adamson, E. D. (1996). *Development* **122,** 3331–3342.

Xie, H-Q., and Hu, V. W. (1994). *Exp. Cell. Res.* **214,** 172–176.

Xu, J., Thompson, K. L., Shephard, L. B., Hudson, L. G., and Gill, G. N. (1993). *J. Biol. Chem.* **268,** 16065–16073.

Yamada, M., Ikeuchi, T., Aimoto, S., and Hatanaka, H. (1996). *J. Neurosci. Res.* **43,** 355–364.

Yang, S. G., Ahmad, S., Wong, N. C., and Hollenberg, M. D. (1994). *Mol. Pharmacol.* **46,** 256–265.

Yarden, R. I., Lauber, A. H., El-Ashry, D., and Chrysogelos, S. A. (1996). *Endocrinology* **137,** 2739–2747.

Yasui, S., Nagai, A., Oohira, A., Iwashita, M., and Konno, K. (1993). *Ped. Pulmon.* **15,** 251–256.

Ye, J., Taylor-Papadimitriou, J., and Pitha, P. (1996). *Mol. Cell. Biol.* **16,** 6178–6189.

Yeh, J., Osathanondh, R., and Villa-Komaroff, L. (1993). *Am. J. Obstet. Gynecol.* **168,** 1569–1573.

Zhang, J., Hagopian-Donaldson, S., Serbedzija, G., Elsemore, J., Plehn-Dujowich, D., McMahon, A. P., Flavell, R. A., and Williams, T. (1996). *Nature* **381,** 238–241.

Zhao, X. Y., and Hung, M. C. (1992). *Mol. Cell. Biol.* **12,** 2739–2748.

4

The Development and Evolution of Polyembryonic Insects

Michael R. Strand and Miodrag Grbic
Department of Entomology
College of Agriculture and Life Sciences
University of Wisconsin—Madison
Madison, Wisconsin 53706

I. Introduction

An important question in developmental biology is how phylogeny and life history interact to affect the course of evolution. Comparative studies indicate that developmental programs are broadly conserved within higher taxa and that

Current Topics in Developmental Biology, Vol. 35
121

fundamental regulatory pathways are often very similar between phyla. The Hox/HOM gene complex, for example, regulates anterior–posterior patterning in both arthropods and chordates, and may likely be used in pattern formation processes across all bilaterian Metazoa (Davidson *et al.*, 1995). Comparisons between higher taxa also suggest that developmental mechanisms evolve slowly and that modifications in embryogenesis are usually associated with marked changes in adult morphology (Gould, 1977; Buss, 1987; Thomson, 1988).

While it is clear that ancestry plays a significant role in shaping how organisms develop, developmental biologists have also identified several instances where distinct differences in embryonic development occur between closely related species without any concomitant changes in adult form (del Pino and Ellinson, 1983; Scott *et al.*, 1990; Wray and Raff, 1990; Jeffery and Swalla, 1991; R. Raff, 1992; Wray and Bely, 1994). These studies indicate not only that alterations in embryogenesis can occur without major consequences for the adult body plan, but that adaptations in early development may arise in response to changes in life history (Wray, 1995). An important goal now is to assess how phylogenetically widespread such punctuated changes in development might be, what kinds of life history shifts favor alterations in early development, and what mechanisms underlie these changes.

In insects, the regulation of embryogenesis has been intensively studied in models like *Drosophila melanogaster*, yet there have been few attempts to analyze how life history has influenced early development of this species-rich group of organisms. Most insects initiate development by forming a syncytial blastoderm: in advanced species like *Drosophila*, the entire body plan is essentially established upon cellularization of the blastoderm (Sander, 1976; Davidson, 1991; St. Johnston and Nüsslein-Volhard, 1992). While differences in early developmental events are known between primitive and advanced insect orders (Akam and Dawes, 1992; Patel, 1994; Tautz and Sommer, 1995), what role, if any, life history has played in shaping how patterning events are regulated is unknown. If life history does play an important role in regulation of early development, we would hypothesize that departures from the general insect developmental ground plan would most likely arise in groups whose eggs develop under conditions very different from those experienced by insects like *Drosophila*. One such group is the parasitic wasps (Clausen, 1940; Askew, 1971; Godfray, 1994). Within this assemblage has evolved a number of specialized life histories, including *polyembryony*, the development of two or more individuals from a single egg (Ivanova-Kasas, 1972; Strand and Grbic, 1997). Our recent studies indicate that polyembryonic wasps represent a major modification of the general insect developmental ground plan. Here we review the development of polyembryonic insects, examine the phylogenetic and life history circumstances that have led to their evolution, and discuss the implications polyembryony has for understanding the evolution of early development in insects.

II. Polyembryony in the Metazoa

Polyembryony refers to the formation of multiple embryos from a single egg. Sporadic polyembryony such as identical twinning occurs in most animal taxa, whereas obligate polyembryony occurs more rarely. Taxa in which obligate polyembryony is reported include certain groups of parasitic invertebrates (some cestodes, trematodes, and insects), colonial, aquatic invertebrates (oligochaetes, bryzoans), and a few mammals (armadillos) (Bell, 1982; Hughes and Cancino, 1985). Polyembryony was first reported in insects by Marchal (1898) and has since been documented in two orders, the Hymenoptera (bees, wasps, and ants) and Strepsiptera (Ivanova-Kasas, 1972). The most extreme examples of obligate polyembryony in the Metazoa occur in the Hymenoptera, where some species produce more than two thousand embryos from each egg they lay (Silvestri, 1937; Ivanova-Kasas, 1972; Strand and Grbic, 1997). Because so little is known about polyembryony in strepsipterans (Noskiewicz and Poluszynski, 1935), most of our discussion focuses on polyembryony in wasps.

III. Insect Phylogeny and Polyembryony

Before discussing the development of polyembryonic insects, we first must consider general phylogenetic relationships. This will allow us to place into context the relatives of the polyembryonic wasps and, later in this review, to assess the possible direction of evolutionary change that has led to polyembryony. Phylogenetic analyses based on the fossil record and living species indicate that the most primitive winged (pterygote) insects reside in hemimetabolous orders such as the Odonata (dragonflies) and Orthoptera (grasshoppers), whereas the most advanced insects occur in the holometabolous orders (Fig. 1). While not fully resolved, the sister group to the Hymenoptera is most likely the panorpoid complex that includes the Diptera (Kristensen, 1991; Lebandeira and Sepkoski, 1993). If so, the nearest relatives at the ordinal level to the wasps are flies such as the model insect *Drosophila melanogaster*.

The Hymenoptera comprise one of the largest (115,000 described species) and most biologically diverse group of insects (LaSalle and Gauld, 1991). Most hymenopterans are in the suborder Apocrita, which consists of both free-living species, like bees and ants, and parasitic species that lay their eggs in or on the body of other arthropods (usually other insects) (Clausen, 1940; Strand, 1986; Strand and Obrycki, 1996). After hatching, the larvae of parasitic wasps consume the host, pupate, and emerge as free-living adults that disperse to seek mates or new hosts to parasitize. Because they are parasitic only during their immature stages, parasitic wasps are usually referred to as *parasitoids* (Godfray, 1994). Polyembryonic wasps occur in four families of Hymenoptera: the Braconidae,

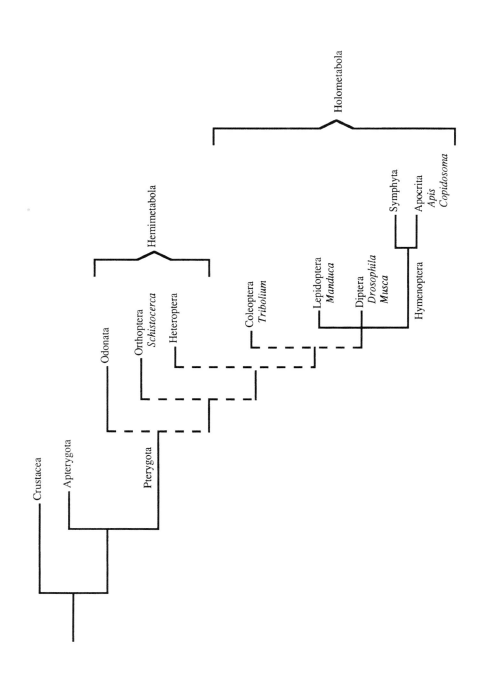

Platygasteridae, Encyrtidae, and Dryinidae (Ivanova-Kasas, 1972; Strand and Grbic, 1997). All wasps in these families are parasitic, and each family contains many more monoembryonic than polyembryonic species (Krombein *et al.*, 1979).

IV. Regulation of Segmentation in Insects

It is also important that we summarize how embryogenesis proceeds in different groups of insects before considering polyembryonic wasps. Our understanding of early development comes from selected species of primitive and advanced insects.

A. Oogenesis and Germ-Band Formation

Insect eggs are formed in ovarioles that consist of a somatic follicular epithelium and germ cells (reviewed by Bünning, 1994). There are two broad categories of ovarioles: *panoistic*, which are usually found in primitive insects like grasshoppers, and *meroistic*, which are found in the more advanced hemimetabolous orders such as the Hemiptera and in most holometabolous insects (see Fig. 1). In panoistic ovarioles, germ cells (cytoblasts) differentiate directly into oocytes, whereas in meroistic ovarioles, cytoblasts undergo a series of divisions to form an oocyte connected to a series of nurse cells. Meroistic ovarioles can be further subdivided into *polytrophic* types, in which each oocyte is connected to its own nurse cells, and *telotrophic* types, in which a nurse complex is connected to multiple oocytes. Nurse cells supply the oocyte with nutrients (yolk), maternal RNAs, and proteins. In the latter stages of oogenesis, the follicular epithelium surrounds the oocyte or oocyte–nurse cell complex and secretes a chorion.

Following oviposition, insect eggs usually initiate development by undergoing syncytial cleavage, whereby several rounds of nuclear division occur without cytokinesis (Schwalm, 1988). Outward migration of some of the nuclei in late cleavage results in the formation of a syncytial blastoderm that then cellularizes. In *Drosophila*, up to 6,000 nuclei reside in a common cytoplasm before cellularization occurs (Turner and Mahowald, 1976; Schwalm, 1988), whereas the tim-

---

Fig. 1 Phylogeny of selected existing orders of insects. Higher categories: subclass Apterygota (wingless primitive insects) and Pterygota (winged insects). The more primitive orders undergo incomplete metamorphosis (hemimetabola), whereas the more advanced orders undergo complete metamorphosis (holometabola). The division of the order Hymenoptera into its two suborders is noted. Selected species discussed in the text are also noted in the figure under their respective ordinal group. The Crustacea are noted as the sister class to the insects. Phylogeny based on the discussions of Kristensen (1991) and Lebandeira and Sepkoski (1993).

ing of cellularization in most other insect embryos is unknown. The cellular blastoderm becomes subdivided into the germ anlage (embryonic rudiment) and the "extraembryonic" blastoderm. Insect embryos can be divided into two basic groups on the basis of how much of the body pattern is formed at the cellular blastoderm stage (reviewed by Sander, 1976). In long germ-band species, the germ anlage gives rise to the entire metameric pattern, whereas in short germ-band species the germ anlage consists only of the future procephalon. All segments posterior to the head region are formed sequentially in an anteroposterior progression during postblastodermal development. Between these two extremes are intermediate germ-band species, in which anterior structures and variable parts of the thoracic region are formed at the cellular blastoderm stage. No clear generalizations can be made about germ-band types and insect phylogeny (Sander et al., 1985). Long germ-band formation occurs predominantly in advanced insects with meroistic ovarioles, but short and intermediate germ-band types occur in insects with both panoistic and meroistic ovaries. At the completion of germ-band extension, all insect embryos look very similar. This stage, therefore, represents the phylotypic stage for insects (Sander, 1983). Germ-band extension is followed by germ-band retraction and segmentation.

B. *Drosophila*: A Paradigm for Long Germ-band Development

Our understanding of how segmentation in insects is regulated derives from the molecular analysis of *Drosophila melanogaster*, a long germ-band dipteran that possesses meroistic, polytrophic ovarioles. The oocyte of *Drosophila* appears to be specified after the first incomplete division of the cytoblast by the asymmetric inheritance of a cytoplasmic structure called the *spectrosome* (Lin and Spradling, 1995). It appears that asymmetric inheritance of the spectrosome predisposes one of the two daughter cells toward giving rise to the oocyte. The next three divisions are also incomplete, resulting in a total of 15 nurse cells connected cytoplasmically to the future oocyte. Embryonic axis formation is initiated by intercellular signaling events between the oocyte and follicular epithelium. These interactions result in specification of the posterior pole of the oocyte (reviewed by Grünert and St. Johnston, 1996). The organization of the oocyte's cytoskeleton allows mRNAs of the maternal-coordinate genes *bicoid* and *nanos* to be asymmetrically distributed to the anterior and posterior of the oocyte, respectively (St. Johnston and Nüsslein-Volhard, 1992). Localization of *nanos* is also coupled with localization of *oskar* mRNA and Vasa protein, components of polar granules that mediate the formation of pole cells (Lasko and Ashburner, 1990; Ephrussi and Lehmann, 1992).

After oviposition, *bicoid* and *nanos* mRNAs are translated, and their proteins diffuse through the syncytium to form an anterior–posterior gradient (St. Johnston and Nüsslein-Volhard, 1992). *Bicoid* protein activates zygotic transcription of the zinc finger containing gap gene *hunchback* (*hb*) in anterioraly positioned nuclei, whereas *nanos* protein inhibits translation of maternal *hb*. Gradients of Hunchback and Bicoid protein regulate other gap genes (*Kruppel, knirps, giant*), in the trunk and head, and a gradient of Caudal protein regulates gap genes in the abdomen (Rivera-Pomar and Jackle, 1996). This results in specific domains of gap gene expression across the syncytial blastoderm. The gap genes then initiate transcriptional regulatory interactions that establish expression domains for pair-rule genes such as *even-skipped* (*eve*), *hairy*, and *fushi tarazu* (*ftz*) (Stanojevic *et al.*, 1991). Pair-rule genes are transcription factors expressed in a characteristic seven-stripe pattern in the *Drosophila* syncytium with double segment periodicity. Many pair-rule genes, like *eve*, are also later expressed in a segmental pattern and in neuroblasts (Doe *et al.*, 1988a, b). At approximately the time the blastoderm cellularizes, the pair-rule genes regulate expression of the segment-polarity genes, which specify the segmental compartments (Ingham, 1988). Segment-polarity genes consist of both transcription factors and signaling molecules that likely reflect a role for cell–cell communication in establishing the final segmental pattern. Finally, expression of the homeotic genes in discrete regions along the anteroposterior axis provides the embryo with region- and segment-specific identities.

Determination of the dorsal–ventral axis also begins during oogenesis through signaling between the oocyte and follicle cells (reviewed by Schüpbach and Roth, 1994). The dorsalizing signal appears to be received by the follicle cells through a receptor encoded by the *torpedo* gene, which encodes a protein very similar in structure to the epidermal growth factor receptor of mammals. Establishment of dorsal and ventral follicle cells specifies embryonic dorsal–ventral polarity through a complex series of interactions that lead to a nuclear concentration gradient of the transcription factor Dorsal. Dorsal then functions to activate a battery of genes required for ventral and ventrolateral cell fates and repression of dorsal-specific genes.

In summary, two types of regulatory processes mediate polarity and patterning of the *Drosophila* embryo. First, cell signaling events establish the polarity and cytoskeletal machinery of the oocyte that localizes specific determinants in the egg. Second, different transcription factors diffuse in the syncytial environment of the newly laid egg and act as morphogens that specify the fate of nuclei.

C. Regulation of Segmentation in Other Insects

The obvious question from the perspective of comparative embryology is whether the *Drosophila* patterning hierarchy universally represents how patterning events are regulated in insects. Nothing is known about how oocyte polarity is established or whether maternal-coordinate genes exist in other species. However, analysis of the patterning cascade in other long germ-band insects like the housefly (*Musca*) and honeybee (*Apis*) suggest that expression of gap, pair-rule, and segment-polarity genes are conserved with *Drosophila* (Fleig, 1990; Fleig *et al.*, 1992; Sommer and Tautz, 1991, 1993). What about intermediate and short germ-band insects that establish part of the body plan after cellularization of the blastoderm? Examination of embryos from advanced orders like the beetles (Coleoptera) and moths (Lepidoptera) indicate that homologues of the *Drosophila* gap genes are expressed in a pattern broadly similar to that of flies (Sommer and Tautz, 1993; Kraft and Jackle, 1994; Wolff *et al.*, 1995). Pair-rule genes likewise are expressed in double segment periodicity, although stripe formation is sequential rather than simultaneous as observed in *Drosophila* (Sommer and Tautz, 1993; Kraft and Jackle, 1994; Patel *et al.*, 1994; Brown *et al.*, 1994). For example, Eve expression in short, intermediate and long germ-band beetles differs only in the number of stripes resolved before gastrulation (Patel *et al.*, 1994). Lastly, the final expression pattern of segment-polarity genes are expressed in a conserved pattern (Brown *et al.*, 1994; Nagy and Carroll, 1994).

The universality of the *Drosophila* patterning hierarchy is called into question, however, by studies of the primitive short germ-band grasshopper, *Schistocerca gregaria* (Patel *et al.*, 1992; Dawes *et al.*, 1994). In contrast to the advanced insects just discussed, grasshoppers possess panoistic ovarioles in which nurse cells are not produced. Therefore, it is unclear how maternal determinants might be localized during oogenesis or whether early elements of the patterning hierarchy as known from *Drosophila* exist. Nothing is known about gap genes in *Schistocerca*, but neither of the pair-rule genes *eve* and *ftz* is expressed in a pair-rule pattern, even though both are expressed in the posterior embryonic primordium and later in development in neuroblasts (Patel *et al.*, 1992; Dawes *et al.*, 1994). The segment-polarity gene *en* is expressed sequentially in the posterior segmental compartments of *Schistocerca* as segments form from the posterior growth zone without pair-rule input (Patel *et al.*, 1992).

These results suggest that segmental patterning in short germ-band species may rely in part on cell–cell interactions, whereas the mechanism involved in *Drosophila* patterning evolved more recently in association with the evolution of long germ-band development (Akam and Dawes, 1992; French, 1996; but see Tautz and Sommer, 1995). One would also predict from these results that the regulatory paradigms established through *Drosophila* would be most conserved in other dipterans and hymenopterans, which represent the closest relatives to *Drosophila*. On the other hand, if the events regulating formation of the insect body plan are influenced by life history, dramatic alterations in early development might occur in advanced insects whose eggs develop under conditions very different from *Drosophila*. As we discuss next, polyembryonic and other parasitic wasps represent just such a departure from the *Drosophila* paradigm.

V. Development of the Polyembryonic Wasp *Copidosoma floridanum*

While insect eggs almost always undergo syncytial cleavage, descriptions of polyembryony from the early 1900s (Marchal, 1898; Silvestri, 1906; Martin, 1914; Patterson, 1921; Leiby, 1922; Silvestri, 1937) suggested to us that polyembryonic wasps undergo holoblastic cleavage. If so, this would be most unexpected, considering that the early development of better-known hymenopterans like the honeybee, *Apis mellifera*, is very similar to that of *Drosophila* (Anderson, 1972; Fleig, 1990). To investigate how development of polyembryonic wasps relates to other insects, we selected *Copidosoma floridanum* (family Encyrtidae) as our model for investigation. We will shortly summarize the general development and segmental patterning of this wasp as reported in several recently completed studies (Baehrecke and Strand, 1990; Strand *et al.*, 1991; Baehrecke *et al.*, 1992; Baehrecke *et al.*, 1993; Grbic *et al.*, 1996a, b; Grbic *et al.*, 1997a, b). In order to follow the discussion, it is essential to understand that the life cycle of *C. floridanum* is intimately linked to developmental events in its lepidopteran host, *Trichoplusia ni*. This relationship is illustrated schematically in Fig. 2.

A. Early Development

Development begins when *C. floridanum* oviposits its egg into the egg stage of the moth *T. ni*. The wasp egg has a clear polarity and is enclosed within a thin chorion that forms a small stalk at the anterior end (Fig. 2). Immediately after oviposition the polar nucleus separates from the pronucleus. The pronucleus migrates to the posterior pole of the egg along with a cytoplasmic structure called the *oosome*, which is an organelle implicated in germ cell determination in Hymenoptera (Silvestri, 1906; Hegner, 1914; Patterson, 1919). Two hours after oviposition the *C. floridanum* egg initiates its first cleavage and forms three cells: two equal-sized blastomeres, from cleavage of the pronucleus, and a polar cell, located anteriorally, that contains the polar nucleus (Fig. 3A). Though polar body nuclei are aborted in most insects, they remain viable and form a polar body cell in *C. floridanum*. The oosome associates with one of the blastomeres. The second cleavage is unequal, with one blastomere forming a large and a small daughter cell and the second blastomere forming two equal-sized blastomeres. The oosome always segregates into the small blastomere. Subsequent cleavages are asynchronous. The nucleus of the polar cell divides without cytokinesis to form a syncytial compartment at the anterior of the egg that migrates as an extraembryonic membrane over the dividing blastomeres (Fig. 3B). The embryo then ruptures out of its chorion and continues development unconstrained by the egg

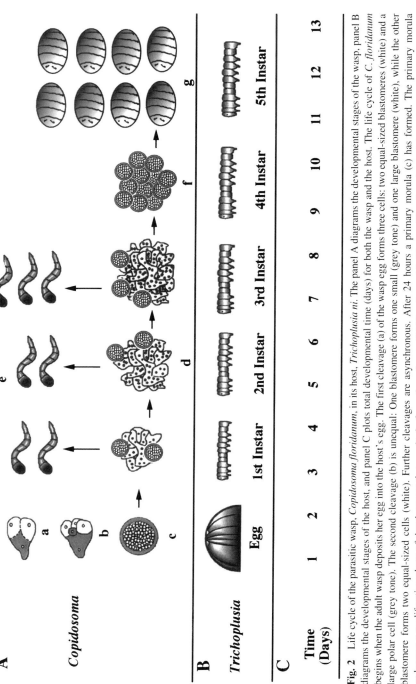

Fig. 2 Life cycle of the parasitic wasp, *Copidosoma floridanum*, in its host, *Trichoplusia ni*. The panel A diagrams the developmental stages of the wasp, panel B diagrams the developmental stages of the host, and panel C plots total developmental time (days) for both the wasp and the host. The life cycle of *C. floridanum* begins when the adult wasp deposits her egg into the host's egg. The first cleavage (a) of the wasp egg forms three cells: two equal-sized blastomeres (white) and a large polar cell (grey tone). The second cleavage (b) is unequal: One blastomere forms one small (grey tone) and one large blastomere (white), while the other blastomere forms two equal-sized cells (white). Further cleavages are asynchronous. After 24 hours a primary morula (c) has formed. The primary morula undergoes a proliferative phase of development that results in formation of up to 2,000 embryos (d). The proliferation phase of development occurs during the host's first through early fourth larval instars (days 3–8 postoviposition). Selected embryos initiate morphogenesis during this period and form precocious larvae (e). During the host's fourth larval instar, development of precocious larvae ceases and the remaining embryos synchronously initiate morphogenesis to form reproductive larvae (f). During the host's fifth larval instar, the reproductive morph embryos undergo germ-band extension, segmentation, and eclose as reproductive larvae (g). Adapted from Grbic *et al.* (1996a).

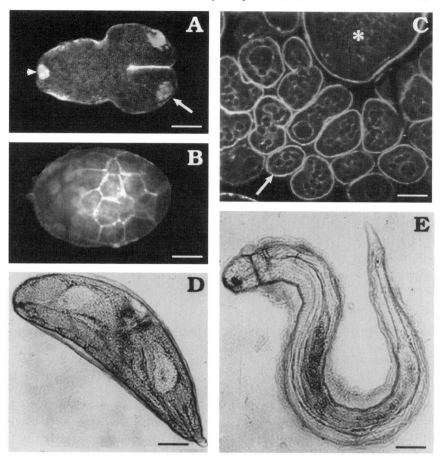

Fig. 3 Confocal and Hoffman images of *C. floridanum* developmental stages. (A) Initiation of egg cleavage (scale bar = 26 μm). Cleavage furrow extends to one-third of the egg, forming two blastomeres. One of the blastomere nuclei is marked by an arrow, the polar nucleus at the anterior of the egg is marked by an arrowhead. (B) Growth of the polar region around the embryonic nuclei that form the primary morula (scale bar = 26 μm). (C) Low-magnification view of a polymorula from a third instar host (scale bar = 22 μm). Numerous proliferating morulae (arrow) are individually surrounded by the enveloping membrane. An embryo undergoing morphogenesis to form a precocious larva is marked by the asterisk (*). (D) A first-instar reproductive larva (scale bar = 40 μm). (E) A precocious larva (scale bar = 35 μm). Adapted from Strand and Grbic (1997); Grbic *et al.* (1997b).

shell. Approximately 24 hours after oviposition, the embryo is enveloped by the polar body–derived extraembryonic membrane. We refer to this structure as the *primary morula*. The inner cells of the primary morula are the source of all embryos that subsequently develop.

B. Proliferation and Development of the Polymorula

After the host egg hatches, the cells of the primary morula continue to divide inside the host larva to form several separate embryonic masses that consist of embryonic cells surrounded by the extraembryonic membrane. We call these embryonic masses *proliferating morulae* and the entire mass a *polymorula* (Figs. 2, 3C). All morulae formed during proliferation result from the invagination of the extraembryonic membrane and subsequent partitioning of embryonic cells. In effect, the extraembryonic membrane creates an increasing number of "chambers," each of which contains clusters of embryonic cells. The process of membrane invagination and cell partitioning repeats itself an indeterminate number of times during the first through fourth instars of the host caterpillar. Nuclear staining of embryonic cells reveals a random pattern of cell division in each morula without any apparent domains of mitotic activity across the polymorula. Partitioning results in progressively fewer embryonic cells per chamber. For example, from 5 to 10 morulae are present in a first-instar host larva, with each morula consisting of several hundred rounded cells. In contrast, more than a thousand morulae are present in a fourth-instar host larva, with each morula consisting of approximately 20 cells.

C. Morphogenesis and Development of Larval Castes

The proliferation of embryos just described has no counterpart in embryogenesis of other nonpolyembryonic metazoans. Equally unusual, however, is that the embryos produced during this process give rise to two morphologically different types of larvae (Figs. 2, 3D, E) (Silvestri, 1906; Patterson, 1921; Cruz, 1981; Grbic *et al.*, 1992). A small number of embryos differentiate during the host's first through fourth instars and develop into what we call *precocious* larvae (Figs. 2, 3E). One or two precocious larvae develop when the host is a first-instar larva, and from 2 to 10 precocious larvae develop, respectively, during the host's second through fourth instars. Thus, the number of precocious larvae progressively increases as the host caterpillar grows, resulting in as many as 200 precocious larvae being present when the host molts to its final (fifth) larval instar. At the end of the host's fourth instar, the majority of embryos (>1000) begin to differentiate and develop into reproductive larvae (Figs. 2, 3D). Reproductive larvae eclose late in the host's fifth instar, consume the host, pupate, and emerge as adult wasps. In contrast, precocious larvae never pupate, and they die from desiccation after the host is consumed by their reproductive siblings.

1. Reproductive Larvae

As already mentioned, each embryo that develops into a reproductive larva consists initially of about 20 cells surrounded by the enveloping membrane. As

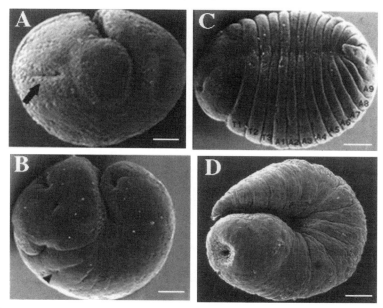

Fig. 4 Scanning electron microscopy images of reproductive and precocious morph embryos. (A) Fully extended germ-band stage of the reproductive morph. The embryo is extensively coiled, so the tip of its tail touches the head region. The labial furrow is marked by a black arrow. (B) Thoracic segmentation of the reproductive morph: A trachial pit is marked with a dark arrowhead. (C) The completely segmented reproductive morph embryo: Three thoracic (T1–T3) and nine abdominal (A1–A9) segments are marked. (D) The segmented precocious morph embryo. Each scale bar = 36 μm. Adapted from Grbic *et al.* (1997b).

cells divide, a solid blastula forms that is comprised of tightly interdigitated cells (Fig. 4A). The only exception is a cluster of cells at the posterior of the primordium (the posterior cluster) that remain rounded. Following formation of a dorsal furrow, embryos undergo germ-band extension along their anterior–posterior axis, resulting in the head region touching the tip of the tail (Fig. 4B). Cells in the cephalic region become rounded and bulged out laterally, separating the head lobe from the overlaying future thorax. Gastrulation follows head lobe formation. When the germ band fully extends, segmentation initiates in the gnathal region, followed by invagination of the stomodeum anteriorally and the proctodeum posteriorally. After full extension of the germ band, embryos that form reproductive larvae undergo germ-band retraction. The process of germ-band shortening changes the shape of the embryo from coiled to "flat," and clearly separates the head from the tail. As condensation continues, the thoracic and abdominal segments form sequentially in a rapid anterior-to-posterior progression. A complete set of segments ultimately forms that corresponds to three thoracic and nine abdominal segments, plus the telson (Fig. 4C).

2. Precocious Larvae

Embryos that develop into precocious larvae begin morphogenesis by forming a solid blastula in a manner identical to reproductive larvae. Germ-band extension follows gastrulation in a manner very similar to the reproductive morph, with the exception that (1) cells of the posterior cluster are absent, and (2) the germ band never shortens (Grbic and Strand, unpublished). Thoracic and abdominal segments then form in a rapid anteroposterior progression resulting in a completely segmented embryo. Because the germ band never shortens, precocious embryos remain coiled, with the head contacting the tip of the tail, and are more slender than reproductive larvae (Fig. 4D).

D. Does *C. floridanum* Development Proceed in a Cellularized Environment?

To determine whether *C. floridanum* embryogenesis proceeds in a cellularized environment, Grbic *et al.* (1996a) conducted cell injection studies from first cleavage through morphogenesis. These studies revealed that *C. floridanum* blastomeres become dye uncoupled during early cleavage. The small blastomere produced after the second cleavage is the first cell to become uncoupled, followed by other blastomeres during subsequent cleavages. Injection of fluorescently labeled dextran tracers indicated that molecules ≥3kDa likely do not pass between embryonic cells, or between embryonic cells and the membrane surrounding each embryo, by the time the wasp embryo reaches the primary morula stage. The transcription factors that initiate the *Drosophila* segmentation cascade are all much larger molecules than the tracers used in this study. Therefore, these results suggest that *Drosophila*-like transcription factors cannot diffuse freely in the cellularized environment of the *C. floridanum* embryo and that axis formation in this species occurs in the absence of a syncytial stage.

E. Segmental Gene Expression in *C. floridanum*

To analyze whether homologous elements of the *Drosophila* segmentation cascade are expressed during embryogenesis of *C. floridanum*, we have also used antibodies that recognize conserved epitopes of selected components of the *Drosophila* patterning hierarchy.

1. Asymmetric Cleavage and Putative Pole Cell Formation

Although, like *Drosophila*, *C. floridanum* possesses polytrophic meroistic ovarioles, it seems unlikely that axial polarity for the thousands of embryos that ultimately develop would be established during oogenesis. However, an antibody generated by Paul Lasko against the Vasa protein stains the oosome in *C. floridanum* eggs (Grbic and Strand, unpublished) (Fig. 5). Vasa, a DEAD box helicase, marks polar granules in the posterior of the early syncytium of *Drosophila*. Polar granules are taken up later by pole cells that form the germ line (Lasko and Ashburner, 1989) (see Section IV.B). After the first cleavage of the *C. floridanum* egg, the oosome stained by anti-Vasa clearly segregates to one of the blastomeres, and during the second cleavage this organelle invariably segregates to the small blastomere. The oosome then dissociates, as evidenced by the uniform staining of the cytoplasm (Fig. 5). In the primary morula, up to six cells contain the putative Vasa homologue.

It is possible that in *C. floridanum* this antibody cross-reacts with some other DEAD box helicase that is not a germ line marker. However, four lines of evidence suggest this antibody recognizes a component involved in germ cell specification of *C. floridanum*. First, the antibody specifically stains the oosome, which has been implicated previously in germ cell determination in hymenopterans (see Section V.A). Second, staining in *C. floridanum* is cytoplasmic, as it is in *Drosophila* (Lasko and Ashburner, 1990). Third, the small blastomere at the four-cell stage to which the putative Vasa homologue segregates is the first cell to become dye uncoupled from the rest of the embryo (Grbic *et al.*, 1996a). This is a characteristic of germ cells in other organisms. Finally, the cell that inherits the oosome in *C. floridanum* exhibits a different cleavage pattern from other blastomeres. In *Drosophila*, migration of pole plasm to the egg's posterior is associated with localization of the posterior determinant *nanos* mediated by *oscar* (Ephrussi and Lehmann, 1992). Although our results in *C. floridanum* should be interpreted with caution, it is tempting to speculate that a similar localization system may function in this wasp, which could be important in germ cell determination and axial patterning.

2. Changes in Pair-Rule Patterning and Conservation of Downstream Events

Grbic *et al.* (1996a) used antibodies to the pair-rule protein Eve (Patel *et al.*, 1992), the segment-polarity protein En (Patel *et al.*, 1989), and homeotic proteins Ubx/AbdA (Kelsh *et al.*, 1994) to analyze segmentation in *C. floridanum*. None of these antigens are detected during early cleavage and embryo proliferation, indicating that these genes have not been coopted for novel functions during the proliferative phase of embryogenesis. However, expression of the Eve antigen is detected in pregastrulation embryos from the posterior end to 55% of the length of the newly formed primordium. This broad band of expression then rapidly resolves, anteriorally–posteriorally after gastrulation, into 15 stripes. No pair-rule expression pattern occurs in *C. floridanum* as seen in other advanced, long germ-band insects like *Drosophila*. Eve expression is followed by appearance of the En protein, beginning in the posterior one to two cells of the mandibular and labial segments, followed by nearly simultaneous formation of the maxillary, thoracic, and abdominal stripes. Expression of *C. floridanum* Ubx/Abd-A occurs in a conserved pattern during postgastrulation, from the posterior metathorax to the eighth abdominal segment.

VI. *Copidosoma floridanum*: A Major Departure from the Insect Developmental Ground Plan

Embryonic development of *C. floridanum* clearly differs from any insect described in the literature. Its eggs lack yolk, cellularization begins early in development, and polar body cells form an extraembryonic membrane that partitions embryonic cells during the proliferative phase of development. Proliferation of multiple embryos has no counterpart among monoembryonic insects, and during morphogenesis the embryonic primoridia of *C. floridanum* form compact blastulae rather than a "blastoderm" on the surface of a yolk mass. Finally, no other group of insects forms two morphologically different larvae from a common egg.

   Polyembryonic development in *C. floridanum* also poses two mechanistic problems for the operation of the segmentation gene cascade described in *Drosophila*. First, axial polarity of the *C. floridanum* egg does not correspond to future embryonic polarity. Embryonic patterning occurs after an extended period of proliferation during which the original anterior–posterior axis of the egg is lost. This process likely excludes the possibility that egg axial polarity participates in the specification of future embryonic axes as it does in *Drosophila*. Second, given that cleavage is holoblastic, transcriptional regulatory factors, such as gap and pair-rule gene products, would not be able to diffuse as occurs during early patterning in *Drosophila*, making it unclear how the segmental patterning cascade is initiated. Below, we discuss factors regulating four key events during polyembryony: polarity, embryo proliferation, segmentation, and caste determination.

A. How Embryonic Polarity Is Established in *C. floridanum*

Although holoblastic cleavage is atypical for insects, the embryos of many other animals cellularize very early in embryogenesis (Davidson, 1990, 1991). In considering animals other than insects, two patterns for establishing polarity in a cellularized environment are observed: autonomous specification of lineage founder cells at one or both poles of the original egg axis (many invertebrates), and prespecification of one egg axis and general regions of the anlagen, with subsequent specification mediated conditionally by diffusable intercellular morphogens (vertebrates) (reviewed by Kessler and Melton, 1994).

Taking into account how *C. floridanum* develops and the paradigms established for *Drosophila*, we can envision at the cellular level two alternatives for establishing axial polarity in polyembryonic wasps (Fig. 6). The first would be that the syncytial, extraembryonic membrane that surrounds each embryo specifies axis formation in a manner analogous to the role follicle cells play in establishing polarity of oocytes in *Drosophila* (see Section IV). This difference in the timing of axis specification relative to the stage of development can be viewed as an example of heterochrony. Morphogens produced in a region-specific manner by the extraembryonic membrane would then trigger the cascade of transcription factors that specify the anterior–posterior axis. If the extraembryonic membrane is involved in inducing embryonic polarity, we would expect two trends: (1) Each primordium in the polymorula, or domains of the polymorula, would exhibit a common polarity and orientation; and (2) primordia mislocalized to a common chamber of the enveloping membrane would exhibit the same polarity and orientation. However, this is not what occurs in *C. floridanum*. Embryos within the polymorula are randomly oriented with respect to one another (Baehrecke *et al.*, 1992; Grbic *et al.*, 1996a), and mislocalization of two or more embryos to a common chamber develop random to one another, as evidenced by the molecular marker *Distalless* (*Dll*), which identifies early anterior head structure in *C. floridanum* and other insects (Panganiban *et al.*, 1995; Grbic *et al.*, 1996b).

The apparent absence of any polarity to the polymorula argues that axial polarity is regulated by a second alternative: Each embryonic primordium establishes its own intrinsic polarity by cell–cell interaction. Intuitively, it seems unlikely that the thousands of embryos produced by *C. floridanum* could be assembled as a mosaic of prespecified cells as occurs in invertebrates like *C. elegans* (Strome, 1989; Wood, 1991). However, partitioning of selected cells with an organizing capacity might occur. Cell lineage(s) with an organizing capacity for the rest of the embryo are known in both invertebrates (Henry and Martindale, 1987; Itow *et al.*, 1991; Ransick and Davidson, 1993) and vertebrates (Waddington, 1933; Spemann, 1938; Storey *et al.*, 1992). One candidate cell lineage in *C. floridanum* would be the putative pole cell lineage that inherits the oosome and stains with anti-Vasa (see Section V.E). Although lineage studies remain incomplete, it is tempting to speculate that the small blastomere produced after the second cleavage is a germ cell precursor and that the cells of the posterior cluster are derived from this lineage. Polar granules are present in diverse taxa (Turner and Mahowald, 1976; Strome and Wood, 1983), and in insects localization of polar granules to the posterior pole is coupled with localization of posterior determinants like *nanos* (St. Johnston and Nüslein-Volhard, 1992; Kimble, 1994; see Section IV.B). The role of polar cells in axial patterning of *C. floridanum* is currently under investigation.

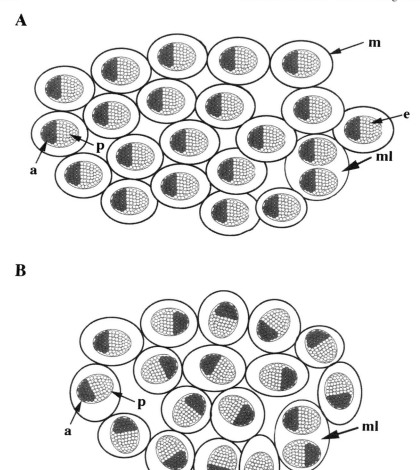

Fig. 6 Model illustrating extrinsic versus intrinsic axial patterning of *C. floridanum* embryos. The *C. floridanum* polymorula consists of multiple embryos. The polymorula is held together by the enveloping membrane (*m*). Each embryo (*e*) in the polymorula resides in a chamber created by the enveloping membrane. At the onset of morphogenesis, most chambers contain a single wasp embryo, although occasionally two embryos become mislocalized (*ml*) to a common chamber, which results in the formation of a duplicated larva (Grbic *et al.*, 1996b). Panel A illustrates an extrinsic model of embryonic axial polarity. A putative signal imposing embryonic polarity comes from the extra-embryonic membrane, causing single and mislocalized embryonic primordia to have their anterior (*a*) and posterior (*p*) regions facing in a common orientation. Panel B illustrates an intrinsic model of axial polarity. Individual and mislocalized embryonic primordia have an inherent polarity. Axis formation in each primordium occurs independent of one another, resulting in the anterior (*a*) and posterior (*p*) regions of each embryonic primordium to be oriented randomly in the polymorula.

B. Regulation of Proliferation

We know that the proliferative phase of development in *C. floridanum* is when thousands of embryos are formed. At the cellular level, this phase of development clearly differs from the nuclear proliferation events that occur during syncytial cleavage in other insects. In *C. floridanum*, the partitioning of embryonic cells by the enveloping membrane appears completely random, without any polarity or spatial organization, yet results in the organization and formation of thousands of embryonic primordia. The molecular events regulating proliferation are unknown. However, two lines of evidence suggest that the embryonic cells partitioned by the enveloping membrane are largely undifferentiated (Grbic *et al.*, 1997a). First, the morphology of these cells is rounded, which is typical for undifferentiated cells. Second, we do not detect the expression of several genes (*eve, en, dll*, and homeotic genes) known to be involved in the initiation of morphogenetic programs and differentiation. All of these proteins are homeobox containing transcription factors that mediate processes leading to terminal differentiation. Experimental evidence from other systems indicate that individual blastomeres isolated early in embryogenesis have the potential to form complete embryos in sea urchins and frogs (Driesch, 1894; Spemann, 1938). This suggests that if blastomeres are prevented from receiving the cell–cell signaling information that normally restricts their fate, they have the potential to form a whole organism. A similar genetic system, whereby cells are set aside from signals to differentiate and therefore remain capable of proliferating, could function in the production of imaginal cells in the larvae of a variety of species (Davidson *et al.*, 1995). The development of *C. floridanum* is consistent with such a scenario.

As we discuss shortly, all embryos initiate morphogenesis after a specific period of time postoviposition, indicating that the timing at which individual embryos initiate morphogenesis is largely autonomous (Baehrecke *et al.*, 1993; Grbic *et al.*, 1997a). How proliferation and morphogenesis might be linked is unknown, but related processes offer some clues. For example, cell proliferation and patterning in imaginal discs is mediated in part by signaling molecules such as Decapentaplegic (DPP), a homologue of the transforming growth factor-β (TGF-β), and Wingless (WG) (Rulifson *et al.*, 1996; Kim *et al.*, 1996). Proliferation in discs is thought to be controlled by gradients of either DPP or WG or through the interaction of these signaling molecules with genes that control cell metabolism and growth (Edgar and Lehner, 1996). Differentiation of precursor cells into oligodendrocytes in rats is also mediated autonomously, possibly by counting the number of cell divisions that occur in response to platelet-derived growth factor (Raff *et al.*, 1988). Results like this suggest that analogous pathways in *C. floridanum* might have a role in terminating cell proliferation and initiating morphogenesis.

C. Regulation of Segmental Patterning

The cellularized environment of the *C. floridanum* embryo violates the most important paradigm of *Drosophila* patterning: Initiation of the segmenation cascade begins with diffusion of transcription factors in a syncytial environment. Even though *C. floridanum* lacks a syncytial stage, it still undergoes long germband development (Grbic *et al.*, 1996a). Germ-band extension arises from elongation of the entire embryonic primordium (see Fig. 4) rather than from a posterior "growth zone" as seen in short germ-band insects. The very rapid anteroposterior expression of Eve and En stripes and the expression of Ubx/Abd-A across the entire putative abdomen of *C. floridanum* are also consistent with segment formation in long germ-band species.

Currently, we have no information on maternal-coordinate or gap gene homologues in *C. floridanum*. Early cellularization of the *C. floridanum* egg, however, makes it unlikely that protein gradients could form by diffusion as occurs during syncytial cleavage in other long germ-band species. For short germ-band insects that develop in a partially cellularized environment, Tautz and Sommer (1995) suggested that the *Drosophila* gap to pair-rule mechanism could operate if cells are connected by junctions that allow for the diffusion of segmentation gene products. Another possibility suggested for short germ-band species is that maternal transcriptional factors could be sequestered in the posterior pole of the embryo during the early syncytial stages (Patel, 1994; Tautz and Sommer, 1995). As cells divide, they could exit the posterior growth zone with different concentrations of maternal factors and create a concentration gradient. Neither of these scenarios likely operates in *C. floridanum*, since embryonic cells become dye uncoupled at the very beginning of embryogenesis and no syncytial stage ever exists preceding germ-band formation. Moreover, the massive proliferation of embryonic cells that occurs prior to morphogenesis of individual embryos makes the activation of downstream genes by dilution of localized maternal factors unlikely.

In our view, there are two options for how segmental patterning is initiated during early development of *C. floridanum*: Either the *Drosophila* paradigm applies in a novel way, or *C. floridanum* does not use a gap to pair-rule segmentation mechanism. If we assume that gap gene homologues exist, one means of creating a diffusion gradient in a cellularized environment would be specifically to translocate transcription factors across cell membranes, as demonstrated for a portion of the Antennapedia protein (Joliot *et al.*, 1991). Another is suggested by the structure of gap genes like *knirps* and *tailless* (Patel, 1994). Both belong to the steroid receptor superfamily that contains other members that act as membrane-bound receptors for small ligands capable of diffusing between cells (Koelle *et al.*, 1991). Gradients could be established by binding a diffusible ligand rather than by diffusion of gap gene products themselves. Finally, it is possible that patterning is initiated by intercellular signaling, using ligand and receptor-type molecules that activate gap gene–like factors to establish gradients through the cellularized embryo. Candidate signaling molecules that might interact with gap genes are currently unknown. If a gap to pair-rule mechanism is not operative in *C. floridanum*, the only clue as to how segmentation might be regulated comes from the structure of the segment-polarity genes in *Drosophila* (Ingham and Martinez Arias, 1992). Some of these genes encode secreted molecules or cell surface receptors that function in cell–cell signaling and thus have the potential to establish segment boundaries.

As mentioned previously, pair-rule patterning occurs in *Drosophila, Tribolium, Manduca*, and other advanced insects (Lawrence *et al.*, 1987; Patel *et al.*, 1994; Kraft and Jackle, 1994), whereas it is not observed in the more primitive grasshopper (Patel *et al.*, 1992; Dawes *et al.*, 1994) or *C. floridanum* (Grbic *et al.*, 1996a). If one looks solely at grasshoppers, one interpretation for this trend would be that pair-rule patterning was not used by the ancestor to the primitive orthopterans and arose later in the more advanced insect orders. Yet this clearly is not supported by the absence of a pair-rule pattern for the Eve antigen in *C. floridanum*. Alternative interpretations for the absence of a pair-rule phase in *C. floridanum* would be that (1) pair-rule patterning simply differs between the hymenopterans and other advanced insects, (2) a pair-rule pattern has been lost specifically in *C. floridanum*, or (3) it reflects when insect embryos cellularize. As we discuss in Section VII, pair-rule patterning does occur in wasps that undergo syncytial cleavage, whereas no pair-rule phase is observed in wasps that exhibit total cleavage. Thus, early cellularization of *C. floridanum*, as in grasshoppers, may be the most important feature associated with the absence of a pair-rule phase.

D. Caste Formation in Polyembryonic Wasps

Precocious and reproductive larvae in *C. floridanum* represent two morphologically distinct castes. Caste formation is a type of alternative polyphenism whereby individuals form different phenotypes in response to some cue (Nijhout, 1994). Among insects, termites (Isoptera) and many ants, bees, and wasps (Hymenoptera) form complex societies structured by sterile workers and reproductives (Brian, 1979). Castes have also evolved in aphids (Homoptera) and thrips (Thysanoptera), where, again, sterile castes have evolved (Crespi, 1992; Stern and Foster, 1996). In *C. floridanum* and other polyembryonic encyrtids, precocious larvae form a sterile, soldier caste whose function is to manipulate the sex ratio of reproductive larvae and to protect the brood from other parasites that attack the same caterpillar (Cruz, 1981; Strand *et al.*, 1990a; Grbic *et al.*, 1992; Ode and Strand, 1995). In contrast, reproductive larvae develop into adult wasps that reproduce by mating and locating new host eggs to parasitize.

Caste determination in insects is thought to be regulated by the interaction between environmental factors and endocrine physiology (Wheeler, 1986). Studies on honeybees, for example, suggest that extrinsic control of caste determination is mediated by the effect of diet on juvenile hormone (JH) titer (de Wilde and Beetsma, 1982; Robinson, 1992). The JHs are a unique set of insect hormones that have been implicated in regulating an array of physiological events in insects (Riddiford, 1985). In honeybees, JH titers are higher in reproductive larvae than in worker larvae. Topical application of JH to workers at a specific period in development results in the expression of reproductive characters, suggesting that JH influences caste determination (Wirtz and Beetsma, 1972). Maternal factors such as queen pheromones (Winston and Slessor, 1992) and season can also influence caste determination (Brian, 1979; Wheeler, 1986). How environmental and endocrine factors influence the molecular events that ultimately dictate morphology is unknown for any species.

Polyembryonic wasps differ from other caste-forming insects in that both castes develop from the same egg, coexist in the same environment (the host), and develop in close proximity to one another in the polymorula. Nonetheless, Fig. 2 indicates that precocious and reproductive larvae develop in different host stages, suggesting that host environment could mediate whether embryos develop into precocious or reproductive larvae. One feature of the host environment that differs between instars is endocrine physiology. Molting and metamorphosis of *T. ni*, as for all insects, is regulated primarily by two hormones, ecdysone and JH (Riddiford, 1985). A rise in the ecdysteroid titer stimulates insects to molt, whereas the JH titer influences the type of molt that occurs. *T. ni* larvae parasitized by *C. floridanum* molt and develop almost identical to normal, unparasitized caterpillars. The obvious exception is that *C. floridanum* reproductive larvae consume parasitized larva at the end of its fifth larval instar, whereas unparasitized *T. ni* molt to a pupa after the fifth instar. As expected from these patterns, ecdysteroid and JH titers also fluctuate very similarly in parasitized and unparasitized larvae (Strand *et al.*, 1990b; Strand *et al.*, 1991). Rises in the ecdysteroid titer stimulate larvae to molt from one instar to another. During the first through third instars, the JH titer remains elevated. However, in the fourth instar the JH titer begins to fall, and it remains low during most of the fifth instar.

Detailed studies of *C. floridanum* indicate that development of both precocious and reproductive larvae is tightly synchronized with the molting cycle of the host (Strand *et al.*, 1991; Baehrecke *et al.*, 1993; Grbic *et al.*, 1997a). Within each host instar, embryogenesis of precocious larvae proceeds as follows: Blastoderm formation begins following ecdysis of the host to a new instar; morphogenesis, as marked by dorsal fold formation, begins when the host ecdysteroid titer begins to rise; and larval eclosion (i.e., emergence from the chamber formed by the enveloping membrane) occurs immediately prior to when the host molts to its next instar. This pattern is similar for reproductive larvae that initiate morphogenesis when the host molts to its fifth instar. Indeed, the primary difference in how precocious and reproductive larvae develop is that discrete numbers of precocious larvae eclose during each host instar, whereas all reproductive larvae eclose simultaneously in the host's final instar.

Given that host JH titers are elevated in early larval instars and decline in the final instar, we initially hypothesized that development of precocious and reproductive larvae could be mediated by host endocrine state. If so, we would expect that elevating the JH titer would perturb either the absolute number or the proportion of larvae that developed into precocious and reproductive larvae. However, we found that the application of JH or JH analogues to the host does not affect caste formation (Strand et al., 1990b; Strand et al., 1991; Grbic et al., 1997a), suggesting that the host JH titer is not the primary factor regulating whether a given embryo develops into a precocious or a reproductive larva.

In contrast, transplantation experiments in which morula-stage embryos from one host instar were implanted into recipient hosts of a different stage suggest that embryos forming both castes must be exposed to an exogenous pulse of ecdysone to complete embryogenesis (Baehrecke et al., 1993). Indeed, both precocious and reproductive larvae develop successfully in novel host stages, such as pupae, so long as they are competent to initiate morphogenesis. In vitro culture experiments support this conclusion and argue that caste formation in C. floridanum does not depend on any factor associated with a specific stage of the host (Grbic et al., 1997a). What differs between the castes is when embryos become competent to initiate morphogenesis and respond to ecdysone. Embryos forming reproductive larvae become competent 9 days after the wasp egg is laid, which coincides precisely with when the host molts to its fifth instar (see Fig. 2). In contrast, some embryos become competent to develop into precocious larvae as early as 2 days postoviposition, which coincides with the hatching of the host first instar from its egg. That other embryos develop into precocious larvae during the host's second through fourth instars suggests that they become competent at later periods (i.e., 3–9 days postoviposition).

What regulates the acquisition of competency in C. floridanum embryos and why only certain precocious morph embryos initiate morphogenesis within each host instar is unclear. It is generally thought that competence to respond to an inductive signal occurs via synthesis of a receptor for the inducing molecule, via synthesis of a molecule that allows a receptor to function, or by repression of an inhibitor. Baehrecke et al., (1993) suggested that the acquisition of competency is associated with the expression of an ecdysone receptor. More recent in vitro studies, however, indicate that embryogenesis of precocious larvae proceeds to the germ-band extension stage without the addition of ecdysone but that completion of embryogenesis requires a pulse of ecdysone (Grbic et al., 1997a). These results suggest that a rise in the host ecdysteroid titer is not the primary factor mediating the acquisition of competency or the initiation of morphogenesis. We suggest that ecdysone is essential for the completion of embryogenesis because it induces cuticle formation and germ-band retraction. Similar events have also been suggested to be mediated by ecdysteroids during embryogenesis of other insects (Lagueux *et al.*, 1979; Lanot *et al.*, 1989).

This returns us to the question of what mediates competency in *C. floridanum*. As discussed previously, competence by imaginal discs to initiate morphogenesis involves the disc anlagen achieving a fixed size via individual cells occupying specific positional values in a morphogenetic, positional field (Mindek, 1972; Edgar and Lehner, 1996). Competency in *C. floridanum* could similarly be mediated by events associated with proliferation of embryos in the polymorula. In summary, these results indicate that host endocrine factors do mediate aspects of *C. floridanum* embryogenesis but not caste determination. The alternative hypothesis is that caste determination is mediated by intrinsic changes that occur relatively early in embryonic development. Experiments that examine whether specific embryonic lineages exist for precocious and reproductive larvae are currently in progress.

VII. Polyembryony in Other Hymenoptera

Thus far, we have discussed how polyembryonic insects develop through our model *C. floridanum*. This wasp is in the family Encyrtidae, but polyembryonic species are also known in three other families of wasps: the Braconidae, the Platygasteridae, and the Dryinidae. An analysis of basic life history characters in a recent comparative review (Strand and Grbic, 1996) showed that polyembryonic wasps in all families have similar life histories. The most notable differences between polyembryonic encyrtids and polyembryonic species in other families include: (1) only encyrtids produce precocious larvae; (2) brood sizes for the other families are much smaller (2–100 offspring) than those produced by encyrtids like *C. floridanum*; and (3) different insects serve as hosts for wasps in each family. We now consider briefly whether similarities in embryonic development also exist between polyembryonic taxa. Although no molecular or embryological manipulation studies are available outside *C. floridanum*, descriptions of embryogenesis in polyembryonic braconids, platygasterids, and dryinids provide some comparative insight (Kornhauser, 1919; Parker, 1931; Daniel, 1932; Silvestri, 1921; Leiby and Hill, 1923; Leiby, 1924; Leiby and Hill, 1924; Hill and Emery, 1937; Nenon, 1978). Table 1 summarizes in brief form the external characteristics of embryos for each polyembryonic family.

One must always be cautious when making comparisons solely on the basis of external characters, yet Table 1 reveals an important evolutionary pattern: Early cleavage, embryo proliferation, and morphogenesis appear very similar in each group. All polyembryonic wasps lay small, yolkless eggs that appear to undergo holoblastic cleavage. Polar bodies form an enveloping membrane, and proliferation occurs when the extraembryonic membrane partitions rounded, loosely aggregated embryonic cells into an increasing number of embryos. Like *C.*

Table 1 External Characters of Polyembryonic Insects from Four Families of Hymenoptera*

Encyrtidae

Polyembryonic genera:	*Copidosoma, Copidosomopsis, Ageniaspis*
Hosts:	Lepidoptera. Wasps oviposit in egg stage of host and progeny complete development in last larval instar of host.
Egg cleavage:	Holoblastic. Polar bodies form an extraembryonic membrane that surrounds the primary morula.
Embryo Proliferation:	Proceeds by invagination of the extraembryonic membrane and partitioning of embryonic cells into multiple morulae.
Germ-band type:	Long germ band.
Caste formation:	Two larval morphs produced during polyembryony of copidosomatine encyrtids. Only one larval morph reported from *Ageniaspis*.
Brood sizes:	Varies with species, from approximately 20 to more than 3,000 wasp progeny per host.
Selected references:	Silvestri (1906); Patterson (1921); Leiby (1922); Nenon (1978); Koscielski and Koscielska (1985); Baehrecke *et al.* (1993); Grbic *et al.* (1996a).

Braconidae

Polyembryonic genera:	*Macrocentrus, Amicroplus.*
Hosts:	Lepidoptera. Wasps oviposit in any larval stage of host, but progeny always complete development in last larval instar of host.
Egg cleavage:	Appears to be holoblastic. An extraembryonic membrane, possibly of polar body origin, surrounds the primary morula.
Embryo proliferation:	Proceeds by invagination of the extraembryonic membrane and partitioning of embryonic cells.
Germ-band type:	Unknown.
Caste formation:	Only one larval morph produced during polyembryony.
Brood sizes:	From 2–40 wasp progeny per host.
Selected references:	Daniel (1932); Parker (1931).

Platygasteridae

Polyembryonic genera:	*Platygaster, Polygnotus.*
Hosts:	Diptera. Wasps oviposit in egg stage of host, and progeny always complete development in last larval instar of host.
Egg cleavage:	Appears to be holoblastic. An extraembryonic membrane is produced by a polar body–derived cell that surrounds the primary morula.
Embryo proliferation:	Proceeds by invagination of the extraembryonic membrane that partitions embryonic cells into multiple morulae.
Germ-band type:	Appear to be long germ band.
Caste formation:	Only one larval morph produced during polyembryony.
Brood sizes:	From 2–15 wasp progeny per host.
Selected references:	Leiby (1924); Leiby and Hill (1924); Hill and Emery (1937).

Dryinidae

Polyembryonic genera:	*Aphelopus.*
Hosts:	Heteroptera. Wasps oviposit in first or second nymphal stage of the host. Progeny complete development in last nymphal stage of host.
Egg cleavage:	Unknown.
Embryo proliferation:	Unknown.
Germ-band type:	Unknown.
Caste formation:	Only one larval morph produced during polyembryony.
Brood sizes:	Approximately 50 wasp progeny per host.
Selected References:	Kornhauser (1919).

*See Figs. 1 and 7 for phylogenetic relationships of these wasp families and their associated hosts.

floridanum, germ-band extension arises in polyembryonic platygasterids by elongation of the entire embryonic primordium, underscoring that at least some other polyembryonic wasps are likely long germ-band–type insects, despite the apparent absence of any syncytial stage. This conservation in the external characteristics of polyembryonic wasps is remarkable when one considers how different early development is in these species relative to other insects, including hymenopterans like the honeybee (Anderson, 1972). Such a punctuated alteration in early development suggests that some shift in life history has been essential in the evolution of polyembryony in the Hymenoptera. It thus becomes instructive to consider what that change could be and how the embryological processes regulating polyembryony might have derived from their nearest ancestors.

VIII. The Evolution of Polyembryony

In considering the evolution of polyembryony, we first need to assess the direction of life history evolution within the Hymenoptera. This requires more detailed phylogenetic information than was presented in Fig. 1. As mentioned previously, most hymenopterans are in the suborder Apocrita, which consists of free-living and parasitic species. Less obvious from the previous discussion is that the free-living apocritans, like honeybees and ants, are restricted to a single group (Aculeata), whereas the remaining superfamilies are parasitic (Fig. 7). Parasitic wasps exhibit two different developmental strategies (Strand, 1986). Some species develop as ectoparasitoids. Ectoparasitoids lay their eggs externally on a host, and their larvae feed by rasping a hole through the host's cuticle. Other species develop as endoparasitoids by injecting their eggs into the hemocoel of the host, where their progeny feed on blood and other tissues.

In what direction has a free-living and parasitic habit evolved in the Hymenoptera? Phylogenetic analyses based on molecular and morphological data place the ectoparasitic Orussoidea as the sister group to the Apocrita (Gibson, 1985; Whitfield, 1992; Dowton and Austin, 1994) (Fig. 7). This is consistent with the observation that most of the apocritan superfamilies contain ectoparasitic basal groups. It also suggests that: (1) all apocritans, including the free-living species, likely evolved from an ectoparasitic ancestor; and (2) endoparasitism has evolved independently at least eight times from different ectoparasitic ancestors. When we look specifically at polyembryony, Fig. 7 reveals that all of the families that contain polyembryonic species are in different taxa (Ichneumonomorpha, Aculeata, and Prototrumorpha), and the nearest relatives to each polyembryonic genus are monoembryonic endoparasitoids. Thus, polyembryony itself appears to have evolved multiple times from different lineages.

A. Embryological Adaptations and Insect Life History

It is unlikely that the correlation between endoparasitism and the evolution of polyembryony is coincidental, considering that no polyembryonic insects, in-

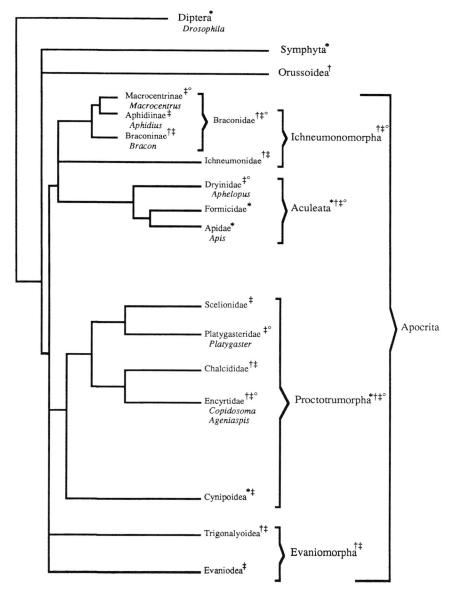

Fig. 7 Phylogeny for the order Hymenoptera. The Diptera are designated as the ordinal-level sister group to the Hymenoptera. The hymenopteran suborders are designated as the Symphyta (sawflies) and the Apocrita (ants, wasps, bees). Selected superfamilies and families within the major subdivisions of the Apocrita are presented. Selected genera discussed in the text are also noted. The following designations are used to indicate the life history of the species within each assemblage: *contains free-living species of terrestrial insects; †contains ectoparasitic species; ‡contains endoparasitic species; °contains endoparasitic species that are also polyembryonic. Although we designate the Diptera as free-living, selected families in this order do contain species that develop as parasitoids (see Clausen, 1940). The phylogeny presented here is a consensus derived from the cladagrams proposed by different authors (Rasnitsyn, 1988; Whitfield, 1992; Dowton and Austin, 1994).

cluding strepsipterans, are free-living or ectoparasitic. That polyembryony has arisen multiple times in the Hymenoptera from monoembryonic ancestors further shows that the types of regulatory changes required for polyembryony can occur relatively easily following the evolution of endoparasitism. So how might a shift in life history to endoparasitism affect insect embryogenesis? Functionally, free-living insects, like *Drosophila* and the honeybee, develop in a terrestrial environment independent of the parent. Adaptations for survival include a rigid chorion to protect the embryo from desiccation and an abundant yolk source to supply the nutrients required for development. As noted by Anderson (1972), the eggs of most free-living Diptera and Hymenoptera are of a similar morphology and develop totally at the expense of nutrients prepackaged during oogenesis, and embryos exhibit no change in volume during development. Ectoparasitic wasps face the same environmental conditions as most other free-living insects, and not surprisingly the structure of their eggs appears similar to that of the honeybee (Clausen, 1940; Bronskill, 1959, 1964). In contrast, the eggs of endoparasitoids develop in a nutrient-rich, aquatic environment (the host), where protection from desiccation and prepackaging of a yolk source are no longer required. Unconstrained by a finite source of nutrition and a chorion, the potential exists for embryos to increase significantly in volume during embryogenesis.

B. Consequences of Endoparasitism for Early Development

If the shift to endoparasitism is important in the alterations observed in early development of polyembryonic wasps, we would expect similar alterations also to be found in monoembryonic endoparasitoids. To examine the consequences of an endoparasitic life history for insect embryogenesis, we focused on two wasps in the Braconidae: a monophyletic assemblage that contains both ecto- and endoparasitic species and is the sister group to the free-living apocritans like the honeybee (Grbic and Strand, 1997) (Fig. 7). *Bracon hebetor* is an ectoparasitoid that lays its eggs on the bodies of caterpillars, whereas *Aphidius ervi* is an endoparasitoid that lays its eggs in the hemocoel of aphids (Tremblay and Caltagirone, 1973; Ode *et al.*, 1996). As we would predict from life history, *B. hebetor* lays yolky eggs with a rigid chorion. Embryogenesis also proceeds very similarly to that in *Drosophila* and the honeybee. Development begins by syncytial cleavage, with the body plan becoming morphologically visible following cellularization of the blastoderm and gastrulation. Formation of the segmented germ band thereafter occurs nearly simultaneously in a classically long germ-band fashion. The similarity in these events with those in other long germ-band species are corroborated molecularly by the expression patterns observed for representative pair-rule, segment-polarity, and homeotic genes (Grbic and Strand, 1997) (Fig. 8). As in *Drosophila*, an antigen for the pair-rule gene *eve* is detected in *Bracon* as a broad domain prior to cellularization of the blastoderm

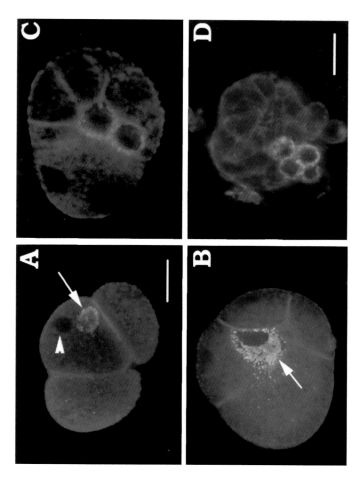

Figure 5 Confocal images of the *C. floridanum* egg and primary morula stained with *Drosophila* anti-Vasa (Lasko and Ashburner, 1989). (A) After the first cleavage (two-cell stage), Vasa is expressed only in the oosome (arrow) inherited by one blastomere (scale bar = 20 μm). The blastomere's nucleus (arrowhead) is unstained. (B) After the second cleavage (four-cell stage), the oosome segregates into the small blastomere. The oosome then dissociates, resulting in a uniform staining pattern with anti-Vasa of the small blastomere's cytoplasm (arrow). (C) After the third cleavage (eight-cell stage), an anti-Vasa protein is detected in two cells that are the progeny of the small blastomere. (D) In the primary morula stage (approximately 256 cells), Vasa is express in four embryonic cells (scale bar = 36 μm). Scale bar in (B) and (C) is the same as in (A).

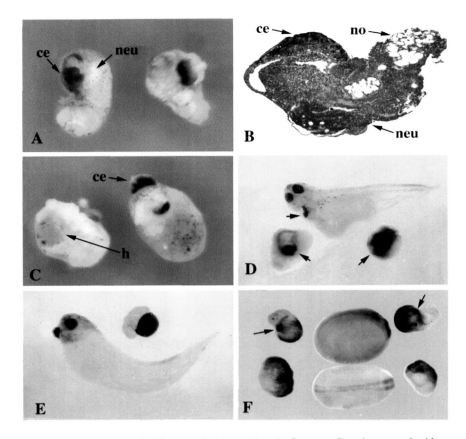

Figure 6.6 Expression of different molecular markers in Spemann Organizer treated with or without suramin. (A) Isolated Spemann organizer cultured for 3 days (*ce* cement gland, *neu* neural structures). (B) Histological section of Spemann organizer shown in panel (A) The explant has differentiated into notochord (*no*), brain structures (*neu*) and cement gland (*ce*). (C) Spemann organizer treated with suramin (150 µM, 3 hr). The explant has differentiated into beating heart (*h* heart tubule) (Grunz, 1992). (D) Spemann organizer treated with suramin (150 µM, 3 hr). Whole-mount *in situ* hybridization with a troponin-marker (Tonissen *et al.*, 1994). All explants show the expression of the heart-specific marker. (E) Spemann organizer treated with suramin (150 µM, 3 hr). Whole-mount *in situ* hybridization with a cement gland-specific marker (Aberger, Lepperdinger, Richter, Grunz, submitted). With exception of cement gland all neural structures are suppressed. (F) Whole-mount *in situ* hybridization of isolated Spemann organizer with a Xsox-3 marker (Penzel, Oschwald, Grunz, submitted). (Culture of the explants until control embryos have have reached stage 20). Untreated Spemann organizer shows the signal (arrows), while in organizer treated with suramin (150 µM, 3 hr) the neural-specific transcription factor is inhibited (explants in the lower part of the figure).

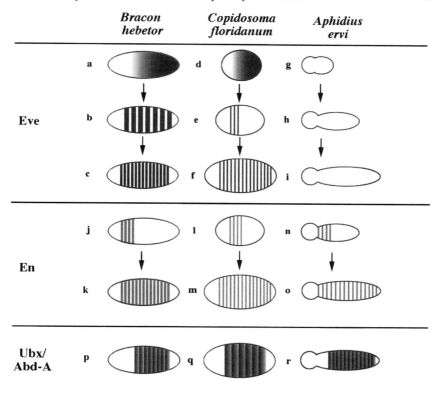

Fig. 8 Expression patterns of the pair-rule protein Even-skipped (Eve) (black), the segment-polarity protein Engrailed (En) (grey), and the homeotic protein Ultrabithorax and Abdominal-A (Ubx/ Abd-A) (shaded) in *Bracon hebetor, Copidosoma floridanum*, and *Aphidius ervi*. In *B. hebetor*, Eve is first expressed in a broad domain in the late syncytial blastoderm (a) followed by a split into pair-rule stripes (b). Pair-rule stripes then split into a segmental pattern in a brief anteroposterior progression (c). In *C. floridanum*, Eve is first expressed from the posterior end to 55% of the length of the cellularized embryonic primordium (d). At the onset of gastrulation, the broad domain of Eve expression resolves directly into segmental stripes in an anteroposterior progression (e). A complete segmental pattern of 15 Eve stripes is observed when the germ band is fully extended (f). In *A. ervi*, no pair-rule or segmental pattern of Eve expression is detected in the cellularized primordium or during germ-band extension (g–i). However, Eve antigen is detected in the dorsolateral mesoderm and neuroblasts of *A. ervi* postgastrulation (not shown). In *B. hebetor* (j, k) and *C. floridanum* (l, m), En is expressed segmentally in a rapid anteroposterior progression during germ-band extension to form a mature pattern of segmentally iterated stripes. As with other short germ-band species, En stripes in *A. ervi* appear sequentially as the germ band extends from the posterior (n), but the mature pattern of segmentally iterated stripes localize to the posterior compartment (o) of each segment, as observed in *B. hebetor* and *C. floridanum*. In all three species, Ubx/Abd-A is expressed in the posterior thorax and abdomen in the retracted germ-band stage (p–r). Note that this figure is highly schematic, since periods of expression for each protein represent a dynamic process. Adapted from Grbic *et al.* (1996a) and Grbic and Strand (1997b).

followed by a pair-rule pattern and later a segmentally reiterated pattern. Homologues of the segment-polarity and homeotic genes *En* and *Ubx/Abd-A* are likewise expressed in a temporal manner characteristic of other canonical long germband insects.

In contrast, the eggs laid by *Aphidius ervi* are enclosed in a thin chorion and appear to be totally devoid of yolk. Cell injection experiments confirm that the *A. ervi* embryo undergoes holoblastic cleavage and that individual blastomeres become dye uncoupled by the 16-cell stage. As in *C. floridanum*, a morula-stage embryo ruptures from the chorion enveloped by membrane of polar body origin. Following formation of a compacted blastula, however, the germ band extends along the anterior–posterior axis in the manner of a short germ-band species like the grasshopper. During this process, no periodic or segmental expression of *eve* antigen is detected, even though Eve expression occurs later in several bilaterally paired neuroblasts in a pattern conserved in all insects (Patel *et al.*, 1992; Patel, 1994). En stripes appear sequentially as the embryo initiates germ-band extension that localizes to the posterior compartment of each segment. Ubx/Abd-A expression in *A. ervi* is also restricted to the abdomen in a pattern conserved in other insects.

The differences in early development between *B. hebetor* and *A. ervi* are as great as any described to date for insects in the comparative developmental literature. Yet, unlike grasshoppers and flies that reside in phylogenetically distant orders, these wasps occur in a monophyletic group of advanced insects. The differences in their early development, therefore, are not due to vastly different phylogenetic histories; rather, we suggest, the shift from an essentially free-living, terrestrial existence (*Bracon*) to development within another organism (*Aphidius*) has favored adaptations in *A. ervi* for survival in a new environment. In particular, the loss of yolk and a chorion in *A. ervi* would appear to be a key alteration in the shift from syncytial to holoblastic cleavage and corresponding alterations in the expression of genes regulating anterior–posterior axis formation.

Although embryogenesis has been examined in relatively few species of Hymenoptera, the descriptive evidence that is available indicates that early cellularization in *A. ervi* and *C. floridanum* are not unique. In each of the four superfamilies in which polyembryony has arisen, parallel changes described for specific monoembryonic endoparasitoids include a reduction in egg size, a thin chorion from which the developing embryo ruptures, the presence of little or no yolk, and the development of an extraembryonic membrane that surrounds the embryo and facilitates uptake of nutrients from the host environment (Leiby, 1922; Daniel, 1932; Tremblay and Calvert, 1972; Ivanova-Kasas, 1972; Tremblay and Caltagirone, 1973; Koscielski *et al.*, 1978; Koscielski and Koscielska, 1985; Strand *et al.*, 1986). At this time, we do not know whether long or short germ-band development is more pervasive among endoparasitic wasps. However, the presence of both short and long germ-band species underscore that other factors besides phylogeny (French, 1996) or rate of embryological development (Sander *et al.*, 1985) are involved in how germ-band types evolve.

C. Evolutionarily Novel Mechanisms Required for Polyembryony

We conclude from our comparative studies of *C. floridanum* and other parasitic wasps that syncytial cleavage is not a requisite for patterning in insects and that substantial changes in early development have no consequences for adult morphology. The evolutionary change from syncytial to holoblastic cleavage has occurred many times in the Hymenoptera, yet the adult morphological characters defining the family Braconidae, for example, are embodied in both *B. hebetor* and *A. ervi*. We conceive the loss of yolk and a rigid chorion as underlying the change to holoblastic cleavage in the endoparasitic Hymenoptera. Free of the limitations set by a chorion, the embryos of endoparasitoids could increase in volume by using host resources absorbed through specialized membranes. The Development of extraembryonic membranes for attachment to tissues and nutrient absorption, and subsequent increases in embryo volume, are in fact characteristics more typically attributed to mammalian embryos (Davidson, 1990; Cross *et al.*, 1994). Yet, if one considers the environmental conditions in which the embryos of endoparasitoids develop, they have more in common with mammalian embryos than they do with other insects (Grbic *et al.*, 1997b).

We conceive that the finite resources and constraints on volume inherent in the architecture of the terrestrial insect egg make the evolution of polyembryony in most orders and families of insects unlikely. However, once the basic changes in early development of monoembryonic endoparasitoids occurred, the evolution of polyembryony could rapidly follow. The extraembryonic membranes already present in monoembryonic endoparasitoids appear to be one preadaptation for polyembryony, since this structure partitions undifferentiated blastomeres in each family of wasps in which polyembryony is reported. Among the genetic regulatory changes required for polyembryony would be the uncoupling of pattern formation processes from early cleavage events in the egg (see Section VI.B). Freed from the constraint of early specification of cell fate, embryo proliferation could proceed by partitioning of blastomeres over the course of the life cycle of the host. That most of the embryos produced by polyembronic wasps synchronously initiate morphogenesis when the host attains its largest net size reflects a pattern observed commonly in organisms that exhibit both asexual and sexual phases of development. Asexual development occurs when resources are abundant and stable, but shifts to sexual reproduction occur when resource quality declines and dispersal becomes necessary. Finally, a host represents a valuable yet defensible resource. Such ecological circumstances present the opportunity for further specialization whereby precocious larvae defend the resource from competitors while reproductive larvae develop to produce reproductively competent adults (Alexander *et al.*, 1991).

IX. Conclusions

Weisblat *et al.* (1994) noted that developmental changes during evolution can occur either as terminal variations appended to highly conserved early events or as modifications of early events leading to relatively well-conserved endpoints. In the case of *C. floridanum* and the other wasps we have examined, the latter stages of the segmentation hierarchy are clearly conserved with the phylotypic stage for insects, yet dramatic differences are seen in the early stages of development. We conclude that polyembryony in the Hymenoptera arose from monoembryonic ancestors that were already endoparasitic. However, parasitism per se does not appear to be the key feature in the evolution of polyembryony. Instead, development in a nutrient-rich, aquatic environment appears to be a much more important consideration in the evolution of this developmental strategy. Not surprisingly, most known polyembryonic species develop either as endoparasites or, like mammals, as organisms whose embryonic phase is spent in the nutrient-rich environment of the parent.

Our comparison of different wasps also underscores that life history can dramatically influence the development of insects. Model systems like *Drosophila* have unquestionably proven their value in analyzing developmental processes at the molecular level and in identifying fundamental genetic pathways that are likely shared across the Metazoa. However, model organisms have their own idiosyncrasies, and overreliance upon them can lead to an overly deterministic view of development. Moreover, comparing model organisms with other species that share both a similar ancestry and a similar life history can further lead to mistaken notions about the universality of the model. Consider the existing comparative literature on insect development. Although some differences exist in the timing of embryological events between *Drosophila* and species like *Tribolium, Manduca*, and *Musca*, the overall segmentation hierarchy of these insects is well described by the fly paradigm (Tautz and Sommer, 1995). Embryogenesis of the honeybee and *Bracon* also appears to proceed very similarly to that of *Drosophila*. Yet, if we consider the life history and oogenesis of these species, we would almost predict a priori that early development proceeds similarly in these advanced insects. Each species possesses meroistic, polytrophic ovarioles, and their eggs develop in very similar environments.

We suggest that the more interesting comparisons for assessing the universality of the paradigms developed from *Drosophila* are to be found by targeting groups in which there would be strong reasons for suspecting differences in early development to exist. One such group would be members of phylogenetically primitive orders, like the Orthoptera, where oogenesis proceeds in a fundamentally different way from that in *Drosophila*. Another would be selected taxa of advanced insects in the Hymenoptera, Diptera, and Coleoptera, where tremendous variation in embryonic life history exists between closely related species. To understand the processes regulating insect development, comparative studies are needed not only between species with life histories and, likely, developmental programs similar to *Drosophila* (i.e., *Tribolium, Musca, Manduca, Apis, Bracon*), but also between species that are considerably different (*Schistocerca, Copidosoma*, and *Aphidius*). It is the only way we can understand how generalizable the detailed knowledge gained through a model like *Drosophila* really is.

Acknowledgments

We thank J. A. Johnson for her assistance in putting this review together. We also thank R. Wharton and J. Whitfield for their advice on the systematics of the Hymenoptera. Funding for the studies on *C. floridanum, B. hebetor*, and *A. ervi* discussed in this paper was provided by grants from the USDA, NSF, Fulbright Foundation, Human Frontiers in Science Program, and Sigma Xi.

References

Akam, M., and Dawes, R. (1992). More than one way to slice an egg. *Curr. Biol.* **8,** 395–398.

Alexander, R. D., Noonan, K. M., and Crespi, B. J. (1991). The evolution of eusociality. *In* "The Biology of the Naked Mole-Rat" (P. W. Sherman, J. U. M. Jarvis, and R. D. Alexander, Eds.), pp. 3–44. Princeton University Press, Princeton, N.J.

Anderson, D. T. (1972). The development of holometabolous insects. *In* "Developmental Systems: Insects" (J. Counce and C. H. Waddington, Eds.), Vol. 1, pp. 165–242. Academic Press, New York.

Askew, R. R. (1970). "Parasitic Insects." Elsevier, New York.

Baehrecke, E. H., and Strand, M. R. (1990). Embryonic morphology and growth of the polyembryonic parasitoid *Copidosoma floridanum* (Ashmead) (Hymenoptera: Encyrtidae). *Int. J. Insect Morph. Embryol.* **19,** 165–175.

Baehrecke, E. H., Grbic, M., and Strand, M. R. (1992). Serosa ontogeny in two embryonic morphs of *Copidosoma floridanum*, the influence of host hormones. *J. Exp. Zool.* **262,** 30–39.

Baehrecke, E. H., Aiken, J. M., Dover, J. M., and Strand, M. R. (1993). Ecdysteroid induction of embryonic morphogenesis in a parasitic wasp. *Devel. Biol.* **158,** 275–287.

Bell, G. (1982). "The Masterpiece of Nature: The Evolution and Genetics of Sexuality." Croom Helm, London.

Brian, M. V. (1979). Caste differentiation and division of labor. *In* "Social Insects" (H. R. Hermann, Ed.), Vol. 1, pp. 121–222. Academic Press, New York.

Bronskill, J. F. (1959). Embryology of *Pimpla turionellae* (L.) (Hymenoptera: Ichneumonidae). *Can. J. Zool.* **37,** 655–688.

Bronskill, J. F. (1964). Embryogenesis of *Mesoleius tenthredinis* Morl. (Hymenoptera: Ichneumonidae). *Can. J. Zool.* **42,** 439–453.

Brown, S. J., Parrish, J. K., Denell, R. E. (1994). Genetic control of early embryogenesis in the red flour beetle, *Tribolium castaneum. Amer. Zool.* **34,** 343–352.

Bünning, J. (1994). "The Insect Ovary." Chapman and Hall, Cambridge, U.K.

Buss, L. W. (1987). "The Evolution of Individuality." Princeton University Press, Princeton, N.J.

Clausen, C. P. (1940). "Entomophagous Insects." McGraw Hill, New York.

Counce, S. J. (1968). Development of composite eggs in *Miastor* (Diptera: Cecidomidae). *Nature* **218,** 781–782.

Crespi, B. J. (1992). Eusociality in Australian gall thrips. *Nature* **359,** 724–726.

Cross, J. C., Werb, Z., Fisher, S. J. (1994). Implantation and placenta: Key pieces of the developmental puzzle. *Science* **266**, 1508–1518.

Cruz, Y. P. (1981). A sterile defender morph in a polyembryonic hymenopterous parasite. *Nature* **294**, 446–447.

Daniel, D. M. (1932). *Macrocentrus ancylivorus* Rohwer, a polyembryonic braconid parasite of the oriental fruit moth. *N.Y. State Agr. Expt. Stn. Tech. Bul.* **187**, 101 pp.

Dawes, R., Dawson, I., Falciani, F., Tear, G., Akam, M. (1994). Dax, a locust Hox gene related to fushi-tarazu but showing no pair-rule expression. *Development* **120**, 1561–1572.

Davidson, E. H. (1990). How embryos work: A comparative view of diverse mode of cell fate specification. *Development* **108**, 365–389.

Davidson, E. H. (1991). Spatial mechanisms of gene regulation in metazoan embryos. *Development* **113**, 1–26.

Davidson, E. H., Peterson, K. J., Cameron, R. A. (1995). Origin of bilaterian body plan: Evolution of developmental regulatory mechanisms. *Science* **270**, 1319–1325.

del Pino, E. M., and Ellinson, R. P. (1983). A novel development pattern for frogs: Gastrulation produces an embryonic disc. *Nature* **306**, 589–591.

de Wilde, J., and Beetsma, J. (1982). The physiology of caste development in social insects. *Adv. Insect Physiol.* **16**, 167–246.

Doe, C. Q., Smouse, D., and Goodman, C. S. (1988a). Control of neuronal fate by the *Drosophila* segmentation gene *even-skipped*. *Nature* **333**, 376–378.

Doe, C. Q., Hiromi, Y., Gehring, W. J., and Goodman, C. S. (1988b). Expression and function of the segmentation gene *fushi tarazu* during *Drosophila* neurogenesis. *Science* **239**, 170–175.

Dowton, M., and Austin, A. D. (1994). Molecular phylogeny of the insect order Hymenoptera: Apocritan relationships. *Proc. Natl. Acad. Sci. USA* **91**, 9911–9915.

Driesch, H. (1894). "Analytische Theorie de Organischen Entwicklung." W. Engelmann. Leipzig.

Edgar, B. A., and Lehner, C. F. (1996). Developmental control of cell cycle regulators: A fly's perspective. *Science* **274**, 1646–1652.

Ephrussi, A., and Lehmann, R. (1992). Induction of germ cell formation by *oskar*. *Nature* **358**, 387–392.

Fleig, R. (1990). *Engrailed* expression and body segmentation in the honeybee *Apis mellifera*. *Roux's Arch. Dev. Biol.* **198**, 467–473.

Fleig, R., Walldorf, U., Gehring, W. J., and Sander, K. (1992). Development of the *Deformed* protein pattern in the embryo of the honeybee *Apis mellifera* L. (Hymenoptera). *Roux's Arch. Dev. Biol.* **198**, 467–473.

French, V. (1996). Segmentation (and *eve*) in very odd insect embryos. *BioEssays* **18**, 435–438.

Gardiner, R. C., and Rossant, J. (1976). Determination during embryogenesis in mammals. *Ciba Found. Sympos.* **40**, 5–18.

Gibson, G. A. P. (1985). Some pro- and mesothoracic structures important for phylogenetic analysis of Hymenoptera, with a review of the terms used for the structures. *Can. Ent.* **117**, 1395–1443.

Godfray, H. C. J. (1994). "Parasitoids: Behavioral and Evolutionary Ecology." Princeton University Press, Princeton, N.J.

Gould, S. J. (1977). "Ontogeny and Phylogeny." Belknap Press, Cambridge, Mass.

Grbic, M., and Strand M. (1997). Life history strategy correlates with reorganization of early development in parasitic wasps. submitted.

Grbic, M., Ode, P. J., and Strand, M. R. (1992). Sibling rivalry and brood sex ratios in polyembronic wasps. *Nature* **360**, 254–256.

Grbic, M., Nagy, L. M., Carroll, S. B., and Strand, M. (1996a). Polyembryonic development: Insect pattern formation in a cellularized environment. *Development* **122**, 795–804.

Grbic, M., Nagy, L. M., and Strand M. (1996b). Pattern duplication in the polyembryonic wasp *Copidosoma floridanum*. *Dev. Genes Evol.* **206**, 281–287.

Grbic, M., Rivers, D., and Strand, M. R. (1997a). Caste formation in the polyembryonic wasp *Copidosoma floridanum*: in vivo and in vitro analysis. *J. Insect Physiol.* In press.

Grbic, M., Nagy, L. M., and Strand, M. (1997b). Embryogenesis of polyembryonic insects: A major departure from the insect developmental ground plan. submitted.

Grünert, S., and St. Johnston, D. (1996). RNA localization and the development of asymmetry during the Drosophila oogenesis. *Curr. Opin. Genet. Dev.* **6**, 395–402.

Hegner, R. W. (1914). Studies on germ cells. *J. Morph.* **26**, 495–561.

Henry, J. J., and Martindale, M. Q. (1987). The organizing role of the D quadrant as revealed through the phenomenon of twinning in the polycheate *Chaetopterus variopedatus*. *Roux's Arch. devl. Biol.* **196**, 499–510.

Hill, C. C., and Emery, W. T. (1937). The biology of *Platygaster herrickii*, a parasite of the Hessian fly. *J. Agric. Res.* **55**, 199–213.

Hughes, R. N., and Cancino, J. M. (1985). An ecological overview of cloning in metazoa. In "Population Biology and Evolution of Clonal Organisms" (J. B. C. Jackson, L. W. Buss, and R. E. Cook, Eds.), pp. 153–186. Yale University Press, New Haven, Conn.

Hülskamp, M., and Tautz, D. (1991). Gap genes and gradients—The logic behind the gaps. *BioEssays* **13**, 261–268.

Ingham, P. W. (1988). The molecular genetics of embryonic pattern formation in *Drosophila*. *Nature* **335**, 25–33.

Ingham, P. W., and Martinez Arias, A. (1992). Boundaries and fields in early embryos. *Cell* **68**, 221–235.

Itow, T., Kenmochi, S., and Mochizuki, T. (1991). Induction of secondary embryos by intra- and interspecific grafts of center cells under the blastopore in horseshoe crabs. *Dev. Growth Diff.* **33**(3), 251.

Ivanova-Kasas, O. M. (1972). Polyembryony in insects. In "Developmental Systems, Insects" (S. J. Counce and C. H. Waddington, Eds.), Vol. 1, pp. 243–271. Academic Press, New York.

Jeffery, W. R., and Swalla, B. J. (1991). An evolutionary change in the muscle lineage of an anural ascidian embryo is restored by the interspecific hybridization with a urodele ascidian. *Dev. Biol.* **145**, 328–337.

Joliot, A., Pernelle, C., Deagostini-Bazin, H., and Prochiantz, A. (1991). Antennapedia homeobox peptide regulates neural morphogenesis. *Proc. Natl. Acad. Sci. USA* **88**, 1864–1868.

Kelsh, R., Weinzierl, R. O. J., White, R. A. H., and Akam, M. (1994). Homeotic gene expression in the locust *Schistocerca*: An antibody that detects conserved epitopes in Ultrabithorax and abdominal-A proteins. *Devel. Genetics* **15**, 19–31.

Kessler, D., and S. Melton, D. A. (1994). Vertebrate embryonic induction: Mesodermal and neural patterning. *Science* **266**, 596–604.

Kim, J., Sebring, A., Esch, J. J., Kraus, M. E., Vorwerk, K., Magee, J., and Carroll, S. B. (1996). Integration of positional signals and regulation of wing formation and identity by *Drosophila vestigial* gene. *Nature* **382**, 133–138.

Kimble, J. (1994). An ancient molecular mechanism for establishing embryonic polarity? *Science* **266**, 577–578.

Koelle, M. R., Talbot, W. S., Segraves, W. A., Bender, M. T., Cherbas, P., and Hogness, D. S. (1991). The *Drosophila* EcR gene encodes an ecdysone receptor, a new member of the steroid receptor superfamily. *Cell* **67**, 59–77.

Kornhauser, S. I. (1919). The sexual characteristics of the membracid, *Thelia bimaculata* (Fabr.). I. External changes induced by *Aphelopus theliae* (Gahan). *J. Morph.* **32**, 531–635.

Koscielski, B., and Koscielska, M. K. (1985). Ultrastructural studies on the polyembryony in *Ageniaspis fuscicollis* Dalm. (Chalcidoidea, Hymenoptera). *Zooogica Polo.* **32**, 203–215.

Koscielski, B., Koscielska, M. K., and Szroeder, J. (1978). Ultrastructure of the polygerm of *Ageniaspis fuscicollis* Dalm. (Chalcidoidea, Hymenoptera). *Zoomorphologie* **89**, 279–288.

Kraft, R., and Jackle, H. (1994). *Drosophila* mode of metamerization in the embryogenesis of the lepidopteran insect *Manduca sexta. Proc. Natl. Acad. Sci. USA* **91,** 6634–6638.

Kristensen, N. P. (1991). Phylogeny of insect orders. *Annu. Rev. Entomol.* **26,** 135–157.

Krombein, K. V., Hurd, P. D., Smith, D. R., and Burks, B. D. (1979). "Catalog of Hymenoptera in America North of Mexico." Smith. Inst. Press, Washington, D.C.

Lagueux, M., Hetru, C., Goltzene, F., Kappler, C., and Hoffman, J. A. (1979). Ecdysone titre and metabolism in relation to cuticulogenesis in embryo of *Locusta migratoria. J. Insect Physiol.* **25,** 709–723.

Lanot, R., Dorn, A., Gunster, B., Thiebold, J., Lagueux, M., and Hoffman, J. A. (1989). Function of ecdysteroids in oocyte maturation and embryonic development of insects. *In* "Ecdysone, From Chemistry to Mode of Action" (J. Koolman, Ed.), pp. 262–270. Georg Thieme Verlag, Stuttgart, Germany.

LaSalle, J., and Gauld, I. D. (1991). Parasitic Hymenoptera and the biodiversity crisis. *Redia* **74,** 540–544.

Lasko, P. F., and Ashburner, M. (1990). Posterior localization of vasa protein correlates with, but is not sufficient for, pole cell development. *Genes Dev.* **4,** 905–921.

Lawrence, P., Johnston, P., Macdonald, P., and Struhl, G. (1987). Borders of parasegments in *Drosophila* embryos are delimited by the *fushi-tarazu* and *even-skipped* genes. *Nature* **328,** 440–442.

Lebandeira, C. C., and Sepkoski, J. J., Jr. (1993). Insect diversity in the fossil record. *Science* **261,** 310–315.

Leiby, R. W. (1922). The polyembryonic development of *Copidosoma gelechiae* with notes on its biology. *J. Morph.* **37,** 195–285.

Leiby, R. W. (1924). The polyembryonic development of *Platygaster vernalis. J. Agric. Res.* **28,** 829–839.

Leiby, R. W., and Hill, C. C. (1923). The twinning and monembryonic development of *Platygaster hiemalis*, a parasite of the Hessian fly. *J. Agric. Res.* **19,** 337–350.

Leiby, R. W., and Hill, C. C. (1924). The polyembryonic development of *Platygaster vernalis. J. Agric. Res.* **28,** 829–839.

Lin, H., and Spradling, A. C. (1995). Fusome asymmetry and oocyte determination in *Drosophila. Dev. Genet.* **16,** 6–12.

Marchal, P. (1898). La dissociation de l'oeuf en un cycle evolutif chez l'*Encyrtus fuscicollis* (Hymenoptere). *Comp. Rend. Acad. Sci. Paris* **126,** 662–664.

Martin, F. (1914). Zur entwicklungsgeschichte des polyembryonalen chalcidiers *Ageniaspis* (Encyrtus) *fuscicollis. Ztschr. f. Wiss. Zool.* **110,** 419–479.

Mindek, G. (1972). Metamorphosis of imaginal discs of *Drosophila melanogaster. Roux's Arch. Entw-Mech. Org.* **169,** 353–356.

Nagy, L. M., and Carroll, S. (1994). Conservation of *wingless* patterning functions in the short-germ embryos of *Tribolium castaneum. Nature* **367,** 460–463.

Nenon, J. P. (1978). La polyembryonie de *Ageniaspis fuscicollis* Thoms. (Hymenoptere, Chalcidien, Encyrtidae). *Bull. Biol. Fr. Belg.* **112,** 13–107.

Nijhout, F. (1994). "Insect Endocrinology". Princeton University Press, Princeton, N.J.

Noskiewicz, J., and Poluszynski, I. (1935). Embryologische untersuchungen an strepsipteren. *Zoologica Polo.* **1,** 53–92.

Ode, P. J., and Strand, M. R. (1995). Progeny and sex allocation decisions of the polyembryonic wasp *Copidosoma floridanum. J. Anim. Ecol.* **64,** 213–224.

Ode, P. J., Antolin, M. F., and Strand, M. R. (1996). Sexual asymmetries in competitive abilities and sex allocation decisions by the parasitic wasp *Bracon hebetor. J. Anim. Ecol.* **65,** 690–700.

Panganiban, G., Sebring, A., Nagy, L., and Carroll, S. B. (1995). The development of crustacean limbs and the evolution of arthropods. *Science* **270,** 1363–1366.

Parker, H. L. (1931). *Macrocentrus gifuensis* Ashmead, a polyembryonic braconid parasite of the european corn borer. *U.S. Dept. Agr. Tech. Bull.* **230**, 62 pp.

Patel, N. H. (1994). Developmental evolution: Insights from studies of insect segmentation. *Science* **266**, 581–590.

Patel, N. H., Martin-Blanco, E., Coleman, D. G., Poole, S. J., Ellis, M. C., Kornberg, T. B., and Goodman, C. S. (1989). Expression of *engrailed* proteins in arthropods, annelids, and chordates. *Cell* **58**, 955–968.

Patel, N. H., Ball, E., and Goodman, C. S. (1992). Changing role of *even-skipped* during the evolution of insect pattern formation. *Nature* **357**, 339–342.

Patel, N. H., Condron, B. G., and Zin, K. (1994). Pair-rule expression patterns of *even-skipped* are found in both short- and long-germ insects. *Nature* **367**, 429–434.

Patterson, J. T. (1919). Polyembryony and sex. *J. Heredity* **10**, 344–352.

Patterson, J. T. (1921). The development of *Paracopidosomopsis*. *J. Morph.* **36**, 1–69.

Raff, M. C., Lillien, L. E., Richardson, W. D., Burne, J. F., and Noble, M. D. (1988). Platelet-derived growth factor from astrocytes drives the clock that times oligodentrocyte development in culture. *Nature* **333**, 562–565.

Raff, R. (1992). Direct-developing sea urchins and the evolutionary reorganization of early development. *BioEssays* **14**, 211–218.

Ransick, A., and Davidson, E. H. (1993). A complete second gut induced by transplanted micromeres in the sea urchin embryo. *Science* **259**, 1134–1138.

Rasnitsyn, A. P. (1988). An outline of the evolution of the hymenopterous insects (Order Vespida). *Oriental Insects* **22**, 115–145.

Riddiford, L. M. (1985). Hormone action at the cellular level. *In* "Comprehensive Insect Physiology, Biochemistry, and Pharmacology" (G. A. Kerbut, and L. I. Gilbert Eds.), Vol. 8, pp. 37–84. Pergamon Press, Oxford.

Rivera-Pomar, R., and Jackle, R. (1996). From gradients to stripes in *Drosophila* embryogenesis: Filling the gaps. *TIGS* **12**, 478–483.

Robinson, G. E. (1992). Regulation of division of labor in insect societies. *Annu. Rev. Entomol.* **37**, 637–665.

Rulifson, E. J., Michelli, C. A., Axelrod, J. D., Perrimon, N., and Blair, S. S. (1996). *Wingless* refines its own expression domain on the *Drosophila* wing margin. *Nature* **384**, 72–74.

St. Johnston, D., and Nüsslein-Volhard, C. (1992). The origin of pattern and polarity in the *Drosophila* embryo. *Cell* **68**, 201–219.

Sander, K. (1976). Specification of the basic body pattern in insect embryogenesis. *Adv. Insect Physiol.* **12**, 125–238.

Sander, K. (1983). The evolution of patterning mechanisms: Gleanings from insect embryogenesis and spermatogenesis. *In* "Development and Evolution" (B. P. Goodwin, Ed.), pp. 137–159. Cambridge University Press, Cambridge, UK.

Sander, K., Gutzeit, H. O., and Jackle, R. (1985). Insect embryogenesis: Morphology, physiology, genetical, and molecular aspects. *In* "Comprehensive Insect Physiology, Biochemistry, and Pharmacology" (G. A. Kerbut, and L. I. Gilbert Eds.), Vol. 5, pp. 319–385. Pergamon Press, Oxford.

Schüpbach, T., and Roth, S. (1994). Dorsoventral patterning in *Drosophila* oogenesis. *Curr. Opin. Genet. Dev.* **4**, 502–507.

Schwalm, F. (1988). "Insect Morphogenesis." Karger, Basel.

Schweisguth, F., Vincent, A., and Lepesant, J. A. (1991). Genetic analysis of the cellularization of the *Drosophila* embryo. *Biol. Cell* **72**, 15–23.

Scott, L. B., Lennarz, W. J., Raff, R. A., and Wray, G. A. (1990). The "lecitotrophic" sea urchin *Heliocidaris erytrogama* lacks typical yolk platelets and yolk proteins. *Dev. Biol.* **138**, 188–193.

Silvestri, F. (1906). Contribuzioni alla conoscenza biologica degli Imenotteri parassiti. Biologia del *Litomastix truncatellus* (Dalm.) (2 nota preliminare). *Ann. Regia Sc. Super. Agric. Portici* **6**, 3–51.

Silvestri, F. (1921). Contribuzioni alla conoscenza biologica degli Imenotteri parassiti. V. sviluppo del *Platygaster dryomyiae* Silv. *Ann. Regia Sc. Super. Agric. Portici* **11**, 299–326.

Silvestri, F. (1937). Insect polyembryony and its general biological aspects. *Bull. Mus. Comp. Zool. Harvard Univ.* **81**, 468–496.

Sommer, R., and Tautz, D. (1991). Segmentation gene expression in the housefly *Musca domestica. Development* **113**, 419–430.

Sommer, R., and Tautz, D. (1993). Involvement of an orthologue of the pair-rule gene *hairy* in segment formation of the short-germ-band embryo of *Tribolium* (Coleoptera). *Nature* **361**, 448–450.

Spemann, H. (1938). "Embryonic Development and Induction." Yale University press, New Haven, Conn.

Stanojevic, D., Small, S., and Levine, M. (1991). Regulators of a segmentation stripe by overlapping activators and repressors in the *Drosophila* embryo. *Science* **254**, 1385–1387.

Stern, D. L., and Foster, W. A. (1996). The evolution of soldiers in aphids. *Biol. Rev.* **71**, 27–79.

Storey, K., Crossley, J. M., De Robertis, E., Norris, W. E., and Stern, C. D. (1992). Neural induction and regionalization in the chick embryo. *Development* **114**, 729–741.

Strand, M. R. (1986). The physiological interactions of parasitoids with their hosts and their influence on reproductive strategies. *In* "Insect Parasitoids" (J. Waage and D. Greathead, Eds.), pp. 97–136. Academic Press, London.

Strand, M. R., and Grbic, M. (1996). Development and life history of polyembryonic parasitoids. In "Parasites: Effect on Host Endocrinology and Behaviour" (N. E. Beckage, Ed.). Chapman and Hall, New York, In press.

Strand, M. R., and Obrycki, J. J. (1996). Host specificity of insect parasitoids and predators. *Bioscience* **46**, 422–429.

Strand, M. R., Meola, S. M., and Vinson, S. B. (1986). Correlating pathological symptoms in *Heliothis virescens* eggs with development of the parasitoid *Telenomus heliothidis. J. Insect Physiol.* **32**, 389–402.

Strand, M. R., Johnson, J. A., and Culin, J. D. (1990a). Intrinsic interspecific competition between the polyembryonic parasitoid *Copidosoma floridanum* and solitary endoparasitoid *Microplitis demolitor* in *Pseudoplusia includens. Entomol. Exp. Appl.* **55**, 275–284.

Strand, M. R., Johnson, J. A., and Dover, B. A. (1990b). Ecdysteroid and juvenile hormone esterase profiles of *Trichoplusia ni* parasitized by the polyembryonic wasp *Copidosoma floridanum. Arch. Insect Biochem. Physiol.* **13**, 41–51.

Strand, M. R., Goodman, W. G., and Baehrecke, E. H. (1991). The juvenile hormone titer of *Trichoplusia ni* and its potential role in embryogenesis of the polyembryonic wasp *Copidosoma floridanum. Insect Biochem.* **21**, 205–214.

Strome, S. (1989). Generation of cell diversity during early embryogenesis in the nematode *Caenorhabditis elegans. Int. Rev. Cytol.* **114**, 81–123.

Strome, S., and Wood, W. B. (1983). Generation of asymmetry and segregation of germ-like granules in early *Caenorhabditis elegans* embryos. *Cell* **35**, 15–25.

Tautz, D., and Sommer, R. (1995). Evolution of the segmentation genes in insects. *TIGS* **1**, 23–27.

Thomson, K. S. (1988). "Morphogenesis and Evolution." Cambridge University Press, Cambridge, U.K.

Tremblay, E., and Caltagirone, L. E. (1973). Fate of polar bodies in insects. *Annu. Rev. Entomol.* **18**, 421–444.

Tremblay, E., and Calvert, D. (1972). New cases of polar nuclei utilization in insects. *Ann. Soc. Ent. Fr.* **8**, 495–498.

Turner, F. R., and Mahowald, A. P. (1976). Scanning electron microscopy of *Drosophila* embryogenesis. I. The structure of the egg envelopes and formation of the cellular blastoderm. *Devp. Bio.* **50,** 95–108.

Waddington, C. H. (1933). Induction by the primitive strak and its derivatives in the chick. *J. Exp. Biol.* **10,** 38–46.

Weisblat, D. A., Wedeen, C. J., and Kostriken, R. G. (1994). Evolution of developmental mechanisms: Spatial and temporal modes of rostrocaudal patterning. *Curr. Topics Devel. Biol.* **29,** 101–134.

Wheeler, D. E. (1986). Developmental and physiological determinants of caste in social Hymenoptera: Evolutionary implications. *Am. Nat.* **128,** 13–34.

Whitfield, J. B. (1992). Phylogeny of the non-aculeate Apocrita and the evolution of parasitism in the Hymenoptera. *J. Hymenop. Res.* **1,** 125–139.

Winston, M. L., and Slessor, K. N. (1992). The essence of royalty: Honey bee queen pheromone. *Amer. Sci.* **80,** 374–385.

Wirtz, P., and Beetsma, J. (1972). Induction of caste differentiation in the honey bee (*Apis mellifera* L.) by juvenile hormone. *Entomol. Exp. Appl.* **15,** 517–520.

Wolff, C., Sommer, R., Schroder, R., Glaser, G., and Tautz, D. (1995). Conserved and divergent expression aspects of the *Drosophila* segmentation gene *hunchback* in the short-germ embryo of the beetle *Tribolium. Development* **121,** 4227–4236.

Wood, W. B. (1991). Evidence for reversal of handedness in *C. elegans* embryos for early cell interactions determining cell fates. *Nature* **349,** 536–538.

Wray, G. A. (1995). Punctuated evolution of embryos. *Science* **267,** 1115–1116.

Wray, G. A., and Bely, A. E. (1994). The evolution of echinoderms is driven by several distinct factors. *Development Supplement,* 97–106.

Wray, G. A., and Raff, R. A. (1990). Novel origins of lineage founder cells in the direct developing sea urchin *Heliocidaris erytrogama. Dev. Biol.* **141,** 41–54.

5

β-Catenin is a Target for Extracellular Signals Controlling Cadherin Function: The Neurocan-GalNAcPTase Connection

Jack Lilien and Janne Balsamo
Department of Biological Sciences
Wayne State University
Detroit, Michigan 48230

Stanley Hoffman
Department of Medicine
Division of Rheumatology and Immunology
Medical University of South Carolina
Charleston, South Carolina 29425-2229

Carol Eisenberg
Department of Cell Biology
Medical University of South Carolina
Charleston, South Carolina 29425-2229

I. Introduction

The stereotypical movement of cells and tissues relative to each other is a dominant feature of early vertebrate development. Over the last 15 years, a vast array of cell–cell and cell–extracellular matrix interactions that play causal roles in establishing tissues and guiding morphogenetic movements has been described. The adhesion molecules that participate in these interactions belong to a limited number of families: the cadherins (reviewed by: Grunwald, 1996; Kuhl

Current Topics in Developmental Biology, Vol. 35
161

and Wedlich, 1996; Marrs and Nelson, 1996; Munro and Blaschuk, 1996; Take-ichi, 1995), N-CAM family members (reviewed by Goridis and Brunet, 1992; Siu, 1996), and integrins (reviewed by Brodt and Dedhar, 1996; Schwartz *et al.*, 1995). However, within each family there are many subtly different forms. This structural diversity is reflected in unique temporal and spatial distribution patterns that determine the course of developmental processes and ultimately the diversity of tissues and organs. Furthermore, many adhesion molecules are expressed simultaneously by the same cells, suggesting that their coordinate function must also be an important factor in guiding morphogenetic rearrangements.

It has recently become apparent that the function of many adhesion molecules is dependent on their association with the cytoskeleton; thus, controlling this association also has the potential to control morphogenetic processes and adds another dimension to the already complex pattern of temporal and spatial expression. In this review we will describe how the association of cadherin with the actin-containing cytoskeleton is rapidly uncoupled by a signal pathway triggered by the interaction between a cell surface glycosyltransferase and its proteoglycan ligand, neurocan. The signals initiated by this unique receptor/ligand pair ultimately act to regulate the binding of a protein tyrosine phosphatase (PTP1B) to the cytoplasmic domain of N-cadherin, where it controls the tyrosine phosphorylation state of β-catenin, one of the cytoplasmic proteins that connect cadherins to the actin-containing cytoskeleton. The phosphorylation state of β-catenin, in turn, regulates the integrity of the cadherin–actin connection and cadherin function.

We will also examine how other receptor–ligand interactions affect the phosphorylation state of β-catenin, and finally we will integrate the role of β-catenin in modulating cadherin function with the key roles that it plays in the Wingless signaling pathway, its recently described transcriptional activation activity, and its association with the APC tumor suppressor protein, important pathways that regulate development and neoplasia.

II. Cadherins Are Associated with a Cell Surface Glycosyltransferase

The cadherin family members and their developmental roles have been reviewed extensively (see references in the first section). We will focus on N- and E-cadherin, two of the so called class I, or classic cadherins. The adhesive function of N- and E-cadherin is dependent on their association with the cytoskeleton (reviewed by Aberle *et al.*, 1996a; Gumbiner and McCrea, 1993; Huber *et al.*, 1996a; Wheelock *et al.*, 1996). This association is mediated by β-catenin bound to the cytoplasmic domain of cadherin; β-catenin in turn binds to α-catenin, which forms the link to actin either directly (Rimm *et al.*, 1995) or through α-actinin (Knudsen *et al.*, 1995). The β-catenin homologue γ-catenin, or plakoglobin, forms similar but separate complexes containing cadherin and α-catenin (Butz and Kemler, 1994; Näthke *et al.*, 1994). A complex homologous to the vertebrate cadherin/β-catenin/α-catenin complex has been structurally and functionally characterized in *Drosophila*; the *Drosophila* homologue of E-cadherin, the product of the *shotgun* gene, is essential for the formation of adherens junctions (Tepass *et al.*, 1996; Uemura *et al.*, 1996) and is associated with the β- and γ-catenin homologue Armadillo (Oda *et al.*, 1994). To complete the picture, Armadillo has been shown to combine with the *Drosophila* homologue of α-catenin (Oda *et al.*, 1993).

We have found that in chick embryo cells both the N- and E-cadherin/catenin complexes are also associated with an *N*-acetylgalactosaminylphosphotransferase (GalNAcPTase), an association that appears to modulate cadherin function (Balsamo and Lilien, 1982; Balsamo and Lilien, 1990). The enzyme is a 220kD glycoprotein (Balsamo *et al.*, 1986) associated with the plasma membrane via a glycophosphatidylinositol linkage (Balsamo and Lilien, 1993). Microsequencing of peptide fragments has revealed that it is closely related to the human *N*-acetylglucosaminyltransferase (Balsamo and Lilien, unpublished). Like N- and E-cadherin, the GalNAcPTase at the cell surface is protected from tryptic digestion by Ca^{2+} ions (Geller and Lilien, 1983). We have also found that several invertebrates, *C. elegans*, sea urchin, and *Drosophila*, possess an immunologically cross-reactive species at the appropriate molecular mass of 220kD (Balsamo and Lilien, unpublished).

Polyclonal and select monoclonal antibodies directed against the GalNAcPTase inhibit both homophilic cadherin-mediated adhesion, assayed as binding of single cells to a substrate of isolated cadherin or to anti-N-cadherin antibody (Balsamo *et al.*, 1991; Bauer *et al.*, 1992), and outgrowth of neurites from neural retina cells on an N-cadherin substrate (Gaya-Gonzales *et al.*, 1991). After treatment of cells with phosphoinositol-specific phospholipase C (PIPLC), anti-GalNAcPTase antibodies are no longer able to inhibit cadherin-mediated adhesion or neurite outgrowth, further implicating the GalNAcPTase as a regulator of cadherin-mediated adhesion. That the inhibition of cadherin-mediated adhesion by select anti-GalNAcPTase antibodies is not due to direct interference with homophilic cadherin–cadherin interaction is indicated by three lines of evidence: (1) The monoclonal anti-GalNAcPTase antibody 7A2 has the same number of binding sites as the inhibitory monoclonal anti-GalNAcPTase antibody 1B11, but does not inhibit cadherin-mediated adhesion. Thus 1B11 recognizes a specific "active site" on the GalNAcPTase. (2) Binding of inhibitory anti-GalNAcPTase antibodies does not interfere with the binding of function-blocking anticadherin antibody to the cell surface. (3) Fab fragments of polyclonal antibodies are also inhibitory (Balsamo *et al.*, 1991).

III. Neurocan Is an Endogenous Ligand for the GalNAcPTase

Our working hypothesis was that the antiadhesive and the antineurite outgrowth activity of anti-GalNAcPTase antibodies mimicked the activity of an endogenous ligand for the GalNAcPTase, and that the interaction of the GalNAcPTase/ligand pair *in situ* would have the potential to control morphogenetic processes. Proteoglycans appeared to be candidate ligands for the GalNAcPTase, since several proteoglycans inhibit adhesion *in vitro*, and both keratan sulfate and chondroitin sulfate proteoglycans have been reported to be inhibitors of neuronal outgrowth *in vitro* (Cole and McCabe, 1991; Guo *et al.*, 1993; Oohira *et al.*, 1991). Furthermore, both types of proteoglycans are present at barriers to neuronal outgrowth *in situ* (Geisert and Bidanset, 1993; Oakley *et al.*, 1994; Pindzola *et al.*, 1993; for reviews see: Margolis and Margolis, 1994; Schwab *et al.*, 1993). We reasoned that the inability of neurites to penetrate such barriers is due to loss of cadherin function, and possibly the function of other adhesion molecules (Gaya-Gonzales *et al.*, 1991), on neurites bearing the GalNAcPTase.

We assayed several brain-derived proteoglycans for their ability to inhibit cadherin-mediated adhesion and to bind directly to the GalNAcPTase. We found that one proteoglycan with a 250kD core protein (250kD PG), when added to retina cells, resulted in inhibition of both homophilic cadherin-mediated adhesion and neurite outgrowth. The inhibitory activity of the 250kD PG was not dependent on the chondroitin sulfate side chains: Removal of these chains with chondroitin ABC lyase had no effect on the inhibitory activity (Balsamo *et al.*, 1995). This is in contrast to the neurite outgrowth inhibitory activity of other chondroitin sulfate proteoglycans, which has been shown to depend on intact chondroitin sulfate chains (Meiners *et al.*, 1995). In fact, isolated chondroitin sulfate can be a potent inhibitor of neurite outgrowth for some neurons (Dou and Levine, 1994).

To characterize the 250kD PG and ultimately determine the molecular basis for its interaction with the GalNAcPTase, we purified the core protein and obtained two peptide sequences as a prelude to obtaining full-length cDNA clones. One of these sequences was from the N-terminus and does not show any sequence similarity to published sequences. A second, internal sequence is very closely related to the hyaluronic acid–binding domain of the aggrecan/versican/neurocan family (Li *et al.*, 1997). Antibodies prepared against the unique N-terminal peptide sequence were used to isolate the proteoglycan to ensure that the sequences were derived from the active component. These immunoaffinity-purified preparations were fully active. Oligonucleotides based on these sequences were then used to probe a chick brain cDNA library from which full-length clones were obtained (Li *et al.*, 1997). The sequence of the full-length clone is very closely related to rat and mouse neurocan.

Neurocan is a member of the aggrecan/versican/brevican/BEHAB family of hyaluronic acid–binding, chondroitin sulfate proteoglycans (Jaworski *et al.*,

1994; Rauch *et al.*, 1992). All three neurocans thus far sequenced (rat, mouse, and chick) have an N-terminal Ig-like domain (Rauch *et al.*, 1995; Rauch *et al.*, 1992) (Fig. 1). In the case of chick neurocan this is preceded by the unique N-terminal sequence. The Ig-like domain is followed by the hyaluronic acid–binding domain. The central region shows little overall sequence similarity, less than 10% (Li *et al.*, 1997). This central region is followed by an EGF-like repeat, a lectin-like domain, and finally a complement regulatory-like domain. The conserved N- and C-terminal regions—approximately 70% sequence identity between chick and rat (Li *et al.*, 1997)—appear as two globular domains separated by an extended linear region (Retzler *et al.*, 1996) composed of the unique central domain that bears all of the chondroitin sulfate side chains. Except for the HA-binding domain, little is known about the function of these domains.

Neurocan has been demonstrated to bind to three members of the N-CAM family of adhesion molecules, N-CAM, NgCAM/L1 (Friedlander *et al.*, 1994), and TAG-1/axonin-1 (Milev *et al.*, 1996), as well as the extracellular matrix glycoprotein tenascin (Grumet *et al.*, 1994). Binding to NgCAM/L1 and N-CAM is reduced by enzymatic removal of chondroitin sulfate side chains, whereas binding to TAG-1/axonin-1 and Tenascin is unaffected (Milev *et al.*, 1996). This indicates that different characteristics of the molecule are responsible for each type of interaction. Our data also indicate that chick neurocan binds directly to the GalNAcPTase and that binding does not require chondroitin sulfate side chains (Balsamo *et al.*, 1995). Furthermore, binding of neurocan to the Gal-NAcPTase is inhibited by the anti-GalNAcPTase antibody 1B11, which inhibits cadherin-mediated adhesion, but not by the noninhibitory antibody 7A2 (Balsamo *et al.*, 1995), suggesting that the binding sites for neurocan and 1B11 may overlap. This strongly suggests the existence of a specific site on the Gal-NAcPTase that is critical for initiating the cascade of events that results in inhibition of cadherin-mediated adhesion, and further demonstrates the specificity of the ligand/receptor interaction.

The modular structure, multiple binding partners, and functional ramifications of neurocan all suggest that it is a molecule that functions in many contexts with many developmental roles. It has the potential to alter cell–cell interactions through *direct* interactions with adhesion molecules or through an *indirect* mechanism involving binding to cell surface receptors, such as the GalNAcPTase, initiating a signal cascade that alters the function of adhesion molecules involved in both cell–cell and cell–extracellular matrix adhesion (Balsamo *et al.*, 1995; Ernst *et al.*, 1995).

IV. The Neurocan–GalNAcPTase Interaction Regulates Both Neural and Nonneural Development

We are just beginning to appreciate how the diversity of molecular interactions involving neurocan may affect development. It is already clear that the temporal

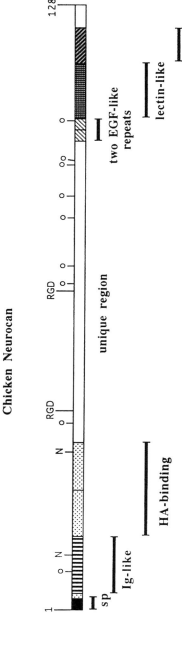

Chicken Neurocan

Fig. 1 Diagram depicting the domain structure of neurocan. *N* indicates the position of potential N-asparagine–linked oligosaccharides. *O* indicates the position of potential O-linked glycosaminoglycan chains; *RGD* indicates the presence of the consensus integrin-binding sequence, Arginine-Glycine-Aspartic Acid; *sp* refers to the signal peptide. The amino acids are labeled 1 through 1287. See text for further details.

and spatial control of neurocan and its receptors, when combined with the temporal and spatial regulation of adhesion molecules, has the potential to affect
profoundly the course of development.

We have recently shown that neurocan, like the anti-GalNAcPTase monoclonal antibody 1B11, inhibits neurite outgrowth by retinal cells *in vitro* (Balsamo *et
al.*, 1997). This is consistent with the distribution of neurocan in the developing
chick retina (Sweatt, Hoffman, Balsamo, and Lilien, unpublished). Neurocan
appears at 7 days of development (E7) in the inner plexiform layer and continues
to be associated with this layer with diminishing intensity until posthatching. It
also appears transiently in the outer plexiform layer between E14 and E20. This
distribution suggests that neurocan is associated with axonal processes, a sugges-
tion consistent with our demonstration that the neurocan/GalNAcPTase interactive pair regulates neurite outgrowth.

Although neurocan has been described as a neural chondroitin sulfate proteoglycan (CSPG) (Rauch *et al.*, 1992), in early chick embryos it is present at a
variety of nonneural sites. For example, at stage 10 it is expressed at such sites as
the heart, the heart-forming fields, the posterior half of the most anterior sclerotomes, the basement membranes surrounding the next group of somites, and
the lateral plate mesoderm (Fig. 2). The distribution of neurocan in the heart-
forming fields is particularly noteworthy (Fig. 2, f). It is expressed in discrete
groups of cells that are migrating from each side of the embryo to extend the
heart tube along the midline of the embryo. This punctate distribution of neurocan may play a critical role in the differentiation of myocardial and endocardial
cells and in the closure of the heart tube.

During the epithelial–mesenchymal transformation in the early heart that leads
to the separation of the myocardial and endocardial lineages, cell–cell adhesion
mediated by N-cadherin is lost prior to down-regulation of N-cadherin expression. Because neurocan (Fig. 2) and the GalNAcPTase (not shown) are expressed
in the early heart, we evaluated whether neurocan/GalNAcPTase interactions
might inhibit N-cadherin function in the early heart (as in retinal cells) and
thereby promote epithelial–mesenchymal transformation. These experiments
took advantage of the availability of a cell line (QCE-6) derived from precardiac
mesoderm that, like precardiac mesoderm, can be induced to differentiate along
either the myocardial or endocardial lineage (Eisenberg and Bader, 1996). When
treated with retinoic acid/bFGF/TGFβ2/TGFβ3/VEGF, QCE6 cells transform
from epithelium to mesenchyme and differentiate along the endocardial lineage,
in that they become competent to migrate into a collagen gel and express the
QH1 antigen, fibrillin, and tenascin.

Just as neurocan and anti-GalNAcPTase monoclonal antibody 1B11 inhibit
N-cadherin-mediated retinal cell adhesion, while anti-GalNAcPTase monoclonal
antibody 7A2 has no effect (Balsamo *et al.*, 1991), both neurocan and 1B11
promote the epithelial–mesenchymal transformation of QCE-6 cells, while 7A2
has no effect (Table 1). Similar to their effects on retinal cell adhesion, other

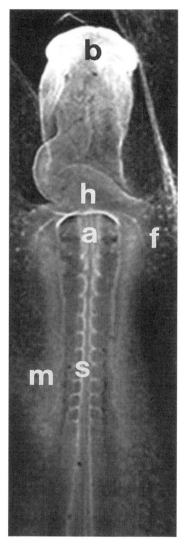

Fig. 2 Neurocan staining in a stage-10 embryo. The embryo was stained in whole mount and observed by confocal microscopy. *b*, brain; *h*, heart; *f*, heart-forming field, note the discrete punctate pattern of staining; *a*, anterior sclerotomes; *s*, somites midway along the anterior–posterior axis; and *m*, lateral plate mesoderm.

CSPGs have little (phosphacan) or no (versican, aggrecan) ability to promote epithelial–mesenchymal transformation.

The ability of neurocan to promote epithelial–mesenchymal transformation

Table 1   Regulation of Epithelial–Mesenchymal
Transformation in QCE-6 Cells

Addition	Migrating Cells
None	0
Mix of growth factors	60
Neurocan, 100 ng/ml	36
Phosphacan, 100 ng/ml	5
Versican, 10 μg/ml	0
Aggrecan, 10 μg/ml	0
1B11, 10 μg/ml	60
7A2, 10 μg/ml	0
NCD-2, 10 μg/ml	0*

*Cell separation, but no migration
Note: QCE-6 cells were plated on top of collagen
gels and then incubated for 48 h in the presence of
the indicated additions. Migration into the gel was
quantified by counting the number of long, thin cells
that migrated to a depth of at least 40 μm.
(Additional migrating cells were found at or near the
surface of the gel, but these cells are not long and
thin.) In each case five fields, each containing 300
total cells, were counted.

involves more than the loss of N-cadherin–mediated adhesion. Anti-N-cadherin
monoclonal antibody NCD-2 also inhibits adhesion, but does not promote migration into the collagen gel (Table 1). Moreover, in addition to promoting deadhe-
sion and migration, the addition of neurocan induces the expression of tenascin,
which serves as a marker for mesenchymal differentiation. In summary, these
results indicate that epithelial–mesenchymal transformation in QCE-6 cells can
be induced by either growth factors or neurocan, the latter most likely involving
binding to the GalNAcPTase.

V. The Neurocan–GalNAcPTase Interaction Controls the Tyrosine Phosphorylation of β-Catenin and the Association of Cadherin with the Cytoskeleton

The interaction of neurocan with the cell surface GalNAcPTase results in loss of
N-cadherin function. This loss of function correlates with enhanced phosphorylation of tyrosine residues on β-catenin and with dissociation of N-cadherin from
the actin-containing cytoskeleton (Balsamo *et al.*, 1995; Balsamo *et al.*, 1996).
Furthermore, we have shown that the N-cadherin-bound β-catenin pool is not
tyrosine phosphorylated, while the pool of β-catenin not bound to N-cadherin

contains phosphorylated tyrosine residues (Balsamo *et al.*, 1996). Enhanced tyrosine phosphorylation of β-catenin has also been reported to correlate with loss of E-cadherin function (Behrens *et al.*, 1993; Hamaguchi *et al.*, 1993; Matsuyoshi *et al.*, 1992; Shibamoto *et al.*, 1994).

Control of phosphorylation of tyrosine residues on β-catenin may in principle be regulated by phosphorylation or dephosphorylation. Both mechanisms appear to be operative under differing conditions. We have discovered that the level of phosphorylated tyrosine residues on β-catenin in chick retina cells depends on the association of a protein tyrosine phosphatase—PTP1B—with the cytoplasmic domain of N-cadherin (Balsamo *et al.*, 1996). Agents that inhibit the activity of the PTP1B result in the accumulation of phosphorylated tyrosine residues on β-catenin, uncoupling of cadherin from its association with the cytoskeleton, and loss of N-cadherin function. Based on these observations, we have proposed that the PTP1B acts as a regulatory switch modulating cadherin-mediated cell–cell adhesion, and that GalNAcPTase/neurocan interactions alter the state of this switch (Balsamo *et al.*, 1996).

We have demonstrated in coprecipitation and overlay experiments that only tyrosine-phosphorylated PTP1B interacts with N-cadherin (Balsamo *et al.*, 1996), a requirement that implicates a tyrosine kinase as a critical regulatory component of this signaling pathway. Indeed, we have shown that binding of neurocan to the cell surface GalNAcPTase results in inhibition of a tyrosine kinase activity that coprecipitates with N-cadherin, and in the release of PTP1B from its association with N-cadherin (Balsamo *et al.*, 1995). Whether this kinase is the same one that is required for tyrosine phosphorylation of β-catenin remains to be determined. Figure 3 is a diagrammatic representation of the presumed events controlling the binding of the PTP1B to cadherin, the resultant alterations in the tyrosine phosphorylation of β-catenin, and how this affects cadherin function.

More recently we have created a dominant-negative form of the PTP1B by introducing a point mutation in the catalytic site, changing cysteine 215 to a serine. This mutation totally eliminates phosphatase activity. We have introduced both the wild-type and mutant forms of PTP1B into L-cells previously transfected to constitutively express N-cadherin. These cells have all the components essential to reconstruct the cadherin–cytoskeletal connection, including endogenous PTP1B, and are able to form cadherin-mediated adhesions. To distinguish endogenous PTP1B from PTP1B introduced via transfection we have introduced a c-*myc* tag in frame at the N-terminus of the protein. Cells expressing the mutant PTP1B are unable to form cadherin-mediated adhesions, while those transfected with the wild-type enzyme show slightly increased cadherin-mediated adhesion. Both the mutant and wild-type PTP1B are bound to the cytoplasmic domain of N-cadherin and are tyrosine phosphorylated. Furthermore, among the cells expressing the dominant-negative PTP1B, β-catenin is tyrosine phosphorylated and N-cadherin is uncoupled from its association with the cytoskeleton (Leung *et al.*,

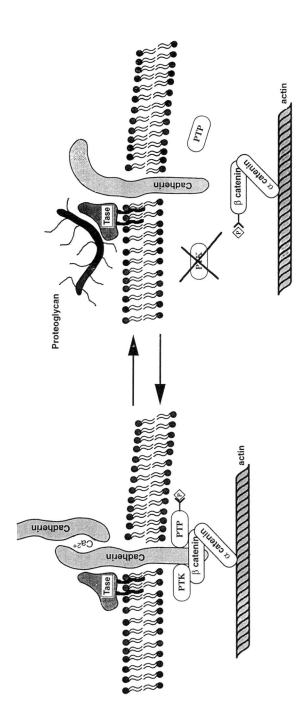

Functional Cadherin

Non-functional Cadherin

Fig. 3 Diagram suggesting the molecular interactions prior to and following binding of neurocan to the GalNAcPTase that results in loss of N-cadherin-mediated adhesion. On the left, fully functional N-cadherin requires N-cadherin-bound PTP1B. The position of the protein tyrosine kinase—PTK—is arbitrary; it is associated with the cadherin complex, but its binding partners are unknown. On the right, following binding of neurocan, the PTK is either released from the complex or inactivated. Consequently, the PTP1B cannot be tyrosine phosphorylated and therefore cannot bind to N-cadherin. This results in tyrosine: phosphorylated β-catenin and loss of the N-cadherin–cytoskeleton association. The fate of the tyrosine phosphorylated β-catenin/α-catenin/actin complex is not known.

1997). Thus, the PTP1B acts as a regulatory switch, controlling cadherin function through its ability to dephosphorylate tyrosine residues on β-catenin.

The precise domains involved in the interaction of the PTP1B with cadherin have not been determined. Cadherin does not contain an SH2 or a PID domain, suggesting that the interaction is not directly mediated by the PTP1B peptides containing the phosphorylated tyrosine residues. One possible binding site on cadherins is a conserved 52-amino-acid sequence at the C-terminus (860–912), which shares a high degree of similarity with the SH3 src-homology domain. PTP1B does contain a short, proline-rich sequence close to the C-terminus (residues 301–313) that is a potential binding domain for SH3 domains (Pawson, 1995).

VI. Transduction of the Neurocan–GalNAcPTase Signal Across the Plasma Membrane

While we have begun to identify the kinase and phosphatase intermediates in the signaling pathway initiated by the neurocan/GalNAcPTase interaction, we have yet to define the most proximate steps. There are several possibilities for transduction of the signal from the exterior to the interior of the cell: The signal may be transduced (1) through cadherin via its association with the transferase, (2) directly via the GPI anchor, or (3) by a combination of both (Fig. 4). The GalNAcPTase is able to transfer N-acetylgalactosamine from UDP-N-acetylgalactosamine to the terminus of O-linked oligosaccharide chains on N-cadherin. There is one consensus sequence for attachment of glycosaminoglycans conserved in N- and E-cadherin, just distal to the transmembrane domain. Glycosylation per se, or a lectinlike interaction based on enzyme/acceptor binding without catalysis, could result in cadherin assuming an altered conformation. This change in conformation, in turn, could be translated into loss of association of a tyrosine kinase with the cadherin complex and consequently loss of binding of the PTP1B to the cytoplasmic tail of cadherin.

Alternatively, or in conjunction with a signal through cadherin, the signal

---

$\longrightarrow$

Fig. 4 Potential pathways for transduction of the signal initiated on binding of neurocan to the GalNAcPTase. Panel A suggests the relationship between the components of the functional cadherin complex (see also Fig. 3). The diamond indicates the potential for an enzyme–acceptor interaction or a lectinlike interaction between N-cadherin oligosaccharide chains and the GalNAcPTase. The filled triangles in the inner plasma membrane leaflet are meant to suggest the position of caveolin. Panel B indicates one possible pathway for the signal mediated by the GPI anchor of the GalNAcPTase. Panel C indicates a second potential pathway involving relay of the signal through cadherin.

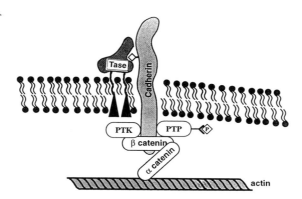

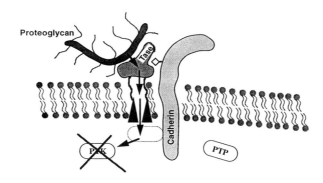

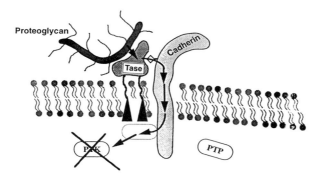

initiated by the neurocan/GalNAcPTase interaction may propagate through the GPI anchor on the GalNAcPTase. GPI-linked proteins have been reported to be associated with nonreceptor tyrosine kinases of the *src* family, and the GPI tail is essential for this association. For example, in leukocytes, the GPI-linked molecules CD48, CD55, and CD59 coimmunoprecipitate with a tyrosine kinase activity, whereas their integral membrane variants do not (Stefanova *et al.*, 1991).

Caveolin may provide one type of link between GPI-anchored proteins and cytoplasmic signaling and cytoskeletal components. Caveolin is found concentrated in membrane microdomains—caveolae—which are enriched in GPI-anchored proteins as well as in components involved in signaling (Lisanti *et al.*, 1995; Stahl and Mueller, 1995). In fact, caveolin itself is an *src* substrate. It has been proposed that caveolin, through its insertion into the inner leaflet of the plasma membrane, may function to anchor or communicate with GPI-tailed proteins (see Fig. 4 and Sargiacomo *et al.*, 1993). Thus caveolin has the potential to transduce signals directly from GPI-anchored proteins to the cytoskeleton.

The similarity between the interaction between the GalNAcPTase and cadherin and that between uPAR, the urokinase receptor, and integrin is striking. uPAR is a GPI-anchored polypeptide that associates with integrin through a cis interaction (Petty and Todd III, 1996). The two molecules colocalize at focal contacts (Pollanen *et al.*, 1987), cocap (Xue *et al.*, 1994), and copurify from leukocytes (Kindzelskii *et al.*, 1996). When complexed with integrin, uPAR mediates cell adhesion to vitronectin and inhibits β1-integrin-mediated adhesion to fibronectin. When the uPAR/integrin complex is disrupted with a peptide competitor, integrin reverts to its initial specificity for adhesion to fibronectin (Wei *et al.*, 1996). Further reinforcing the striking similarity between the uPAR/integrin interaction and the GalNAcPTase/cadherin interaction, carbohydrate competitors disrupt the interaction between uPAR and integrin (Xue *et al.*, 1994), indicating a lectinlike interaction, as we suggested earlier for the GalNAcPTase–cadherin interaction.

Cells engineered with a dominant-negative mutant form of integrin, which prevents wild-type integrin association with actin, have inactive (nonadhesive) uPAR (Wei *et al.*, 1996). Thus uPAR and integrin reciprocally control or modulate each other's adhesive interactions. It was suggested that both the integrin and the uPAR adhesion processes are controlled through the association between integrin and the cytoskeleton (Wei *et al.*, 1996), integrin providing the link between uPAR and the cytoskeleton, allowing signals received by uPAR to be translated into changes in adhesive preference.

uPAR's GPI anchor also appears to play an important role in the reciprocal interaction: Transmembrane forms of uPAR do not bind to vitronectin (Wei *et al.*, 1996). This suggests a role for the GPI anchor in uPAR's ligand specificity. Furthermore, the uPAR/integrin complex isolated from neutral detergent extracts is enriched for caveolin (Wei *et al.*, 1996), and it may be this association that brings uPAR into proximity with signaling molecules, such as kinases or phosphatases, necessary for altered ligand specificity.

VII. Multiple Kinases and Phosphatases Control Cadherin Function

The correlation between increased phosphorylation of β-catenin on tyrosine residues and loss of cadherin function, as well as its interaction with the cytoskeleton, suggests that the establishment and maintenance of cell–cell adhesions require a dynamic equilibrium between tyrosine phosphorylation and dephosphorylation on β-catenin. The levels of phosphorylated tyrosine residues must be regulated by the concerted action of protein tyrosine kinases (PTKs) and protein tyrosine phosphatases (PTPs).

We have already described the binding of PTP1B to the cytoplasmic tail of N-cadherin and its role in maintaining fully functional N-cadherin by controlling the level of tyrosine phosphorylation of β-catenin. Additionally, two distinct receptor tyrosine phosphatases, LAR-PTP and hPTPk, have been reported to associate directly with β-catenin. LAR-PTP is found associated with the β-catenin/cadherin complex in PC12 cells (Kypta *et al.*, 1996) and appears to interact directly with the amino-terminal domain of β-catenin. Tyrosine phosphorylation of LAR-PTP itself decreases the levels associated with β-catenin (Debant *et al.*, 1996). LAR-PTP has also been reported to colocalize with a LAR-interacting protein (LIP.1) at focal adhesions, suggesting a role in regulating cell–matrix interactions (Serra-Pages *et al.*, 1995). hPTPk was identified in a human mammary tumor cell line and localized at adherens junctions. It shows sequence similarity with the intracellular domain of cadherins, and appears to interact directly with both β- and γ-catenin, probably via the Armadillo motifs (Fuchs *et al.*, 1996). Another receptor tyrosine phosphatase, RPTPμ also shows sequence similarity with the cytoplasmic domain of cadherin, but appears to associate directly with a domain near the transmembrane region of the cadherin molecule (Brady-Kalnay *et al.*, 1995). RPTPμ has been shown to mediate homophilic adhesive interactions (Brady-Kalnay *et al.*, 1993), which may trigger either association with cadherin or activation and removal of phosphate from β-catenin.

Several PTKs have been shown to affect the levels of phosphorylated tyrosine residues on β-catenin. Activation of the epidermal growth factor receptor (EGFR) or the scatter factor/hepatocyte growth factor receptor, c-*met*, result in tyrosine phosphorylation of β-catenin, with concomitant loss of cadherin function (Shibamoto *et al.*, 1994; see Fig. 5). The cytoplasmic domain of the EGF receptor tyrosine kinase associates directly with, and can phosphorylate, β-catenin (Hoschuetzky *et al.*, 1994). Furthermore, activation of overexpressed EGFR results in phosphorylation of β-catenin and loss of E-cadherin-mediated adhesion. Additionally, the solubility of E-cadherin is altered, presumably due to altered attachment of E-cadherin to the cytoskeleton (Fujii *et al.*, 1996). The closely related transmembrane tyrosine kinase, c-*erb*B-2 (Yamamoto *et al.*, 1986), has also been reported to associate directly with β-catenin (Kanai *et al.*, 1995; Ochiai *et al.*, 1994) and is thus potentially able to directly phosphorylate β-catenin and down-

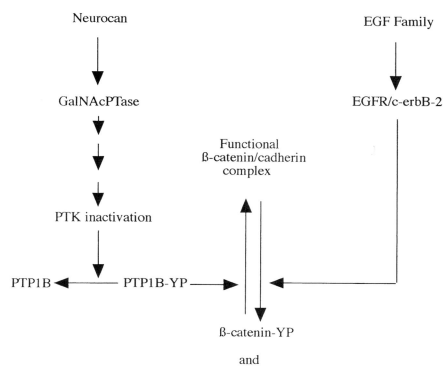

Fig. 5 Diagram summarizing two pathways affecting the tyrosine phosphorylation of β-catenin and the association of cadherin with the cytoskeleton. On the left, binding of neurocan to the Gal-NAcPTase results in inactivation or loss of a protein tyrosine kinase—PTK—from the N-cadherin complex. Consequently, the PTP1B cannot be phosphorylated and therefore cannot bind to N-cadherin. On the right, receptor protein tyrosine kinases can directly phosphorylate β-catenin. Both pathways lead to accumulation of tyrosine-phosphorylated β-catenin and loss of the N-cadherin–cytoskeleton association.

regulate cadherin function. In this context, when the association between c-*erb*B-2 and endogenous β-catenin is disrupted by introduction of a dominant-negative form of β-catenin lacking the cadherin-binding domain but retaining the c-*erb*B-2 binding domain, invasive behavior is reduced (Shibata *et al.*, 1996). These results suggest that reduced tyrosine phosphorylation of cadherin-associated β-catenin reduces invasiveness, possibly through strengthening of cadherin-mediated cell–cell adhesion. The transmembrane tyrosine kinase SF/HGF-receptor, c-*met*, may act similarly. As the name implies, interaction of SF/HGF with its receptor does appear to be one trigger for cells to assume a migratory phenotype that is characteristic of loss of cell–cell adhesion. Indeed, the interaction of SF/HGF with c-*met* is essential for the migration of myogenic precursors into the limb (Bladt *et al.*, 1995). In each of these cases, the increased tyrosine phosphorylation of β-catenin may overwhelm the ability of the PTP1B or other phosphatases to dephosphorylate β-catenin, resulting in loss or reduction of cadherin-mediated adhesion and assumption of a more migratory phenotype.

Transmembrane tyrosine kinases in the EPH family have recently been implicated in the patterning of retinal axons on the tectum (Cheng *et al.*, 1995; Drescher *et al.*, 1995; Holash *et al.*, 1995; Worley and Holt, 1996). The mechanisms through which these receptors influence axon targeting are unknown. Given the role of receptor protein tyrosine kinases, RPTKs, in modulating cadherin function, it is possible that one role for these RPTKs in axonal patterning is to modulate the function of cadherins or other adhesion molecules through alterations of their cytoskeletal associations induced by phosphorylation.

Consistent with the effect of neurocan and growth factors on β-catenin phosphorylation and cadherin-mediated adhesion, increasing tyrosine kinase activity via transfection of cells with v-*src* (Behrens *et al.*, 1993; Hamaguchi *et al.*, 1993) or infection of chick embryo fibroblasts with Rous Sarcoma virus (Matsuyoshi *et al.*, 1992) results in tyrosine phosphorylation of β-catenin and loss of cadherin function. Similarly, in the *ras*-transformed human breast cell line MCF-10A, the level of tyrosine phosphorylation on β-catenin is elevated, and the association of E-cadherin with the cytoskeleton is much less stable, than in nontransformed counterparts. Concomitant with the increased phosphorylation of β-catenin, its association with E-cadherin is dramatically reduced (Kinch *et al.*, 1995).

Among these *ras*-transformed cells, the decreased association of β-catenin with E-cadherin is accompanied by increased binding of p120*cas* to E-cadherin (Kinch *et al.*, 1995). p120*cas* is a member of the Armadillo family and was originally characterized as an *in vivo* and *in vitro* substrate for *src* (Reynolds *et al.*, 1992). p120*cas* binds directly to E-cadherin (Shibamoto *et al.*, 1995), and binding to E-cadherin requires tyrosine phosphorylation (Aghib and McCrea, *1995*). Furthermore, loss of E-cadherin function has been correlated with increased binding of tyrosine phosphorylated p120 (Skoudy, 1996). It is notable that some splice variants of p120*cas* migrate very similarly to β-catenin on SDS PAGE (Reynolds *et al.*, 1994). Given the reciprocal relationship of p120*cas* and β-catenin binding to E-cadherin and the similar migration, it is possible that prior reports that tyrosine-phosphorylated β-catenin remains associated with E-cadherin are due to a case of mistaken identity (see Behrens *et al.*, 1993; Hamaguchi *et al.*, 1993; Matsuyoshi *et al.*, 1992; Shibamoto *et al.*, 1994).

VIII. β-Catenin Is at the Center of Several Developmentally Important Pathways

β-Catenin plays a decisive regulatory role in cadherin-mediated adhesion. We have seen that tyrosine phosphorylation of β-catenin is crucial to its role in

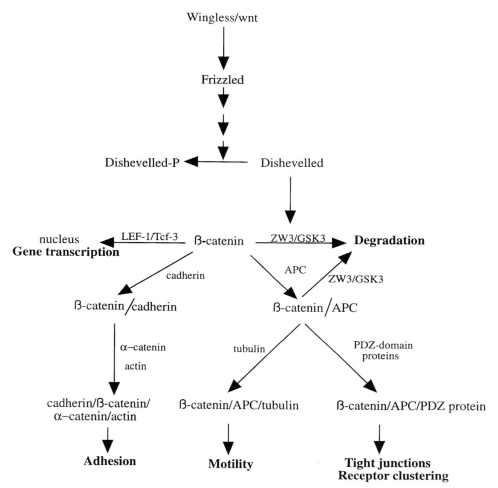

Fig. 6 Diagram indicating the potential role of β-catenin in the Wingless pathway, cadherin-mediated adhesion, and the activity of APC. The figure indicates our present understanding of the potential interactions that involve β-catenin. See text for details.

mediating the association of cadherin with the cytoskeleton, and therefore to cadherin function. β-Catenin also plays a decisive role the in *Wingless/Wnt* signaling pathway, which is central to embryonic patterning, and interacts with the Adenomatous polyposis coli (APC) protein, which may play a role in cell motility and extension (see Fig. 6). These interactions do not appear to regulate cadherin function directly, but may affect the levels of β-catenin and therefore indirectly ramify on cadherin function.

In *Drosophila*, the β- and γ-catenin homologue, Armadillo, not only is a component of the cadherin complex (Oda *et al.*, 1993; Oda *et al.*, 1994) and plays an important role in adherens junction formation (Cox *et al.*, 1996; Tepass *et al.*, 1996; Uemura *et al.*, 1996), but is a component of the Wingless signaling pathway that is involved in specifying many aspects of body patterning (reviewed by Eaton and Cohen, 1996; Gumbiner, 1995; Peifer, 1996). The secreted Wingless protein is bound by the recently characterized Frizzled protein, a receptor protein with seven transmembrane domains (Bhanot *et al.*, 1996). This interaction results in hyperphosphorylation of the cytoplasmic protein Dishevelled and its recruitment to the plasma membrane (Yanagawa *et al.*, 1995). In the absence of phosphorylation, Dishevelled activates Zeste White-3/Shaggy (ZW-3), a serine-threonine kinase that phosphorylates Armadillo (Peifer *et al.*, 1994a) and thereby marks it for degradation. Thus, activation of the Wingless pathway results in increased steady-state levels of Armadillo (Peifer *et al.*, 1994b). Consistent with this, transfection of *wingless* cDNA into a *Drosophila* imaginal disk cell line results in increased levels and stability of the Armadillo protein (van Leeuwen *et al.*, 1994).

Vertebrate homologues of the polypeptide components of the Wingless pathway have been identified. In *Xenopus*, a Wingless-like pathway is involved in dorsal specification. Maternal X*wnt*8 appears to initiate the signal (Miller and Moon, 1996; Yang-Snyder *et al.*, 1996; Yost *et al.*, 1996); the vertebrate homologue of Dishevelled is upstream of β-catenin (Sokol, 1996), as is XGSK (the *Xenopus* homologue of ZW-3, *Xenopus* glycogen synthase kinase). Transcription factors within the Spemann organizer are downstream of β-catenin (Carnac *et al.*, 1996; Wylie *et al.*, 1996). Consistent with this overall scheme, β-catenin is able to mimic the activity present in Nieuwkoop center (Guger and Gumbiner, 1995). β-Catenin's activity undoubtedly depends on its interaction with LEF-1/Tcf-3 transcription factors. This interaction results in translocation to the nucleus and activation of transcription (Behrens *et al.*, 1996; Huber *et al.*, 1996b; Molenaar *et al.*, 1996). LEF-1 also promotes dorsal mesoderm formation (Behrens *et al.*, 1996; Huber *et al.*, 1996b; Molenaar *et al.*, 1996), and nuclei containing β-catenin demarcate dorsalizing centers in both *Xenopus* and zebrafish (Schneider *et al.*, 1996). Furthermore, injection of either β-catenin or LEF-1 mRNAs result in axis duplication in *Xenopus* embryos (Behrens *et al.*, 1996; Huber *et al.*, 1996b). In addition to activation of transcription of downstream targets within the organizer region, the β-catenin/LEF-1 complex binds to the promoter for E-cadherin (Huber *et al.*, 1996b), presumably activating its expression. In *Drosophila*, Armadillo is also shuttled to the nucleus in combination with transcription factors, regulating expression of the *engrailed* gene (van de Wetering *et al.*, 1997; Brunner *et al.*, 1997); activation of the Wingless pathway is essential for the maintenance of *engrailed* expression in adjacent territories (Danielian and McMahon, 1996; Vincent and Lawrence, 1994).

Activation of the Wingless/Wnt pathway and increased levels of β-catenin appears to have two distinct types of effects (see Fig. 6). On the one hand, β-catenin is complexed with transcription factors of the LEF-1 family, affecting transcription as discussed earlier. Increased levels of β-catenin also affect cadherin-mediated adhesion. Transfection and expression of *Wnt-1* in vertebrate cell lines expressing N-cadherin results in accumulation of β-catenin and plakoglobin, increased binding of β-catenin in cadherin complexes, and increased stability of N-cadherin-mediated adhesion (Hinck *et al.*, 1994a and b). Among cells bearing E-cadherin, *Wnt-1* expression leads to increased levels of plakoglobin as well as to its redistribution into areas of cell contact, increased expression of E-cadherin, and increased calcium-dependent adhesion (Bradley *et al.*, 1993).

β-Catenin has another partner, the Adenomatous polyposis coli, or APC protein (see Fig. 6). APC was originally identified via its role in the formation of colon tumors: mutations in APC that result in loss of its carboxy-terminus result in overproduction of epithelial cells and the formation of polyps (Groden *et al.*, 1991). APC is a very large polypeptide—>300kD—related to the Armadillo family of proteins, and its interaction with β-catenin is mediated by Armadillo repeats on each protein (Hülsken *et al.*, 1994b; Polakis, 1995). APC and E-cadherin form separate and distinct complexes with β-catenin (Hülsken, *et al.*, 1994a; Rubinfeld *et al.*, 1996).

APC may also control the level of β-catenin; in contrast to *Wnt* signals, increased expression of APC decreases the steady-state levels of β-catenin. This appears to be dependent on phosphorylation of APC by GSK. Unphosphorylated APC binds β-catenin only weakly; however, the APC/β-catenin complexes recruits GSK, which results in phosphorylation of APC and enhanced binding of β-catenin, possibly at a second site (Rubinfeld *et al.*, 1996). Phosphorylation and enhanced binding results in degradation of β-catenin (Munemitsu *et al.*, 1995; Rubinfeld *et al.*, 1996). The APC and Wingless pathways appear to be inextricably linked through the role of GSK; thus, in the absence of a *Wnt* signal, β-catenin is phosphorylated by GSK and targeted for degradation. Interaction with APC is either essential for or enhances targeted degradation of β-catenin.

The interaction of APC with β-catenin also appears to play an important role in cell motility. APC interacts directly with tubulin (Munemitsu *et al.*, 1994; Rubinfeld *et al.*, 1996; Smith *et al.*, 1994). In motile cells, APC is localized at sites of active migration, and this pattern of localization depends on intact microtubules (Näthke *et al.*, 1996). Furthermore, the role of APC in motility appears to be controlled by β-catenin. Expression of N-terminally deleted β-catenin in MDCK cells blocks tubulogenesis induced by HGF/SF. This is correlated with the accumulation of mutant β-catenin with APC (Pollack *et al.*, 1997). Thus the mutant β-catenin is preventing the normal targeting or turnover of APC, suggesting that APC and β-catenin may control each other's stability. Further solidifying the role of APC in motility, increased expression in the intestinal epithelium of transgenic mice results in disordered migration in intestinal villi (Wong *et al.*,

1996). In contrast, increased expression of E-cadherin slows, but does not alter, the order of migration (Hermiston *et al.*, 1996).

The APC/β-catenin complex also binds to the *Drosophila* Disks Large tumor suppressor protein (DLG), an interaction mediated by the carboxy-terminus of the APC protein (Matsumine *et al.*, 1996). Additionally, APC was found to colocalize with DLG at lateral plasma membrane sites in epithelial tissues and in presynaptic nerve terminals (Matsumine *et al.*, 1996). The mammalian synaptic protein PSD-95/SAP-90 (Cho *et al.*, 1992 Kistner *et al.*, 1993; Woods and Bryant, 1993) and the tight junction proteins ZO-1 and ZO-2 are closely related to DLG (Itoh *et al.*, 1993; Jesaitis and Goodenough, 1994; Willot *et al.*, 1993; Woods and Bryant, 1993). This family of proteins contains a so-called PDZ (*PSD/DLG/ZO*-1) or DHR (discs large homology domain) domain, which binds directly to the *N*-methyl-D-aspartate (NMDA)-type glutamate receptor (Kim *et al.*, 1995; Kornau *et al.*, 1995; Niethammer *et al.*, 1996) and Shaker-type K+ channels (Kim *et al.*, 1995) and may be involved in clustering of these receptors at specific sites (see Sheng, 1996). While the role of APC and β-catenin in these interactions is unknown, it is possible that β-catenin serves as a necessary linker mediating a labile or modulatable attachment of these cell surface receptors to the cytoskeleton, as we have demonstrated for cadherin.

IX. Conclusions

The cadherin, the Wingless/Wnt, and the APC pathways appear to be closely interrelated, each ramifying on the levels of β-catenin, yet each has a distinct cellular function. Cadherin-mediated adhesion is one of the key processes guiding the cell–cell interactions essential to the formation and maintenance of regionally specific tissue organization. β-Catenin has the potential to play an important role in this process as its tyrosine phosphorylation/dephosphorylation can rapidly switch cadherin between functional and nonfunctional states. The neurocan/GalNAcPTase pair is one of several ligand/receptor interactions that can trigger this process. The fine-tuning of many morphogenetic processes may depend on the temporal and spatial distribution of these ligand/receptor pairs. Furthermore, inactivation of cadherin via phosphorylation or retention of phosphate on tyrosine residues can lead to much more rapid changes in morphogenetic potential than regulation at the transcriptional level. Activation of the Wingless/Wnt pathway results in up-regulation of β-catenin and transport to the nucleus, where it participates in transcriptional activation of a specific subset of genes. This is one of the key processes in establishing the embryonic axis. Up-regulation of β-catenin also stabilizes cadherin-mediated adhesion. APC/β-catenin complexes also appear to regulate the level of β-catenin and to have several independent roles involving motility, cellular extension, and receptor localization. Though there is much to be learned about each of the processes in which β-catenin plays an important role, we are beginning to appreciate how inextricably linked seemingly independent cellular processes are.

Acknowledgments

Studies by the authors reported in this chapter were supported by grants from the NSF (JL), the Whitehall Foundation (JL), the Karmanos Cancer Institute (JL), and the U.S. Public Health Service (HL37641, SH; HL55923, CE and Program Project HL52813).

References

Aberle, H., Schwartz, H., and Kemler, R. (1996). Cadherin-catenin complex: Protein interactions and their implications for cadherin function. *J. Cell. Biochem.* **61,** 514–523.

Aghib, D. F., and McCrea, P. D. (1995). The E-cadherin complex contains the src substrate p120. *Exp. Cell Res.* **218,** 359–369.

Balsamo, J., and Lilien, J. (1982). An *N*-acetylgalactosaminyltransferase and its acceptor in embryonic chick neural retina exist in interconvertible particulate forms depending on their cellular location. *J. Biol. Chem.* **257,** 349–354.

Balsamo, J., and Lilien, J. (1990). *N*-cadherin is stably associated with and is an acceptor for a cell surface *N*-acetylgalactosaminylphosphotransferase. *J. Biol. Chem.* **265,** 2923–2928.

Balsamo, J., and Lilien, J. (1993). The retina cell-surface *N*-acetylgalactosaminylphosphotransferase is anchored by a glycophosphatidylinositol. *Biochem.* **32,** 8246–8250.

Balsamo, J., Pratt, R. S., and Lilien, J. (1986). Chick neural retina *N*-acetylgalactosaminyltransferase/acceptor complex: Catalysis involves transfer of *N*-acetylgalactosamine phosphate to endogenous acceptors. *Biochem.* **25,** 5402–5407.

Balsamo, J., Thiboldeaux, R., Swaminathan, N., and Lilien, J. (1991). Antibodies to the retina *N*-acetylgalactosaminylphosphotransferase modulate *N*-cadherin-mediated adhesion and uncouple the *N*-cadherin transferase complex from the actin-containing cytoskeleton. *J. Cell Biol.* **113,** 429–436.

Balsamo, J., Ernst, H., Zanin, M. K., Hoffman, S., and Lilien, J. (1995). The interaction of the retina cell surface *N*-acetylgalactosaminylphosphotransferase with an endogenous proteoglycan ligand results in inhibition of cadherin-mediated adhesion. *J. Cell Biol.* **129,** 1391–1401.

Balsamo, J., Leung, T., Ernst, H., Zanin, M. K. B., Hoffman, S., and Lilien, J. (1996). Regulated binding of a PTP1B-like phosphatase to *N*-cadherin: Control of cadherin-mediated adhesion by dephosphorylation of β-catenin. *J. Cell Biol.* **134,** 801–813.

Balsamo, J., Hoffman, S., and Lilien, J. (1996). Control of cadherin-mediated cell–cell adhesion through regulated association with the cytoskeleton. *J. Braz. Adv. Sci.* **48,** 341–346.

Bauer, G. E., Balsamo, J., and Lilien, J. (1992). Cadherin-mediated adhesion in pancreatic islet cells is modulated by a cell surface *N*-acetylgalactosaminylphosphotransferase. *J. Cell Sci.* **103,** 1235–1241.

Behrens, J., Vakaet, L., Friis, R., Winterhager, E., Van Roy, F., Mareel, M. M., and Birchmeier, W. (1993). Loss of epithelial differentiation and gain of invasiveness correlates with tyrosine phosphorylation of the E-cadherin/beta-catenin complex in cells transformed with a temperature-sensitive v-*src* gene. *J. Cell Biol.* **120,** 757–766.

Behrens, J., von Kries, J. P., Kuhl, M., Bruhn, L., Wedlich, D., Grosschedl, R., and Birchmeier, W. (1996). Functional interaction of beta-catenin with the transcription factor LEF-1. *Nature* **382,** 638–642.

Bhanot, P., Brink, M., Samos, C. H., Hsieh, J. C., Wang, Y. S., Macke, J. P., Andrew, A. D., Nathans, J., and Nusse, R. (1996). A new member of the Frizzled family from *Drosophila* functions as a Wingless receptor. *Nature* **382,** 225–230.

Bladt, F., Riethmacher, D., Isenman, S., Aguzzi, A., and Birchmeier, C. (1995). Essential role for the c-met receptor in the migration of myogenic precursor cells in the limb bud. *Nature* **376,** 768–771.

Bradley, R. S., Cowin, P., and Brown, A. M. (1993). Expression of Wnt-1 in PC12 cells results in modulation of plakoglobin and E-cadherin and increased cellular adhesion. *J. Cell Biol.* **123,** 1857–1865.

Brady-Kalnay, S. M., Flint, A. J., and Tonks, N. K. (1993). Homophilic binding of PTPμ, a receptor-type protein tyrosine phosphatase, can mediate cell–cell aggregation. *J. Cell Biol.* **122,** 961–972.

Brady-Kalnay, S. M., Rimm, D. L., and Tonks, N. K. (1995). Receptor protein tyrosine phosphatase PTPμ associates with cadherins and catenins *in vivo. J. Cell Biol.* **130,** 977–986.

Brodt, P. and Dedhar, S. (1996). The integrins: Mediators of cell–extracellular matrix and intercellular communication. *In* "Cell Adhesion and Invasion in Cancer Metastasis", Pnina Brodt, ed. pp. 35–59, R. G. Landers Co.

Brunner, E., Peter, O., Schweizer, L., Basler, K. (1997). Pangolin encodes a Lef-1 homologue that acts downstream of Armadillo to transduce the Wingless signal in *Drosphila. Nature* **385,** 829–833.

Butz, S., and Kemler, R. (1994). Distinct cadherin-catenin complexes in Ca(2+)-dependent cell–cell adhesion. *FEBS Lett.* **355,** 195–200.

Carnac, G., Kodjabachian, L., Gurdon, J. B., and Lemaire, P. (1996). The homeobox gene siamois is a target of the Wnt dorsalization pathway and triggers organizer activity in the absence of mesoderm. *Develop.* **122,** 3055–3065.

Cheng, H. J., Nakamoto, M., Bergemann, A. D., and Flanagan, J. G. (1995). Complementary gradients in expression and binding of ELF-1 and Mek-4 in development of the topographic retinotectal projection map. *Cell* **82,** 371–381.

Cho, K., Hunt, C. A., and Kennedy, M. B. (1992). The rat brain postsynaptic density fraction contains a homolog of the *Drosophila* disc-large tumor supressor protein. *Neuron* **9,** 929–942.

Cole, G. J., and McCabe, C. F. (1991). Identification of a developmentally regulated keratan sulfate proteoglycan that inhibits cell adhesion and neurite outgrowth. *Neuron* **7,** 1007–1018.

Cox, R. T., Kirkpatrick, C., and Peifer, M. (1996). Armadillo is required for adherens junction assembly, cell polarity, and morphogenesis during *Drosophila* embryogenesis. *J. Cell Biol.* **134,** 133–148.

Danielian, P. S., and McMahon, A. P. (1996). Engrailed-1 as a target of the Wnt-1 signaling pathway in vertebrate midbrain development. *Nature* **383,** 332–334.

Debant, A., Serra-Pages, C., Seipel, K., O'Brien, S., Tang, M., Park, S. H., and Streuli, M. (1996). The multidomain protein TRIO binds the LAR transmembrane tyrosine phosphatase, contains a protein kinase domain, and has separate rac-specific and rho-specific guanine nucleotide exchange factor domains. *Proc. Natl. Acad. Sci.* **93,** 5466–5471.

Dou, C. L., and Levine, J. M. (1994). Inhibition of neurite growth by the NG2 chondroitin sulfate proteoglycan. *J. Neurosci.* **14,** 7616–28.

Drescher, U., Kremoser, C., Handweker, C., Loschinger, J., Noda, M., and Bonhoeffer, F. (1995). *In vitro* guidance of retinal ganglion cell axons by RAGS, a 25kDa tectal protein related to ligands for Eph receptor tyrosine kinases. *Cell* **82,** 359–370.

Eaton, S., and Cohen, S. (1996). Wnt signal transduction: More than one way to skin a (β-) cat? *Trends Cell Biol.* **6,** 287–290.

Eisenberg, C., and Bader, D. M. (1996). Establishment of the mesodermal cell line QCE-6: A model system for cardiac cell differentiation. *Circ. Res.* **78,** 205–216.

Ernst, H., Zanin, M. K. B., Everman, D., and Hoffman, S. (1995). Receptor-mediated adhesive and anti-adhesive functions of chondroitin sulfate proteoglycan preparations from embryonic chick brain. *J. Cell Sci.* **108**, 3807–3816.

Friedlander, D. R., Milev, P., Karthikeyan, L., Margolis, R. K., Margolis, R. U., and Grumet, M. (1994). The neuronal chondroitin sulfate proteoglycan neurocan binds to the neural cell adhesion molecules Ng-CAM/L1/NILE and N-CAM, and inhibits neuronal adhesion and neurite outgrowth. *J. Cell Biol.* **125**, 669–680.

Fuchs, M., Muller, T., Lerch, M. M., and Ulrich, A. (1996). Association of human protein-tyrosine phosphatase k with members of the armadillo family. *J. Biol. Chem.* **271**, 16712–16719.

Fujii, K., Furukawa, F., and Matsuyoshi, N. (1996). Ligand activation of overexpressed epidermal growth factor receptor results in colony dissociation and disturbed E-cadherin function in HSC-1 human cutaneous squamous carcinoma cells. *Exp. Cell Res.* **223**, 50–62.

Gaya-Gonzales, L., Balsamo, J., Swaminathan, N., and Lilien, J. (1991). Antibodies to the retina *N*-acetylgalactosaminylphosphotransferase inhibit neurite outgrowth. *J. Neurosci. Res.* **29**, 474–480.

Geisert, E. E., Jr., and Bidanset, D. J. (1993). A central nervous system keratan sulfate proteoglycan: Localization to boundaries in the neonatal rat brain. *Dev. Brain Res.* **75**, 163–173.

Geller, R. L., and Lilien, J. (1983). Repair of a calcium-dependent adhesive mechanism of embryonic neural retina cells: Kinetic and molecular analysis. *J. Cell Sci.* **60**, 29–49.

Goridis, C., and Brunet, J. F. (1992). NCAM: Structural diversity, function and regulation of expression. *Sem. Cell Biol.* **3**, 189–197.

Groden, J., Thliveris, A., Samowitz, W., Carlson, M., Gelbert, L., Albertsen, H., Joslyn, G., Stevens, J., Spirio, L., Robertson, M., and et al. (1991). Identification and characterization of the familial adenomatous polyposis coli gene. *Cell* **66**, 589–600.

Grumet, M., Milev, P., Sakurai, T., Karthikeyan, L., Bourdon, M., Margolis, R. K., and Margolis, R. U. (1994). Interactions with tenascin and differential effects on cell adhesion of neurocan and phosphacan, two major chondroitin sulfate proteoglycans of nervous tissue. *J. Biol. Chem.* **269**, 12142–12146.

Grunwald, G. B. (1996). Cell adhesion molecules in retina development and pathology. *In* "Progress in Retinal and Eye Research", Vol. 15. pp. 363–392, Elsevier Science, Ltd.

Guger, K. A., and Gumbiner, B. M. (1995). β-Catenin has Wnt-like activity and mimics the Nieuwkoop signaling center in *Xenopus* dorsal-ventral patterning. *Dev. Biol.* **172**, 115–125.

Gumbiner, B. M. (1995). Signal transduction by β-catenin. *Curr. Opin. Cell Biol.* **7**, 634–640.

Gumbiner, B. M., and McCrea, P. (1993). Catenins as mediators of the cytoplasmic function of cadherins. *J. Cell Sci., Suppl.* **17**, 155–158.

Guo, M., Dow, K. E., Kisilevsky, R., and Riopelle, R. J. (1993). Novel neurite growth-inhibitory properties of an astrocyte proteoglycan. *J. Chem. Neuroanat.* **6**, 239–245.

Hamaguchi, M., Matsuyoshi, N., Ohnishi, Y., Gotoh, B., Takeichi, M., and Nagai, Y. (1993). p60v-src Causes tyrosine phosphorylation and inactivation of the *N*-cadherin-catenin cell-adhesion system. *EMBO J.* **12**, 307–314.

Hermiston, M. L., Wong, M. H., and Gordon, J. I. (1996). Forced expression of E-cadherin in the mouse intestinal epithelium slows cell migration and provides evidence for nonautonomous regulation of cell fate in a self-renewing system. *Genes Dev.* **10**, 985–996.

Hinck, L., Näthke, I. S., Papkoff, J., and Nelson, W. J. (1994a). Dynamics of cadherin/catenin complex formation: Novel protein interactions and pathways of complex assembly. *J. Cell Biol.* **125**, 1327–1340.

Hinck, L., Nelson, W. J., and Papkoff, J. (1994b). Wnt-1 modulates cell–cell adhesion in mammalian cells by stabilizing beta-catenin binding to the cell adhesion protein cadherin. *J. Cell Biol.* **124**, 729–741.

Holash, J. A., Pasquale, and B. E. (1995). Polarized expression of the receptor protein tyrosine kinase Cek5 in the developing avian visual system. *Dev. Biol.* **172**, 683–693.

Hoschuetzky, H., Aberle, H., and Kemler, R. (1994). Beta-catenin mediates the interaction of the cadherin-catenin complex with epidermal growth factor receptor. *J. Cell Biol.* **127,** 1375–1380.

Huber, O., Bierkamp, C., and Kemler, R. (1996a). Cadherins and catenins in development. *Curr. Opin. Cell Biol.* **8,** 685–691.

Huber, O., Korn, B., McLaughlin, J., Ohsugi, M., Hermann, B. G., and Kemler, R. (1996b). Nuclear localization of β-catenin by interaction with transcription factor LEF-1. *Mech. Dev.* **59,** 3–10.

Hülsken, J., Birchmeier, W., and Behrens, J. (1994a). E-cadherin and APC compete for the interaction with beta-catenin and the cytoskeleton. *J. Cell Biol.* **127,** 2061–2069.

Hülsken, J., Behrens, J., and Birchmeier, W. (1994b). Tumor-suppressor gene products in cell contacts: The cadherin-APC-armadillo connection. *Curr. Opin. Cell Biol.* **6,** 711–716.

Itoh, M., Nagafuchi, A., Yonemura, S., Kitani-Yasuda, T., Tsukita, S., and Tsukita, S. (1993). The 220-kD protein colocalizing with cadherins in non-epithelial cells is identical to ZO-1, a tight junction-associated protein in epithelial cells: cDNA cloning and immunoelectron microscopy. *J. Cell Biol.* **121,** 491–502.

Jaworski, D. M., Kelly, G. M., and Hockfield, S. (1994). BEHAB, a new member of the proteoglycan tandem repeat family of hyaluronan-binding proteins that is restricted to the brain. *J. Cell Biol.* **125,** 495–509.

Jesaitis, L. A., and Goodenough, D. A. (1994). Molecular characterization and tissue distribution of ZO-2, a tight junction protein homologous to ZO-1 and the *Drosophila* discs-large tumor suppressor protein. *J. Cell Biol.* **124,** 949–961.

Kanai, Y., Ochiai, A., Shibata, T., Oyama, T., Ushijima, S., Akimoto, S., and Hirohashi, S. (1995). c-erbB-2 Gene product directly associates with beta-catenin and plakoglobin. *Biochem. Biophys. Res. Commun.* **208,** 1067–1072.

Kim, E., Niethammer, M., Rostchild, A., Jan, Y. N., and Sheng, M. (1995). Clustering of Shaker-type K+ channels by interaction with a family of membrane-associated guanylate-kinases. *Nature* **378,** 85–88.

Kinch, M. S., Clark, G. J., Der, C. J., and Burridge, K. (1995). Tyrosine phosphorylation regulates the adhesions of *ras*-transformed breast epithelia. *J. Cell Biol.* **130,** 461–471.

Kindzelskii, A. L., Laska, Z. O., Todd III, R. F., and Petty, H. R. (1996). Urokinase-type plasminogen activator receptor reversibly dissociates from complement receptor type 3 (alpha M beta 2' CD11b/CD18) during neutrophil polarization. *J. Immunol.* **156,** 297–309.

Kistner, V., Wenzel, B. M., Veh, R. W., Cases-Langhoff, C., Garner, A. M., Appeltaur, V., Voss, B., Gundelfinger, E. D., Garner, C. C. (1993). SAPGO, a rat presynaptic protein related to the product of the Drosophila tumor suppressor gene dlg-A. *J. Biol. Chem.* **268,** 4580–4583.

Knudsen, K. A., Soler, A. P., Johnson, K. R., and Wheelock, M. J. (1995). Interaction of alpha-actinin with the cadherin/catenin cell–cell adhesion complex via alpha-catenin. *J. Cell Biol.* **130,** 67–77.

Kornau, H. C., Schenker, L. T., Kennedy, M. B., and Seeburg, P. H. (1995). Domain interaction between NMDA receptor subunits and the postsynaptic density protein PSD-95. *Science* **269,** 1737–1740.

Kuhl, M., and Wedlich, D. (1996). *Xenopus* cadherins: Sorting out types and function in embryogenesis. *Dev. Dynam.* **207,** 121–134.

Kypta, R. M., Su, H., and Reichardt, L. F. (1996). Association between a transmembrane protein tyrosine phosphatase and the cadherin-catenin complex. *J. Cell Biol.* **134,** 1519–1529.

Leung, T., Lilien, J., and Balsamo, J. (1997). Cadherin-bound PTP1B is required for adhesion function. In preparation.

Li, H., Leung, T. C., Balsamo, J., Hoffman, S., and Lilien, J. (1997). cDNA cloning, expression and cadherin adhesion inhibitory activity of chick neurocan. In preparation.

Lisanti, M. P., Tang, Z., Scherer, P. E., and Sargiacomo, M. (1995). Caveolae purification and glycosylphosphatidylinositol-linked protein sorting in polarized epithelia. *Meth. Enzymol.* **250**, 655–668.

Margolis, R. U., and Margolis, R. K. (1994). Aggrecan-versican-neurocan family proteoglycans. *Meth. Enzymol.* **2455**, 105–126.

Marrs, J. A., and Nelson, W. J. (1996). Cadherin cell-adhesion molecules in differentiation and embryogenesis. *In* "International Review of Cytology", Vol. *165*. pp. 159–205, Academic Press, Inc.

Matsumine, A., Ogai, A., Senda, T., Okumura, N., Satoh, K., Baeg, G. H., Kawahara, T., Kobayashi, S., Okada, M., Toyoshima, K., and Akiyama, T. (1996). Binding of APC to the human homolog of the Drosophila discs large tumor suppressor protein. *Science* **272**, 1020–1023.

Matsuyoshi, N., Hamaguchi, M., Taniguchi, S., Nagafuchi, A., Tsukita, S., and Takeichi, M. (1992). Cadherin-mediated cell–cell adhesion is perturbed by v-*src* tyrosine phosphorylation in metastatic fibroblasts. *J. Cell Biol.* **118**, 703–714.

Meiners, S., Powell, E. M., and Geller, H. M. (1995). A distinct subset of tenascin/CS-6-PG-rich astrocytes restricts neuronal growth *in vitro. J. Neurosci.* **15**, 8096–8108.

Milev, P., Maurel, P., Häring, M., Margolis, R. K., and Margolis, R. U. (1996). TAG-1/Axonin-1 is a high-affinity ligand of neurocan, phosphocan/protein-tyrosine phosphatase-ζ/β, and N-CAM. *J. Biol. Chem.* **271**, 15716–15723.

Miller, J., and Moon, R. T. (1996). Signal transduction through β-catenin and specification of cell fate during embryogenesis. *Genes Dev.* **10**, 2527–2539.

Molenaar, M., van de Wetering, M., Oosterwegel, M., Peterson-Maduro, J., Godsave, S., Korinek, V., Roose, J., Destree, O., and Clevers, H. (1996). XTcf-3 transcription factor mediates β-catenin-induced axis formation in *Xenopus* embryos. *Cell* **86**, 391–399.

Munemitsu, S., Souza, B., Muller, O., Albert, I., Rubinfeld, B., and Polakis, P. (1994). The APC gene product associates with microtubules *in vivo* and promotes their assembly *in vitro. Cancer Res.* **54**, 3676–3681.

Munemitsu, S., Albert, I., Souza, B., Rubinfeld, B., and Polakis, P. (1995). Regulation of intracellular beta-catenin levels by the adenomatous polyposis coli (APC) tumor-suppressor protein. *Proc. Natl. Acad. Sci.* **92**, 3046–3050.

Munro, S. B. and Blaschuk, O. W. (1996). The structure, function and regulation of cadherins. *In* "Cell Adhesion and Invasion in Cancer Metastasis" Pnina Brodt, ed. pp. 17–33, R. G. Landes Co.

Näthke, I. S., Hinck, L., Swedlow, J. R., Papkoff, J., and Nelson, W. J. (1994). Defining interactions and distributions of cadherin and catenin complexes in polarized epithelial cells. *J. Cell Biol.* **125**, 1341–1352.

Näthke, I. S., Adams, C. L., Polakis, P., Sellin, J. H., and Nelson, W. J. (1996). The adenomatous polyposis coli tumor suppressor protein localizes to plasma membrane sites involved in active cell migration. *J. Cell Biol.* **134**, 165–179.

Niethammer, M., Kim, E., and Sheng, M. (1996). Interaction between the C-terminus of NMDA receptor subunits and multiple members of the PSD-95 family of membrane-associated guanylate kinases. *J. Neurosci.* **16**, 2157–2163.

Oakley, R. A., Lasky, C. J., Erickson, C. A., and Tosney, K. W. (1994). Glycoconjugates mark a transient barrier to neural crest migration in the chicken embryo. *Develop.* **120**, 103–114.

Ochiai, A., Akimoto, S., Kanai, Y., Shibata, T., Oyama, T., and Hirohashi, S. (1994). c-*erb*B-2 gene product associates with catenins in human cancer cells. *Biochem. Biophys. Res. Commun.* **205**, 73–78.

Oda, H., Uemura, T., Shiomi, K., Nagafuchi, A., Tsukita, S., and Takeichi, M. (1993). Identification of a *Drosophila* homologue of alpha-catenin and its association with the armadillo protein. *J. Cell Biol.* **121**, 1133–1140.

Oda, H., Uemura, T., Harada, Y., Iwai, Y., and Takeichi, M. (1994). A *Drosophila* homolog of cadherin associated with armadillo and essential for embryonic cell–cell adhesion. *Dev. Biol.* **165**, 716–726.

Oohira, A., Matsui, F., and Katoh-Semba, R. (1991). Inhibitory effects of brain chondroitin sulfate proteoglycans on neurite outgrowth from PC12D cells. *J. Neurosci.* **11**, 822–827.

Pawson, T. (1995). Protein modules and signalling networks. *Nature* **373**, 573–580.

Peifer, M. (1996). Regulating cell proliferation: As easy as APC. *Science* **272**, 974–975.

Peifer, M., Pai, L. M., and Casey, M. (1994a). Phosphorylation of the *Drosophila* adherens junction protein Armadillo: Roles for wingless signal and zeste-white 3 kinase. *Dev. Biol.* **166**, 543–556.

Peifer, M., Sweeton, D., Casey, M., and Wieschaus, E. (1994b). *Wingless* signal and Zeste-white 3 kinase trigger opposing changes in the intracellular distribution of Armadillo. *Develop.* **120**, 369–380.

Petty, H. R., and Todd III, R. F. (1996). Integrin as promiscuous signal transduction devices. Immunol. *Today* **17**, 209–212.

Pindzola, R. R., Doller, C., and Silver, J. (1993). Putative inhibitory extracellular matrix molecules at the dorsal root entry zone of the spinal chord during development and after root and sciatic nerve injury. *Dev. Biol.* **156**, 34–48.

Polakis, P. (1995). Mutations in the APC gene and their implications for protein structure and function. *Curr. Opin. Genet. Dev.* **5**, 66–71.

Pollack, A. L., Barth, A. I. M., Nelson, W. J. and Mostov, K. E. (1997). Dynamics of β-catenin interactions with APC protein regulate epithelial tubulogenesis. Submitted.

Pollanen, J., Saksela, O., Salonen, E. M., Andreasen, P., Nielsen, L., Dano, K., and Vaheri, A. (1987). Distinct localizations of urokinase-type plasminogen activator and its type 1 inhibitor under cultured human fibroblasts and sarcoma cells. *J. Cell Biol.* **104**, 1085–96.

Rauch, U., Karthikeyan, L., Maurel, P., Margolis, R. U., and Margolis, R. K. (1992). Cloning and primary structure of neurocan, a developmentally regulated, aggregating chondroitin sulfate proteoglycan of brain. *J. Biol. Chem.* **267**, 19536–19547.

Rauch, U., Forsberg, N., Kulbe, G., Arnold-Ammer, I., and Faessler, R. (1995). Amino-acid sequence of mouse neurocan and brevican and their different expression during embryonic development. Unpublished.

Retzler, C., Wiedemann, H., Kulbe, G., and Rauch, U. (1996). Structural and electron microscopical analysis of neurocan and recombinant neurocan fragments. *J. Biol. Chem.* **271**, 17107–17113.

Reynolds, A. B., Herbert, L., Cleveland, J. L., Berg, S. T., and Gaut, J. R. (1992). p120, A novel substrate of protein tyrosine kinase receptors and of p60v-*src*, is related to cadherin-binding factors beta-catenin, plakoglobin and armadillo. *Oncogene* **7**, 2439–2445.

Reynolds, A. B., Daniel, J., McCrea, P. D., Wheelock, M. J., Wu, J., and Zhang, Z. (1994). Identification of a new catenin: The tyrosine kinase substrate p120cas associates with E-cadherin complexes. *Mol. Cell Biol.* **14**, 8333–8342.

Rimm, D. L., Koslov, E. R., Kebriaei, P., Cianci, C. D., and Morrow, J. S. (1995). Alpha 1(E)-catenin is an actin-binding and -bundling protein mediating the attachment of F-actin to the membrane adhesion complex. *Proc. Natl. Acad. Sci.* **92**, 8813–8817.

Rubinfeld, B., Albert, I., Porfiri, E., Fiol, C., Munemitsu, S., and Polakis, P. (1996). Binding of GSK3β to the APC-β-catenin complex and regulation of complex assembly. *Science* **272**, 1023–1026.

Sargiacomo, M., Sudol, M., Tang, Z., and Lisanti, M. P. (1993). Signal transducing molecules and glycosyl-phosphatidylinositol-linked proteins form a caveolin-rich insoluble complex in MDCK cells. *J. Cell Biol.* **122**, 789–807.

Schneider, S., Steinbeisser, H., Warga, R. M., and Housen, P. (1996). β-Catenin translocation into nuclei demarcates the dorsalizing centers in frog and fish embryos. *Mech. Dev.* **57**, 191–198.

Schwab, M. E., Kapfhammer, J. P., and Bandtlow, C. E. (1993). Inhibitors of neurite growth. *Neurosci.* **16**, 565–595.

Schwartz, M. A., Schaller, M. D., and Ginsberg, M. H. (1995). Integrins: Emerging Paradigms of Signal Transduction. *In Ann. Rev. Cell Dev. Biol.*, pp. 549–599.

Serra-Pages, C., Kedersha, N. L., Fazikas, L., Medley, Q., Debant, A., and Streuli, M. (1995). The LAR transmembrane protein tyrosine phosphatase and a coiled-coil LAR-interacting protein co-localize at focal adhesions. *EMBO J.* **14,** 2827–2838.

Sheng, M. (1996). PDZs and receptor/channel clustering: rounding up the latest suspects. *Neuron* **17,** 575–578.

Shibamoto, S., Hayakawa, M., Takeuchi, K., Hori, T., Oku, N., Myiazawa, K., Kitamura, N., Takeichi, M., and Ito, F. (1994). Tyrosine phosphorylation of β-catenin and plakoglobin enhanced by hepatocyte growth factor and epidermal growth factor in human carcinoma cells. *Cell Adhes. Comm.* **1,** 295–305.

Shibamoto, S., Hayakawa, M., Takeuchi, K., Hori, T., Miyazawa, K., Kitamura, N., Johnson, K. R., Wheelock, M. J., Matsuyoshi, N., and Takeichi, M. (1995). Association of p120, a tyrosine kinase substrate, with E- cadherin/catenin complexes. *J. Cell Biol.* **128,** 949–957.

Shibata, T., Ochiai, A., Gotoh, M., Machinami, R., and Hirohashi, S. (1996). Simultaneous expression of cadherin-11 in signet-ring cell carcinoma and stromal cells of diffuse-type gastric cancer. *Cancer Lett.* **99,** 147–153.

Siu, C. H. (1996). Cell-adhesion molecules of the immunoglobulin family. *In* "Cell Adhesion and Invasion in Cancer Metastasis", Pnina Brodt, ed. pp 61–75, R. G. Landes Co.

Skoudy, A., Gomez, S., Fabre, M. and Garcia de Herreros, A. (1996). p120-catenin expression in human colorectal cancer. *Int. J. Cancer* **68,** 14–20.

Smith, K. J., Levy, D. B., Maupin, P., Pollard, T. D., Vogelstein, B., and Kinzler, K. W. (1994). Wild-type but not mutant APC associates with the microtubule cytoskeleton. *Cancer Res.* **54,** 3672–3675.

Sokol, S. Y. (1996). Analysis of Dishevelled signalling pathways during *Xenopus* development. *Curr. Biol.* **6,** 1456–1467.

Stahl, A., and Mueller, B. M. (1995). The urokinase-type plasminogen activator receptor, a GPI-linked protein, is localized in caveolae. *J. Cell. Biol.* **129,** 335–344.

Stefanova, I., Horejsi, V., Ansotegui, I. J., Knapp, W., and Stockinger, H. (1991). GPI-anchored cell-surface molecules complexed to protein tyrosine kinases. *Science* **254,** 1016–1019.

Takeichi, M. (1995). Morphogenetic roles of classic cadherins. *Curr. Opin. Cell Biol.* **7,** 619–627.

Tepass, U., Gruszynski-DeFeo, E., Haag, T. A., Omatyar, L., Torok, T., and Hartenstein, V. (1996). *Shotgun* encodes *Drosophila* E-cadherin and is preferentially required during cell rearrangement in the neurectoderm and other morphogenetically active epithelia. *Genes Dev.* **10,** 672–685.

Uemura, T., Oda, H., Kraut, R., Hayashi, S., Kotaoka, Y., and Takeichi, M. (1996). Zygotic *Drosophila* E-cadherin expression is required for processes of dynamic epithelial cell rearrangement in the *Drosophila* embryo. *Genes Dev.* **10,** 659–671.

van de Wetering, M., Cavallo, R., Doorjes, D., van Beest, M., van Es, J., Loureiro, J., Ypam, A., Hursh, D., Jones, T., Bejsovec, A., Peifer, M., Mortin, M. and Clevers, H. (1997). Armadillo coactivates transcription driven by the product of the Drosophila segment polarity gene dTCF. *Cell* **88,** 789–799.

van Leeuwen, F., Samos, C. H., and Nusse, R. (1994). Biological activity of soluble wingless protein in cultured *Drosophila* imaginal disc cells. *Nature* **368,** 342–344.

Vincent, J. P., and Lawrence, P. A. (1994). *Drosophila wingless* sustains engrailed expression only in adjoining cells: Evidence from mosaic embryos. *Cell* **77,** 909–915.

Wei, Y., Lukashev, M., Simon, D. I., Bodary, S. C., Rosenberg, S., Doyle, M. V., and Chapman, H. A. (1996). Regulation of integrin function by the urokinase receptor. *Science* **273,** 1551–1555.

Wheelock, M. J., Knudsen, K. A., Johnson, K. R. (1996). Membrane-cytoskeleton interactions with cadherin cell adhesion proteins: Role of catenins as linker proteins. *In* "Current Topics in Membranes" **43**, 169–185.

Willot, E., Balda, M. S., Fanning, A. S., Jameson, B., Itallie, C. V. and Anderson, J. M. (1993). The tight junction protein 20-1 is homologous to the *Drosophila* discs-large tumor suppressor protein of septate junctions. *Proc. Natl. Acad. Sci. U.S.A.* **90**, 7834–7838.

Wong, M., Hermiston, M. L., Syder, A. J., and Gordon, J. I. (1996). Forced expression of the tumor suppressor adenomatosis polyposis coli protein induces disordered cell migration in the intestinal epithelium. *Proc. Natl. Acad. Sci.* **93**, 9588–9593.

Woods, D. F., and Bryant, P. J. (1993). Apical junctions and cell signalling in epithelia. *J. Cell Sci., Suppl.* **17**, 171–181.

Worley, T., and Holt, C. (1996). Inhibition of protein tyrosine kinases impairs axon extension in the embyronic optic tract. *J. Neurosci.* **16**, 2294–2306.

Wylie, C., Kofron, M., Payne, C., Anderson, R., Hosobuchi, M., Joseph, E., and Heasman, J. (1996). Maternal β-catenin establishes a dorsal signal in early *Xenopus* embryos. *Develop.* **122**, 2987–2996.

Xue, W., Kindzelskii, A. L., Todd III, R. F., and Petty, H. R. (1994). Physical association of complement receptor type 3 and urokinase-type plasminogen activator receptor in neutrophil membrane. *J. Immunol.* **152**, 4630–4640.

Yamamoto, T., Ikawa, S., Akiyama, T., Semba, K., Nomura, N., Miyama, N., Saito, T., and Toyoshima, K. (1986). Similarity of the protein encoded by the human c-*erb*B2 gene to epidermal growth factor receptor. *Nature* **319**, 230–234.

Yanagawa, S., van Leeuwen, F., Wodarz, A., Klingensmith, J., and Nusse, R. (1995). The dishevelled protein is modified by wingless signaling in *Drosophila*. *Genes Dev.* **9**, 1087–1097.

Yang-Snyder, J., Miller, J. R., Brown, J. D., Lai, C. J., and Moon, R. T. (1996). A frizzled homolog functions in a vertebrate Wnt signalling pathway. *Curr. Biol.* **6**, 1302–1306.

Yost, C., Torres, M., Miller, J. R., Huang, E., Kimelman, D. and Moon, R. T. (1996). The axis-inducing activity, stability, and subcellular distribution of beta-catenin is regulated in Xenopus embryos by glycogen synthase kinase 3. *Genes Dev.* **10**, 1443–1454.

6

Neural Induction in Amphibians

Horst Grunz
Department of Zoophysiology
University GH Essen
45117 Essen, Germany

I. Introduction

Over 100 years ago Wilhelm Roux described the importance of a stream of egg substances that can be observed opposite the entrance point of the sperm (Roux, 1885, 1887). This phenomenon, later analyzed in elegant experiments, is known today as *cortical rotation* (Gerhart *et al.*, 1989). Opposite the sperm entrance point, the zone of Spemann organizer will be established up until the early gastrula. So even prior to the first cleavage, one of the main body axes and the dorsal/ventral polarity will be determined in the vertebrate embryo. However, many important steps have already taken place during oogenesis; these cannot be discussed in this chapter (Pierandrei-Amaldi and Amaldi, 1994). During gastrulation a further symmetry, the anteroposterior axis, will be added.

    The organizer experiment initiated gold rush–like activities in the scientific community to search for the mechanisms responsible for the establishment of the complex three-dimensional body plan at late gastrula from a relatively simple organized egg (Spemann and Mangold, 1924; Nobel Prize for Spemann in 1935).

Current Topics in Developmental Biology, Vol. 35
191

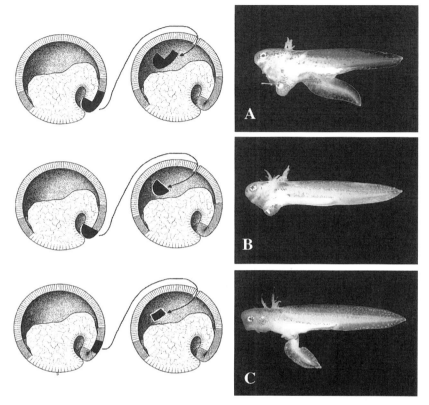

Fig. 1 **A.** The transplantation (Einsteck experiment) of the complete dorsal blastopore lip (Spemann organizer) into the blastocoel of an early host gastrula of Triturus *alpestris* results in the formation of a complete secondary axis. **B.** The transplantation (Einsteck experiment) of the anterior part ("head inducer") into the blastocoel of an early host gastrula of Triturus *alpestris* results in the formation of head structures. **C.** The transplantation (Einsteck experiment) of the posterior part ("trunk inducer") into the blastocoel of an early host gastrula of Triturus *alpestris* results in the formation of tail structures.

One of the most important processes during early embryonic development is neural induction, resulting in the formation of the central nervous system. When a small area above the blastopore of an early amphibian gastrula is grafted to the prospective belly side of a host embryo, a complete secondary embryo will be formed (Fig. 1). Spemann named this territory *organizer* (organisator). The removal of this zone from the early gastrula causes severe defects of the embryo, which lacks the central nervous system. Such embryos will develop into trunk/tail structures only.

Not only was the Spemann organizer experiment a milestone in developmental biology and embryology, but it was the ignition event for many scientists up to today to start the search for factors responsible for neural induction. However, in the last five years the traditional concept of neural induction has been dramatically revised. Formerly, the whole animal cap was considered a uniform and undetermined entity. It was thought to receive its information for neural determination from the dorsal mesoderm (Spemann organizer) only as the instructor (instructive induction). During the last decade (similar to the situation in the 1930s and '40s) it was the expectation and hope of embryologists to find only one neuralizing factor responsible for the primary steps of neural induction. This is true even after a renaissance of amphibian embryology following the introduction of *Xenopus* embryos as a favorite vertebrate model system, on one hand, and molecular techniques, on the other. Just in the last two years, insight into the processes of neural induction reached a new dimension. It could be shown that the ventral side of the amphibian egg, including the ventral marginal zone and the ventral ectoderm, is much more important for the development of the embryo than originally expected (see later). Furthermore, it turned out that there exists not only one organizing center, but a complicated interaction of many genes (including maternal ones) and their products, synthesized and distributed in gradients rather than in strict centers. Because over 15 organizer-specific and a few antagonistic ventral-specific genes have been characterized, the problem of neural induction is far from being solved. Still, this fact makes it one of the most exciting research fields in modern biology.

II. Historical Background

From the very beginning the organizer phenomenon stimulated embryologists to seek a unique neuralizing factor. Organizer tissue devitalized by heating or treatment with various chemicals still possessed inducing activity. Bautzmann and co-workers concluded from such results that the organizer must contain inducing substances. However, the search in the succeeding decades for specific neuralizing inducers was not successful. It could be shown that many different chemicals are able to evoke neural inductions in competent ectoderm, for example, sterols and even kaolin and silicon (reviewed in detail by Saxén and Toivonen, 1962; Nakamura and Toivonen, 1978; Nieuwkoop *et al.*, 1985; Hamburger, 1988; Saxén, 1989; Gilbert and Saxén, 1993; Tiedemann *et al.*, 1996). At that time the responding tissue, the ectoderm, was considered less important. Since the ectoderm differentiates into neural structures after heat shock, high pH, ammonia, urea, etc., Holtfreter postulated that some cells are damaged by so-called "sublethal cytolysis," which in turn neuralizes neighboring, still-vital cells (Holtfreter, 1947). Much later, John and collaborators (1984) showed that neuralizing factors are present in the early gastrula ectoderm. It was concluded from experiments with various unspecific treatments that neuralizing factors are present in the ectoderm in a masked form, which will be released by homogenization and various incubations with different chemicals. These neural inductions of the ectoderm without any influence of the organizer were quite irritating in the 1930s. It was concluded that there may not exist a specific neuralizing factor at all. This confusing situation was the reason that many scientists lost interest in this research field. In retrospect, the results with "autoneuralized" ectoderm fit very well our concept of neural induction and the determination of the dorsal and ventral sides of the embryo. (I will report these very recent findings about the neural default state of the ectoderm in detail later in the chapter.)

Nevertheless, a few groups in Finland, Japan, and Germany continued to search for early embryonic inducers. However, it turned out that amphibian embryos are not suitable to isolate inducing factors by biochemical methods. All amphibian embryonic cells contain a large amount of yolk and lipids, which severely disturb the traditional biochemical isolation and purification steps. Furthermore, the inducing factors are present in low concentrations only. The researchers tried to circumvent this problem by using as the source for the isolation of inducing factors well-differentiated or adult tissues, which are available in large quantities.

The background for this approach was the fact that living tissues—especially devitalized tissues from adult salamander as well as from other animals (guinea pig bone marrow, mouse liver, mouse kidney, etc.)—were able to evoke neural and mesodermal inductions in competent ectoderm (Bautzmann *et al.*, 1932; Holtfreter, 1933a, 1934; Toivonen, 1938; Chuang 1938, 1939). Almost simultaneously in the late '40s and early '50s, several groups therefore started fractionation experiments (Kuusi 1951a,b; Yamada, 1950; Tiedemann and Tiedemann, 1956; Yamada, 1958). However, while the other groups did not follow this line of experiments because of the tremendous technical difficulties, only Tiedemann's group continued to isolate a factor from 11-day-old chicken embryos in highly purified form. He invented the phenol extracting method, which showed that the biological activity could be found in the phenol layer and not in the aqueous phase. This very important result was proof that it was not nucleic acids (as discussed previously) but proteins that must be the expected candidates for morphogenetic and inducing factors.

Because this factor induces in competent ectoderm tissues derived from the vegetal part of the embryo (endomesoderm), Tiedemann named the inducer *vegetalizing* factor. Later, a factor with similar biological activity (XTC-MIF) could be isolated from *Xenopus* tadpole cells (XTC-MIF = XTC-mesoderm inducing factor; J. Smith, 1987). Meanwhile, it is clear that these factors are homologous to *activin*, a member of TGFβ superprotein family (for details see Tiedemann *et al.*, 1995, 1996; Asashima, 1994; Fukui and Asashima, 1994; Grunz, 1996). Besides ventral mesodermal structures, activin is able to induce dorsal mesodermal structures like notochord and somites in competent ectoderm

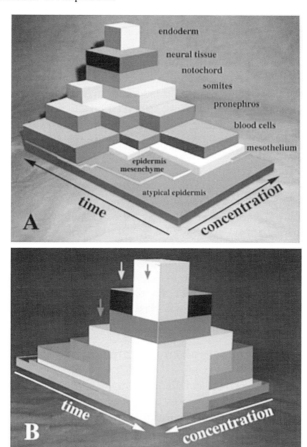

Fig. 2 A. Model showing the results of ectoderm treated with *activin* (vegetalizing factor) for different periods and concentrations (Grunz, 1983) (see also Fig. 4). At short incubation time and low concentration, ectoderm will differentiate into atypical epidermis. Longer incubation periods and higher concentrations will result in the formation of mesenchyme, blood precursor cells, mesothelium, pronephros, muscle, notochord, and neural structurs. Very high concentrations and long incubation times cause the induction of endoderm and heart structures (Ariizumi *et al.*, 1991). The heart structures will be formed because a partial induction of the ectoderm into mesoderm besides endoderm probably takes place, a prerequisite for heart development (Okada, 1954). The simultaneous presence of different cell types and the expression of specific genes in the same explant is caused by secondary cell interactions (Grunz, 1979; Minuth and Grunz, 1980; Green *et al.*, 1994; Wilson and Melton, 1994). **B.** Same model as in A, but the view is from the back. It demonstrates that only under special conditions will induced ectoderm form one cell type only (right arrow). At very high concentrations, with long incubation time, or with treatment of disaggregated cells plus relatively high concentrations of inducer, the ectoderm will form endoderm (Grunz, 1983) or notochord only (Grunz and Tacke, 1989). Lower concentration and incubation time will result in the formation of different structures in the same explant (left and middle arrows).

(Grunz, 1983; Asashima *et al.*, 1990, 1991b; J. Smith *et al.*, 1990; Wilson and Melton, 1994). Important parameters include the concentration of the inducer, the incubation time, the cell number, and the competence of the ectoderm (Grunz, 1979, 1983; Gurdon, 1988; Ariizumi *et al.*, 1991) (Fig. 2). Factors of another protein superfamily (FGF superfamily), like acidic and basic fibroblast growth factor, preferentially induce ventral-mesoderm-derived tissues like mesothel and blood precursor cells (Grunz and Tacke, 1986; Grunz *et al.*, 1987; Knöchel *et al.*, 1987; Slack *et al.*, 1987; Tiedemann *et al.*, 1988; Ding *et al.*, 1992). These approaches show that primarily only mesoderm/endoderm-inducing factors could be received in homogeneous form. However, as we will see later, these factors are correlated directly with the processes of neural induction and determination. So far it had not been possible to isolate and purify neuralizing factors to homogeneity (Born *et al.*, 1989; Janeczek *et al.*, 1984, 1992; Mikhailov and Gorgolyuk, 1987; Mikhailov *et al.*, 1995;) However, there were at least two misconceptions in the past discussion of neural induction. One was the idea that only a single neuralizing factor should be responsible for the primary steps of neural induction. The second was that the reacting tissue (the ectodermal target cells) receives all information from the inducing chordamesoderm only and that it behaves like a naive entity. Both assumptions must be revised on the basis of very recent results from many laboratories.

III. The Search for Neuralizing Factors

Niu kept Spemann organizer in culture medium for several days. In this so-called conditioned medium he placed pieces of competent ectoderm. He reported that this ectoderm differentiated into neuronal cells (Niu and Twitty, 1953). These results could never be confirmed by other authors. Several years ago we disaggregated 40 dorsal blastopore lips (Spemann organizer) into single cells and alongside we placed competent ectoderm in the supernatant. However, under these conditions we did not get an induction of the ectoderm. Apparently the medium was so diluted that the necessary threshold concentrations could not be established. Very recent results show that the dorsal mesoderm indeed secretes a protein (*chordin*), which can stimulate competent ectoderm to differentiate into dorsal mesoderm and neural structures in a concentration of 1 nM (Piccolo *et al.*, 1996). Factors that can trigger competent ectoderm into neural pathways of differentiation (reviewed by Tiedemann *et al.*, 1996) could be isolated from the brain of chicken embryos (Mikhailov and Gorgolyuk, 1989; Mikhailov *et al.*, 1995), neuroblastoma cell lines (Lopashov *et al.*, 1992), *Xenopus* oocytes, and gastrula/neurula stages (Janeczek *et al.*, 1992).

Whether these factors are identical with secreted factors like *chordin* and *noggin* is not known. However, they could act in a similar way to that postulated for *chordin*, by interaction with *BMP-4*-like molecules (Hogan, 1996; Piccolo *et*

VENTRALIZATION and ANTINEURALIZATION

DORSALIZATION and NEURAL INDUCTION

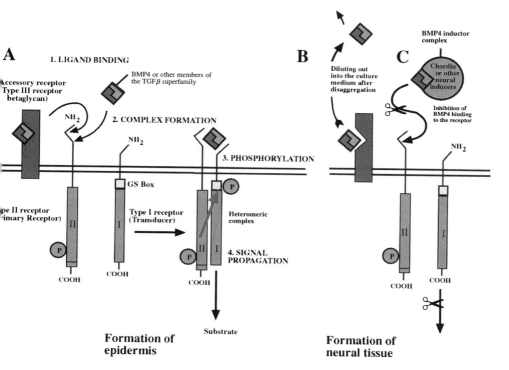

Fig. 3 Interaction of neuralizing factors with *BMP-4*. In this currently favored model, *BMP-4* or other members of TGFβ superfamily (for review see Massagué and Weis-Garcia, 1996) bind not only to the specific receptors (Type I and II), but also to membrane-anchored proteoglycans, important components of the extracellular matrix (ECM). This fact may explain why the removal of ECM components (diluting out *BMP-4*) after disaggregation results in neuralization of *Xenopus* ectoderm (Grunz and Tacke, 1989, 1990). The interaction of neuralizing factors with *BMP-4* prevents its binding to its receptor, which also causes neuralization. For the formation of epidermis, signaling of *BMP-4* dimers via Type I and Type II receptors is needed, which form a heteromeric complex. Type II receptors are thought to have constitutive kinase activity. The transphosphorylation of the cytoplasmic GS (glycine-serine)-rich domain of the Type I receptor by the Type II kinase will propagate the signal intracellularly, resulting in the formation of epidermis.

al., 1996) (Fig. 3). It was shown that *BMP-4* prevents neuralization of disaggregated ectodermal cells (Wilson and Hemmati-Brivanlou, 1995). On the basis of these results and data with a dominant-negative mutant of the *activin* receptor it was postulated that the default status of the ectoderm is neural (see later) (Grunz and Tacke 1989, 1990; Wilson and Hemmati-Brivanlou, 1995; Hemmati-Brivanlou and Melton, 1994). The extracellular matrix (ECM) and specific receptors for *BMP*-like molecules on the plasma membrane of the ectodermal target cells are important factors in the neural induction pathway initiated by secreted factors emanating from the chordamesoderm.

IV. The Responding Tissue

Earlier approaches showed that neural structures could be induced by interactions with targets downstream of the plasma membrane. The treatment of ectoderm with phorbolesters (TPA), which activates the protein kinase C (PKC) causes neural induction in *Triturus* ectoderm (Davids *et al.*, 1987; Davids, 1988) and also in *Xenopus* ectoderm (Otte *et al.*, 1988). Although cyclic nucleotides alone do not induce neural structures (Grunz and Tiedemann, 1977), they act synergistically with TPA (Otte *et al.*, 1989). Of interest is the fact that TPA activates PKC in the dorsal, but not the ventral, ectoderm (Otte *et al.*, 1991). Furthermore, Otte and collaborators observed a translocation of PKC from the cytoplasm to the cell membrane in response to neuralizing signals. These results show that a difference in competence apparently exists between the dorsal and ventral ectoderm to react to inducing signals. While dorsal ectoderm treated with *activin* formed mainly dorsal mesodermal structures (notochord and somites), ventral ectoderm differentiated into mesothelium and blood precursor cells (Sokol and Melton, 1991). Apparently this dorsal–ventral polarity is generated very early in development (Grunz, 1977; Kageura and Yamana, 1983, 1984; London *et al.*, 1988; Gallagher *et al.*, 1991; Li *et al.*, 1996). Recently it was shown that β-catenin, a component of the *Wnt*-signaling pathway, is located mainly in the two dorsal blastomeres of the four-cell stage (Miller and Moon, 1996; Li *et al.*, 1996).

These results indicate that a difference in competence to react to neural inducing signals is already established before gastrulation. It is likely that different factors (dorsalizing and ventralizing molecules) are distributed in the ectoderm in the form of a dorsoventral gradient. These differences, correlated with autonomous cell mechanisms, cause a different predisposition of the dorsal and ventral ectoderm. Therefore, a neural predetermination of the dorsal ectoderm must not necessarily be established at the beginning of gastrulation by planar signals (see later). In any case, the predisposition is very labile and reversible. We could not observe differences in the quality and quantity of neural structures when we separately disaggregated dorsal and ventral ectoderm (Grunz, 1996). Disaggregation and delayed reaggregation causes the neuralization of ectoderm, as shown earlier (Grunz and Tacke, 1989; see also later). Furthermore, pieces of dorsal gastrula ectoderm transplanted into ventral ectoderm, and vice versa, respond as the surrounding cells on *activin*, which indicates that the predisposition is reversible (Chen and Grunz, in press). The autonomous importance of the ectoderm and its influence on the vegetal half were shown by the recombination of animal caps with vegetal parts of early cleavage stages (Kato and Gurdon, 1994; Tiedemann,

1993; Bauer *et al.*, 1996). The differentiation of notochord is much improved when substantial amounts of ectoderm are added to the vegetal part (including the marginal zone) of the embryo.

In another approach, it was shown that animal caps from stages later than stage 7 can stimulate erythrocyte differentiation (Maéno *et al.*, 1992, 1994b). The same group showed that the overexpression of a functionally, defective mutant of *BMP-4* receptor in the animal cap, which blocks the *BMP*-signaling pathway, causes the dorsalization of ventral mesoderm (Maéno *et al.*, 1994a). These results indicate that the animal half contains factors that are very important for the determination of not only the ectoderm per se, but also for the programming of the vegetal part of the embryo. Apparently, *BMP-4*, other members of the TGFβ superfamily, and still-unknown factors play an important role in both processes (Fig. 3).

V. Autoneuralization, or the Hypothesis of the Neural Default Status of the Ectoderm

As mentioned earlier, the ectoderm will form neural structures under the influence of dorsal mesoderm (Spemann organizer). Transfilter experiments have shown that a diffusible neuralizing factor is emanating from the inducing chordamesoderm (Saxén, 1961) (Fig. 4C). A neuralizing factor could also be isolated from the intercellular gap between the inducing chordamesoderm and the overlying neural ectoderm (John *et al.*, 1983). In contrast to the ideas of the 1940s that a masked neuralizing factor must be activated in the ectoderm, for example, by high or low pH or by partial lysis (sublethal cytoloysis) of the ectoderm, it has become generally accepted that neuralizing signals are transmitted via receptors on the plasma membrane (Born *et al.*, 1986). In retrospect, the autoneuralization experiments do not appear to contradict the hypothesis of neural induction, since the ectoderm is the target only downstream of the inducing signals emanating from the inducing chordamesoderm. Members of the neuralizing signaling chain are the specific receptors of the plasma membrane (Massagué and Weis-Garcia, 1996), adenylcyclase, cyclic AMP, G-proteins, PKC, internal Ca^{2+} (Moreau *et al.*, 1994; Drean *et al.*, 1995), and further factors located in the cytosol and the nucleus.

However, in light of recent data, the receptors on the plasma membrane must not be the direct target for neuralizing factors emanating from the chordamesoderm. Instead, the neuralizing factors might be interacting with factors on the plasma membrane or the extracellular matrix, like *BMP-4*, which inhibit the ectoderm from forming neural structures but cause the differentiation of epidermis without external inducers (Fig. 3). The evidence that the neuralization of the ectoderm takes place by a lifting of an inhibition rather than by an instructive induction is based on several approaches. It was shown that the ectoderm will

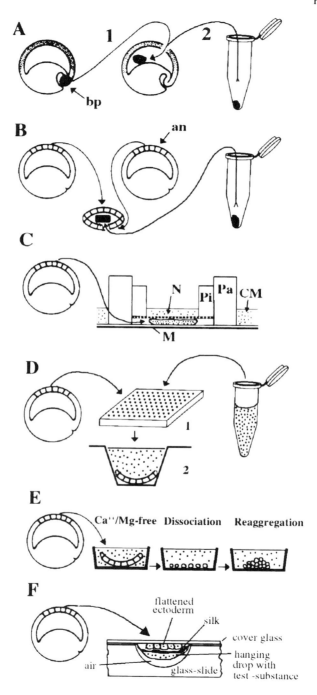

form neural structures or express neural specific markers without external inducers, by the following approaches:

1. Disaggregation and delayed reaggregation of ectoderm
2. Overexpression of dominant-negative mutants of *activin* or *BMP* and their receptors
3. Injection of *follistatin*-mRNA, antagonist of *activin*

In 1987 we started disaggregation experiments with *Xenopus* quite similar to those published in our earlier papers with *Triturus alpestris*. Primarily I used the disaggregation technique to demonstrate that the reaggregation and cell contact depend on an active protein synthesis (Grunz, 1969) (Fig. 4E). Later we used the technique to show that the formation of mesoderm from ectoderm treated with vegetalizing factor (identical with *activin*) depends on secondary cell interactions (Minuth and Grunz, 1980).

Ectoderm treated with vegetalizing factor (homologous to *activin*) disaggregated and immediately reaggregated, differentiated into mesodermal derivatives like notochord and somites. However, disaggregated induced ectoderm kept as single cells for 24 hours prior to reaggregation formed endodermal structures, some blood precursor cells, and atypical epidermis. We concluded from these results that in high concentrations, the vegetalizing factor induces endodermal structures (Fig. 2), which are able to interact with uncommitted ectodermal cells to form mesodermal structure, and in turn, by secondary cell interactions, neural structures. However, these interactions are possible only in intact explants or in reaggregated cell heaps that allow direct contact between the cells (Sargent *et al.*, 1986).

We repeated our experiments in 1988, using *Xenopus* ectoderm and XTC-MIF (crude *activin*) as inducing factor (J. Smith, 1987). Since the factor was not highly purified (it did not induce endoderm even in high concentrations) we could not confirm the data obtained with vegetalizing factor and *Triturus* ecto-

←————————————————————————————————————

Fig. 4 Different test methods to analyze early embryonic induction: **A.** Implantation method (Einsteck experiment after Spemann and Mangold, 1924): (1) Implantation of the blastopore lip (compare with Fig. 1); (2) implantation of a pellet (insoluble factors or so-called heterogeneous inducers). **B.** Sandwich technique for the test of insoluble factors or heterogeneous inducers (Holtfreter, 1933a). *an*, animal cap (ectoderm). **C.** Nucleopore® chamber (Grunz and Tacke, 1986, modified from Sáxen, 1961) for tests of soluble factors, which can act only on the former blastocoelic side of the explant. The Nucleopore® filter prevents the curling up of the ectoderm. **D.** Test plate (1) with flat-bottom wells (Terasaki plate) for the test of soluble factors in small amounts of medium (10 μl) like *FGF* or *activin* (animal cap assay); (2) one well at higher magnification. *N*, Nucleopore membrane; *Pa*, outer Plexiglas ring; *Pi*, inner Plexiglas ring; *CM*, culture medium; *M*, O_2- and CO_2-permeable membrane (Petriperm®, Fa. Heraeus). **E.** Dissociation and reaggregation method (Townes and Holtfreter, 1955; Saxén *et al.*, 1964; Grunz, 1969; Minuth and Grunz, 1980; Gualandris and Duprat, 1981). Disaggregation and delayed reaggregation of ectoderm results in the formation of neural structures (Grunz and Tacke, 1989, 1990; Godsave and Slack, 1989). **F.** Hanging-drop culture for the test of inducing factors (Becker *et al.*, 1959).

derm. However, we got a totally unexpected result: The disaggregated uninduced ectodermal cells (control series), which were kept as single cells in Ca^{2+}/Mg^{2+}-containing medium for 4–5 hours prior to the reaggregation, differentiated into large neural structures (Grunz and Tacke, 1989; Godsave and Slack, 1989). These observations suggest that the extracellular matrix contains factors that may prevent the neuralization of the ectoderm. When we added the supernatant of a large number of disaggregated embryos to the disaggregated single ectodermal cells, we could prevent the neuralization (Grunz and Tacke, 1990). These results indicated that factors must be present in intact animal caps that will be diluted out into the culture medium when the cells are kept disaggregated for a longer period. These factors will prevent intact ectoderm from forming neural structures. On the basis of our results with XTC-MIF–treated ectoderm, it seems unlikely that *activin* is a good candidate as "inhibiting" molecule. The disaggregated cells differentiated either exclusively into notochord (high concentrations of XTC factor) or mainly into neural structures (lower concentrations of XTC-MIF) but never into epidermis.

This result was recently confirmed with recombinant *activin* A (Grunz, 1996). Using basic fibroblast factor (a "weaker") mesodermalizing factor of the FGF superprotein family, we could not prevent the neuralization. However, the results suggest that under *in vitro* conditions bFGF shifts the neural structures from an anterior (archencephalic) to a more posterior (deuterencephalic) pattern (Grunz, 1996). This observation is in agreement with data that FGF is important in the regionalization of the central nervous system (Kengaku and Okamoto, 1995; Doniach, 1995). However, very recent data with transgenic *Xenopus* embryos show that it is unlikely that FGF plays an important role in the regionalization of the central nervous system in the normal embryo (Kroll and Amaya, 1996).

On the other hand, *BMP-4*, a member of the TGFβ superprotein family that induces ventral mesodermal structures (Köster *et al.*, 1991), is able to prevent neural induction of disaggregated cells and stimulate the expression of an epidermal marker (Wilson and Hemmati-Brivanlou, 1995). Recently we could show by Western blotting that the supernatant of disaggregated early gastrula cells, using the same procedure with a Pro-di-concentration/dialysis equipment as in our earlier experiments (Grunz and Tacke, 1990), contains both bone morphogenetic factors, *BMP-2* and *BMP-4* (Knöchel, collaborators, and Grunz, unpublished results). The observation of Wilson and Hemmati-Brivanlou, together with our earlier results, supports the view that the default state of the ectoderm is neural. This hypothesis is also supported by the fact that the injection of *follistatin* mRNA in two-cell stages and the overexpression of dominant-negative mutants of *activin*- or *BMP*-receptors (Hemmati-Brivanlou and Melton, 1992, 1994; Hemmati-Brivanlou *et al.*, 1994; Xu *et al.*, 1995) caused the neuralization of animal caps. However, it turned out that *BMP*-like molecules, in addition to their participation in the stabilization of the epidermal determination of the ectoderm, play an important role in the formation of the dorsal–ventral polarity as antagonists to factors located and expressed preferentially in the dorsal mesoderm (Spemann organizer). In contrast to the traditional opinion, it is now believed that the ventral side (ventral mesoderm) requires active signaling by *BMP*-like factors and is not determined simply by signals emanating from the dorsal side.

Very puzzling was the observation that two closely related homeobox genes that are expressed exclusively on the ventral side of the embryo act, together with *BMP*-like molecules, antagonistically toward genes and their products on the dorsal side (Gawantka *et al.*, 1995; Onichtchouk *et al.*, 1996; Schmidt *et al.*, 1996; Papalopulu and Kintner, 1996; Ladher *et al.*, 1996).

VI. The Old and New Organizer Concepts

The classical Einsteck experiment showed that a small piece above the blastopore of an early *Triturus* gastrula could induce a secondary embryo on the belly side of a host embryo (Spemann and Mangold, 1924). Later on it was found that the anterior part of the dorsal blastopore lip induced head structures, while the posterior part formed trunk/tail structures (Mangold, 1933) (Fig. 1). Apparently there are different steps necessary for the formation of the central nervous system. First comes a primary determination of the neural anlage during early gastrulation. In a second step, the regionalization in the different brain areas during the late-gastrula and early-neurula stages are needed (for reviews see Saxén, 1989; Gurdon, 1992; Gilbert and Saxén, 1993; Tiedemann *et al.*, 1995, 1996; Grunz, 1996). These differences in inducing activity are today discussed as a differential and sequential activation of genes and the interaction of secreted and nonsecreted molecules in the dorsal mesoderm (see later).

VII. The Establishment of the Organizer

Immediately after insemination, a cortical rotation of the peripheral cytoplasm takes place from the vegetal pole to the presumptive dorsal side of the embryo (Ancel and Vintemberger, 1948; Vincent and Gerhart, 1987). This cytoplasmic movement can be prevented by UV irradiation of the vegetal pole of the uncleaved embryos, resulting in larvae consisting of trunk/tail structures lacking the head area (Scharf and Gerhart, 1980; Houliston and Elinson, 1991; Houliston, 1994; Gerhart *et al.*, 1989). This treatment severely inhibits the formation of dorsal mesoderm (preferentially notochord), which is the prerequisite for the induction of neural structures (Fujisue *et al.*, 1993). The formation of the dorsal vegetal zone depends on the formation of microtubuli and the distinct translocation of cytoplasmic components (Kloc and Etkin, 1994, 1995; Zhou and King, 1996; King 1996). The experiment of Nieuwkoop (1969) had already shown that the dorsal vegetal zone (now called the *Nieuwkoop center*) is the prerequisite of the formation of the Spemann organizer. However, it must be pointed out that determination factors for the dorsal mesoderm are located not only in the dorsal vegetal zone during early cleavage stages, but also in the dorsal marginal zone and even in the dorsal animal hemisphere (Grunz, 1977; Cardellini, 1988; Grunz, 1994; Li et al., 1996; Miller and Moon, 1996; Larabell et al., 1997). Since there is growing evidence that most genes and their products (including maternal factors) are distributed in all animals in the form of gradients (Johnston and Nüsslein-Volhard, 1992; Nüsslein-Volhard, 1994; Frohnhöfer and Nüsslein-Volhard, 1986; Nüsslein-Volhard et al., 1987; Haffter and Nüsslein-Volhard, 1996; Green and Smith, 1991; Dawid and Taira, 1994; Harger and Gurdon, 1996; Nüsslein-Volhard, 1996; De Robertis et al., 1991), the Nieuwkoop center must be considered an area of highest concentration of certain factors. How these genes are activated and maternal factors converted into an active state in distinct areas of the embryo is still not very well understood. However, there are indications that factors at the vegetal pole are translocated and converted to their biologically mature form (Kessler and Melton, 1995; Thomsen and Melton, 1993).

Up to the early-gastrula stage the Nieuwkoop area will in turn form the dorsal mesoderm (Spemann organizer). In a further series of events special genes and their products (secreted and nonsecreted proteins) are responsible for the primary steps of neural induction. There are indications that at least in part, dorsal blastopore lip (chordamesoderm) gains its neural inducing activity during the early phase of involution (Grunz, 1993a). This was shown by treatment of the Spemann organizer with *suramin*, which causes a shift of dorsal to ventral mesodermal structures (mainly heart structures) and also inhibits the formation of neural structures (Grunz, 1992, 1993a; Oschwald et al., 1993) (Fig. 6). *Suramin* is thought to prevent the binding of growth factors to their receptors (Goetschy et al., 1996). It is possible that secreted factors like *chordin* and *noggin* during

---

Fig. 5 Model of the signaling pathway of *Wnt* and the crosstalk between *Wnt* and catenin/cadherin (modified from Miller and Moon, 1996; Kühl and Wedlich, 1996; Huber et al., 1996a). In the presence of *Wnt*, disheveled (dsh) antagonizes the function of *Notch* as well as glycogen synthese kinase (GSK-3). The repressed function of GSK-3 causes a decreased phosphorylation of β-catenin and a higher stability of the APC (adenomatous polyposis coli tumor-suppressor protein)/β-catenin complex. The increased cytoplasmic β-catenin will promote the interaction with downstream effectors (X = transcription factors with an HMG box) and the translocation of the X/β-catenin complex to the nucleus. The binding to the DNA will initiate transcription and the change of cell fate (determination of mesenchymal or dorsal mesodermal cell types) (Kemler, 1993; Behrens et al., 1996; Molenaar et al., 1996). Furthermore, cytoplasmic β-catenin can interact with membrane-associated α-catenin and APC, resulting in changes of cell shape, migration, and cell adhesion. In the absence of *Wnt*, GSK-3 will phosphorylate β-catenin, which is unstable in cytoplasm and will rapidly degrade. Because of the low level of cytosolic β-catenin, the interaction with downstream transcription factors is prevented. In the absence of active *Wnt* signaling (left), dorsal genes are thought to be suppressed while *BMP-4* (ventralizing factor) synthesis is increased. *Frizzled* homologues are suggested as *Wnt* receptors or components of a receptor complex (Yang-Snyder et al., 1996).

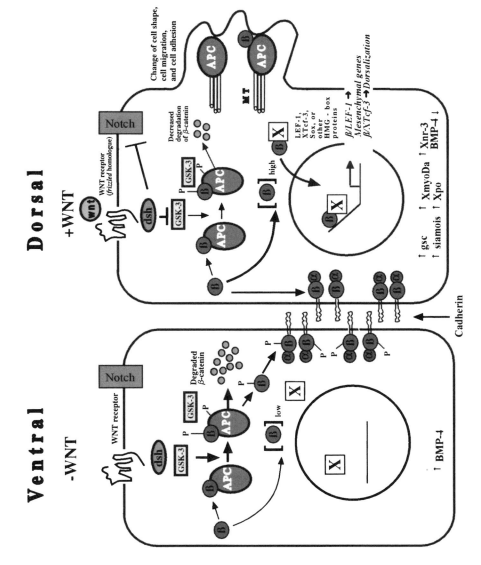

Ventral — WNT

WNT receptor

Notch

dsh ▐ GSK-3

Degraded β-catenin

GSK-3 P
P β APC

P β APC

P β

[β]low

X

X

↑ BMP-4

Dorsal + WNT

wnt

WNT receptor (*frizzled homologue*)

Notch

dsh ▐ GSK-3

Decreased degradation of β-catenin

GSK-3 P
P β APC

β APC

β

[β]high

β X

LEF-1, XTcf-3, Sox, or other HMG - box proteins

β/LEF-1
Mesenchymal genes →
β/XTcf-3 →*Dorsalization*

↑ gsc   ↑ XmyoDa   ↑ Xnr-3
↑ siamois   ↑ Xpo   BMP-4 ↓

Change of cell shape, cell migration, and cell adhesion

APC

β APC

MT

Cadherin

suramin treatment no longer can interact with their receptors, which prevents the dorsalization of mesoderm and in turn neuralization. Its regional inducing ability will be established during neuralization (Saha and Grainger, 1992). Also, ectoderm treated with *activin* A, a member of the TGFβ superprotein family, can function as Spemann organizer (Smith *et al.*, 1990; Asashima *et al.*, 1991b). However, the full range of anteriormost embryonic structures cannot be induced by *activin* (Ariizumi and Asashima, 1995).

VIII. The Organizer, Also a Unique Area in the Molecular Sense

On the basis of the classical Einsteck experiment of Spemann and Mangold (1924), there was no doubt that the dorsal mesoderm (Spemann organizer) is a unique area in the amphibian embryo. The new molecular techniques made it possible to search for peculiarities on the molecular level. The hope still existed of finding one single or a few neural inducer(s) that alone would be responsible as an instructor or evocator (Needham *et al.*, 1934). Indeed the mRNA of *goosecoid*, a gene with a homeobox closely related to the *Drosophila bicoid* and *gooseberry* genes, evoked the formation of a secondary embryo when injected into the ventral blastomeres of early-cleavage stages (Cho *et al.*, 1991; Blumberg *et al.*, 1991). However, it turned out that many additional genes and their products participate in the formation of the dorsalization of the organizer area and its role as neuralizing factors.

Two classes of genes could be defined. Similar to *goosecoid*, further homeobox-containing could be located in the organizer area, such as *Xlim-1* (Taira *et al.*, 1992), *XFD1/XFH1/pintallavis* (S. Knöchel *et al.*, 1992; Dirksen and Jamrich, 1992; Ruiz i Altaba and Jessel, 1992), *Xnot* (Von Dassow *et al.*, 1993), *Xnot 2* (Gont *et al.*, 1996), *Siamois* (Lemaire *et al.*, 1995), *Otx 2* (Pannese *et al.*, 1995; Blitz and Cho, 1995), and *Xanf* (Zariasky *et al.*, 1995). Also, secreted proteins located in the dorsal blastopore area, such as *noggin*, could be identified. However, as one substance alone, *noggin* is able to induce neural structures in rather unphysiologically high concentrations only (W. Smith and Harland, 1992). Still, *Xenopus* genes related to the mouse gene nodal *Xnr1 (Xnr3)*, members of the TFGβ superfamily, act in synergy with *noggin* and induce complete secondary axes in whole embryos (W. Smith *et al.*, 1995; Jones *et al.*, 1995; Lustig *et al.*, 1996).

Follistatin, an antagonist to *activin*, is synthesized in the dorsal mesoderm (Asashima *et al.*, 1991; Hemmati-Brivanlou *et al.*, 1994; Nakamura *et al.*, 1990; Ueno *et al.*, 1987). The injection of *follistatin* mRNA into the two-cell stage results in neuralization of ectoderm that was isolated at the early-gastrula stage (Hemmati-Brivanlou *et al.*, 1994). On the other hand *follistatin* protein cannot induce neural structures in the animal cap assay (Asashima *et al.*, 1991; Grunz,

1996). In contrast to *activin*, microinjection of *chordin* mRNA induces complete secondary embryos (Sasai *et al.*, 1994, 1995; Holley *et al.*, 1995).

Only one gene (*ADMP*) and its product, also a member of the TGFβ protein family and most prominently expressed in the Spemann organizer, have antidorsalizing activity. This antidorsalizing morphogenetic protein (*ADMP*) down-regulates *noggin, goosecoid*, and *follistatin* as well as dorsal markers like NCAM and MyoD. It may modulate the dorsalizing influences of other dorsalizing factors in the Spemann organizer (Moos *et al.*, 1995). Recently the transcript of a gene (*cerberus*) that encodes for a putative secreted protein has been detected in the yolky endomesodermal cells located in the deep layer of the organizer. The signals of whole-mount *in situ* hybridization are found in the leading edge of the most anterior cells and extend into the dorsal half of the floor of the blastocoel at the early-gastrula (stage 10.5). Microinjection of *cerberus* mRNA into a single D4 blastomere of the 32-cell stage results in the formation of an ectotopic head and duplicated hearts and livers (Bouwmeester *et al.*, 1996). Of special interest is the fact that in animal caps *cerberus* cannot induce beating hearts, although a marker for the cardiogenic region, NKX-2.5, is induced (Tonissen *et al.*, 1994). Since functional hearts are formed in intact embryos after *cerberus* injection, additional factors may be required for overt cardiac differentiation (Okada, 1954). Apparently, heart induction requires at least two separate signals, one from the organizer and another from the endoderm (Sater and Jacobson, 1990; Nascone and Mercola, 1995). These results are in agreement with our experiments with *suramin*-treated Spemann organizer and *activin*-treated animal caps. While *suramin*-treated Spemann organizer with enclosed endoderm formed beating hearts (Grunz 1992, 1993a; Fig. 6C,D), animal caps with *activin* and *suramin* treatment formed blood precursor cells only (Oschwald *et al.*, 1993).

IX. Transmission of Neuralizing Signals Between the Inducing Spemann Organizer (Chordamesoderm) and the Ectodermal Target Cells

The spatial and temporal expression of *cerberus* is of great relevance to the discussion about the importance of planar and vertical signals during the early steps of neural induction, since vertical signals may emanate from the leading edge of the anteriormost endomesoderm to the overlaying ectoderm even before gastrulation starts (see the next section, "Planar versus Vertical Signals"). Up to now it has not been clear whether one of these neural inducing factors, like *noggin, cerberus, follistatin*, or *chordin*, is identical with factors isolated from chicken brain, *Xenopus* embryos, and neuroblastoma or retinoblastoma cell lines (Mikhailov and Gorgolyuk, 1987, 1989; Mikhailov *et al.*, 1995; Janeczek *et al.*, 1992; Lopashov *et al.*, 1992). The factor from *Xenopus* is, like several growth factors, part of a large anionic protein complex (reviewed by Tiedemann *et al.*,

1995, 1996). While these factors are able to induce neural structures in competent ectoderm, *follistatin* (Hemmati-Brivanlou *et al.*, 1994) and scatter factor (hepato-cyte growth factor; Stern and Ireland, 1993) were not biologically active in the animal cap assay (Asashima *et al.*, 1991a; Grunz, 1996). Additional factors are needed for neuralization that are not present in the isolated ectoderm but are present in the intact embryo. However, all secreted and isolated neuralizing factors may have in common the ability to prevent *BMP*-like ligands from bind-ing to their receptors on the ectodermal target cells rather than interacting imme-diately with specific receptors and causing neural induction (see later discussion and Fig. 3).

In 1961 Lauri Saxén published a paper describing how, with a special filter chamber, neuralizing signals passed a Millipore filter (Saxén, 1961) (Fig. 4). These results were confirmed later with Nucleopore filter (Toivonen and War-tiovaara, 1976). Even electron microscopy could detect no cell protrusions in the filter pores (channels) between the inducing chordamesoderm and the reacting ectoderm. Forebrain structures were induced when Nucleopore filters with a pore size less than 0.5 μm were used. Larger pore sizes (0.5 μm and up) apparently also allow cell-to-cell contacts between the inducing and reacting tissues, which caused the shift in the more posterior neural structures accompanied by mesoder-mal derivatives (reviewed by Saxén, 1989). However, a prolonged induction is needed for the realization of the more posterior structures.

These transfilter experiments also indicated that neuralization consists of at least two main steps. In the first step neural structures of a more anterior charac-ter are determined that can be overridden by several inducing events. This leads to the regionalization of the central nervous system. These observations are in agreement with the activation-transformation hypothesis of Nieuwkoop. Also, newer reports have distinguished between primary and secondary events in neu-ral induction (Saha and Grainger, 1992). Thus, the labile determination of cement gland (one of the anteriormost structures) can be overridden by the prolonged action of "stronger" neuralizing signals (Sive, 1993; Sive and Bradley, 1996). Furthermore, FGF-like factors are thought to play an important role in the second period of neural induction, the regionalization of the central nervous system (Doniach, 1995), although very recent results with transgenic *Xenopus* indicate that signaling through FGF receptor is not required *in vivo* for neural induction or patterning (Kroll and Amaya, 1996; see also later). However, it is generally accepted that the formation of the central nervous system can be subdivided into two main phases, the primary step of neuralization and the patterning or regional-ization of the presumptive brain areas. These phases had already been formulated during the period from the late 1930s to the late '50s in various hypotheses: The quantitative theory, or the concept of morphogenetic potentials of Dalcq and Pasteels (1938), the double-potency theory of Yamada (1947), the double-gradi-ent hypothesis of Toivonen and Saxén (Toivonen and Saxén, 1955; Saxén and Toivonen, 1961; Tiedemann and Tiedemann, 1964), and the activation-transformation hypothesis of Nieuwkoop (1952) (reviewed in Nakamura and Toivonen, 1978; Nieuwkoop *et al.*, 1985; Saxén, 1989; Gilbert and Saxén, 1993; Tiedemann *et al.*, 1996).

Meanwhile, the two most discussed models—the double-graient model and the two-step (activation-transformation) model—could be merged. In the first step, the ectoderm will be labile determined into anterior neural structures. The regionalization into more caudal structures will then take place via the interaction of chordamesoderm and the already-neural-determined ectoderm. Mesoderm will be formed only in the posteriormost part of the neural plate, where the neural signal is overridden by strong mesodermal signals or where no neural signals are present at all. Although there are indications from the transfilter experiments and isolation from the gap between chordamesoderm and neurectoderm in the neurula (John *et al.*, 1983) that the neuralizing signals are transmitted by diffusible signals, the exact mechanisms are not well understood. However, it is unlikely that signals are traveling over long distances. Two mechanisms are discussed—a transfer from cell to cell, either like a relay mechanism (Reilly and Melton, 1996) or in the form of a gradient (Gurdon *et al.*, 1994, 1996). Electron microscopy showed that the reacting ectoderm must be located in close apposition to the inducing chordamesoderm to get induced into neural structures (Grunz and Staubach, 1979; Tacke and Grunz, 1988). The action of neuralizing factors will be discussed later in the chapter in the context of the new hypothesis of the neural default state of the ectoderm and the importance of ventralizing factors.

X. Planar versus Vertical Signals

Holtfreter (1933b) concluded from his exogastrula experiments with *Axolotl* embryos that neural inducing factors can reach the ectoderm via vertical signals only. Results with *Xenopus* embryos using exogastrulae and Keller sandwiches suggest that planar signals from the dorsal blastopore lip (Spemann organizer) travel to the neighboring neuroectoderm (Doniach, 1992; Doniach *et al.*, 1992; Ruiz i Altaba, 1992). However, it turned out that the *Xenopus* embryo is not very suitable for deciding the question of whether planar signals prior to the beginning of gastrulation are as important as vertical signals after the involution of the endomesoderm during gastrulation. Since the morphological organization and the morphogenetic movements of *Xenopus* are quite different in comparison to urodeles, we used two different species in comparative studies (Grunz *et al.*, 1995; Hollemann *et al.*, in preparation).

There are substantial arguments that planar signals during the early steps of neural induction are of minor importance (Nieuwkoop and Koster, 1995; Grunz *et al.*, 1995; reviewed in Grunz, 1996). On the basis of the recent results of De Robertis and collaborators (Bouwmeester *et al.*, 1996) that a neuralizing gene *cerberus* is expressed at the leading edge of the endomesoderm long before the bottle cells of the external blastopore lip start ingressing, it could be argued that the ectoderm of Keller sandwiches prepared from early-gastrula stages (even stage 10, Nieuwkoop and Faber, 1956) have already been triggered by vertical neuralizing signals emanating from the internal underlying endomesoderm (Bouwmeester *et al.*, 1996). Further experiments with *Triturus alpestris* and *Xenopus* laevis, where we combined the exogastrula method with the Keller sandwich technique, using *Triturus-* and *Xenopus*-specific *Pax 6* markers, support the view that planar signals are less important for the primary steps of neural induction (Hollemann *et al.*, in preparation). On the other hand, horizontal or planar signals apparently play an important role during the period of regionalization and patterning of the central nervous system (Mangold and Spemann, 1927; Grunz 1990; Grunz *et al.*, 1986; Albers, 1987).

XI. The "Antiorganizers"

In the traditional view, Spemann organizer was considered the source, that dorsalizes the lateral and ventral marginal zones in a graded fashion. According to this view, the default state of the vegetal marginal zone is the formation of ventral mesoderm. However, recent reports indicate that the ventral marginal zone is not just an inert zone receiving signals from the dorsal side, but, rather, contains active signals for the specification of the ventral state and even influences the lateral and the dorsal/animal hemispheres (Fig. 7). *Xwnt-8*, a member of the *Wnt* family of secreted molecules, may act as an important factor to specify ventral territories (Christian and Moon, 1993; Hoppler *et al.*, 1996). After midblastula transition, it is expressed preferentially in the ventral and lateral marginal zones (presumptive mesoderm). Loss of function through a dominant-negative *Wnt* ligand has implicated *Xwnt-8* signaling in the expression of MyoD and the formation of skeletal muscle in *Xenopus* embryos (Kato and Gurdon, 1994; Hoppler *et al.*, 1996). On the other hand, the overexpression of *Wnt-8* on the ventral side of early-cleavage stages causes the formation of a secondary axis. *Xwnt-8b* (Cui *et al.*, 1995) is thought to function as an endogenous signal modifying the response of cells to dorsal–ventral axis inducers.

Many articles have reported a close correlation between the *Wnt* signaling pathway (Fig. 5) and cell-to-cell adhesion (reviewed by Kemler, 1993; Huber *et al.*, 1996a; Kühl and Wedlich, 1996; Miller and Moon, 1996; Gumbiner, 1995). An important member of this *Wnt* signaling pathway is β-catenin. On one hand, it participates in the formation of the α, β-catenin/cadherin complex for the formation of cell contacts (Takeichi, 1995). On the other hand, it plays a putative role in the *Wnt* signaling chain from the plasma membrane to the nucleus. When it is overexpressed at the ventral side of *Xenopus* embryos, it leads to the formation of a secondary body axis (Funayama *et al.*, 1995).

The stability of β-catenin in the cytoplasm, resulting in further signaling, depends on its phosphorylation state (Fig. 5). Only the unphosphorylated form will not be degraded. Responsible for the phosphorylation is glycogen synthase kinase-3 (GSK-3) (Yost *et al.*, 1996). Microinjection of rat GSK-3β mRNA into animal ventral blastomeres of eight-cell-stage embryos caused the development of ectopic cement glands with adjacent anterior neural tissues. In contrast, animal dorsal injection of the same dose of GSK-3β mRNA caused eye deficiencies, whereas vegetal injection had no pronounced effects on normal development (Itoh *et al.*, 1995). The overexpression of GSK-3 causes the loss of dorsal axial structures (He *et al.*, 1995). On the other hand, the inhibition of GSK-3β mRNA by lithium results, in extreme cases, in entirely dorsalized *Xenopus* embryos that lack identifiably ventral tissues (Kao *et al.*, 1986; Kao and Elinson, 1988; Klein and Melton, 1996). In the animal cap assay (ectoderm of *Xenopus*), LiCl has no significant inducing activity. However, it should be mentioned that *Triturus* and *Axolotl* ectoderm forms even dorsal mesodermal structures after LiCl treatment (Masui, 1961; Gebhart and Nieuwkoop, 1964; Grunz 1968, 1993b).

In the *Wnt* signaling cascade downstream of β-catenin, the architectural transcription factor LEF-1, a mammalian HMG box factor, has been postulated (Huber *et al.*, 1996b; Behrens *et al.*, 1996). It binds directly to β-catenin and translocates β-catenin to the nucleus, where it alters the DNA-binding properties of LEF-1. In *Xenopus* a related factor, *Xtcf-3*, has been found that also binds to β-catenin and translocates it to the nucleus (Molenaar *et al.*, 1996). A close relationship exists between HMG and SRY box–containing proteins. They have been named Sox, for Sry-type HMG box genes (for review see Baxevanis and Landsman, 1995; Laudet *et al.*, 1993.

The mouse genome contains at least 20 Sox genes subclassed into seven groups, named A to G (Wright *et al.*, 1993). Genes from group B, including Sox-1, Sox-2 and Sox-3, Sox-19, seem to be involved in the determination of the central nervous system (Vriz *et al.*, 1996). We have recently isolated a *Xenopus*-specific HMG box–containing gene, Xsox 3, which is expressed in the presumptive central nervous system (Penzel *et al.*, submitted) (Fig. 5F). Nothing is yet known about its biological function. Whether this gene participates in the control of the process determining the presumptive brain structures is still unknown.

Besides *Xwnt-8*, another peptide growth factor, *BMP-4*, is expressed in the ventral marginal zone. Similar to *Xwnt-8*, it is able to override dorsal mesodermal tissue specification (Köster *et al.*, 1991; Dale *et al.*, 1992; Jones *et al.*, 1992, 1996; Christian and Moon, 1993; Fainsod *et al.*, 1994; Schmidt *et al.*, 1995). Microinjection of a dominant-negative *BMP-4* receptor results in the dorsalization of ventral mesoderm (Graff *et al.*, 1994; Maéno *et al.*, 1994a; Suzuki *et al.*, 1994). A role in dorsoventral pattern formation and in ventralization is correlated with the observation that *BMP-4* is expressed in the ventral and lateral marginal zones of the early gastrula, but not in the organizer (Fainsod *et al.*, 1994; Steinbeisser *et al.*, 1995; Sasai *et al.*, 1995; Schmidt *et al.*, 1995). *BMP-4* is also expressed in the animal hemisphere, where it suppresses neuralization and induces epidermis (Grunz and Tacke 1989, 1990; Wilson and Hemmati-Brivanlou, 1995). These results indicate that *BMP* acts in an antagonistic way to signals located in the dorsal marginal zone. Thus, the neuralizing activity of the secreted protein *chordin* can be suppressed by *BMP-4* (Sasai *et al.*, 1995). *BMP-4* is also an important candidate for dorsal-ventral patterning in Zebrafish (Granato and Nüsslein-Volhard, 1996).

One very exciting observation was that on the ventral side, several homeobox-containing genes are expressed acting downstream of *BMP-4*. In an attempt to identify genes involved in marginal zone patterning, Niehrs and collaborators performed a large-scale screen by whole-mount *in situ* hybridization with digoxigenin-labeled riboprobes made from cDNA clones that are picked randomly from a neural cDNA plasmid library. By this method two novel homeobox genes, *vent-1* and *vent-2*, were identified. Their mRNAs are found mainly in the ventral and ventrolateral marginal zones at the early gastrula. On the dorsal side, an arc of ~160°, comprising more than the organizer proper, is excluded from expression. At stage 11, transcripts can be detected additionally in the ventral part of the animal caps (Gawantka *et al.*, 1995; Onichtchouk *et al.*, 1996). Independently in screens for homeobox genes, *Xbr-1* and *Vox*, which are identical with *vent-2*, were isolated by two other labs (Papalopulu and Kintner, 1996; Schmidt *et al.*, 1996). *Xvent-1 [PV.1]* (Gawantka *et al.*, 1995; Ault *et al.*, 1996) can be induced by *BMP-4* and can interact with *goosecoid* in a negative cross-regulatory loop. *Vent-2 (Vox, Xbr-1)* is initially expressed in the marginal zone and animal cap region of gastrulae, specifically excluding the organizer region. The expression is clearly different from that of *Xvent-1*, whose expression boundary in the marginal zone is in a more lateral position. *Xvent-2* is expressed in regions that also express *BMP-4*. It interacts with *BMP-4* in a positive feedback loop. *Xvent-2* suppresses *goosecoid*, and vice versa, in a cross-regulatory loop. *Xom*, identical with *Xvent-2*, is a further homeobox gene that the authors characterized in dorsal–ventral patterning, also downstream of *BMP-4* (Ladher *et al.*, 1996). Similar to *Xpo*, a gene expressed preferentially in the ventral and ventrolateral marginal zones (Amaya *et al.*, 1993; Sato and Sargent, 1991), which are not affected by *Xvent-1*, *Xom* does not inhibit the general mesodermal marker Xbra (Ladher *et al.*, 1996). On the other hand, injection of *noggin* into one blastomere of the two-cell-stage inhibits the expression of *Xom*. Overexpression of *Xom* causes a phenotype similar to that caused by overexpression of *BMP-4*. However, the effects of *Xom* are not as marked as those of *BMP-4*, since *Xom* is not able to reverse completely the dorsalization effects of *noggin*. Furthermore, much larger amounts of *Xom* mRNA (4 ng), similar to *Vox*, is necessary for ventralization in comparison to *vent-1* (200 pg).

Perhaps other genes are additionally needed for a full ventralization. *Xom* is activated before *BMP-4*. This holds true also for *Vox*. However, it cannot be excluded that *Vox* and *Xom* are activated by low levels of maternal *BMP-4*, for which low levels of maternal transcripts can be detected (Dale *et al.*, 1992). On the other hand, the initial activation could be triggered by *BMP-2*, which has high levels of maternal transcripts (Clement *et al.*, 1995). However, since the overexpression of *BMP-4* causes expression of *Xom* to occur in the dorsal marginal zone as well as in ventral tissues, similar to the results with *vent-1* and *vent-2* (*Vox*), it can be suggested that all these genes function in the *BMP-4* signaling pathway (probably downstream of *BMP-4*).

XII. The Establishment of the Animal–Vegetal, Dorsal– Ventral, and Anterior–Posterior Pattern Depends on the Formation of Distinct Gradients

Summarizing our preceding discussion, there is growing evidence that several gradients (representing maternal factors, homeobox-containing genes, and genes coding for secreted proteins) must be established from the growing oocyte until late gastrula to form the embryonic body axis with the distinct organized central nervous system (Fig. 7). Prior to fertilization, the amphibian egg shows an animal–vegetal polarity. Certain maternal factors, such as *vg1*, are preferentially localized at the vegetal pole (Weeks and Melton, 1987). Sperm entry causes cortical rotation, which shifts factors in direction (Kloc and Etkin, 1994, 1995) from the vegetal pole to the dorsal marginal zone. From this dorsal vegetal zone, also named the Nieuwkoop center, signals are emanating to the presumptive Spemann organizer, which in this dorsal marginal zone activates several transcription factors, such as *goosecoid, Xlim-1, otx-2, XFD-1/XFH1/pintallavis, Xnot, Xnot-2, Xanf-1*, and *Siamois*. Furthermore, secreted proteins like *follistatin, noggin*, and *chordin* are localized in this zone. Also, members of the TGFβ superfamily, *Xnr-1, Xnr-3*, and *ADMP*, are expressed predominantly in the dorsal lip of early gastrulae.

The dorsal gradient represented by the factors just mentioned is counteracted by a ventralizing gradient (Fig. 7). This gradient is established and maintained by *Xwnt-8, BMP-4, vent-1*, and *vent-2* (*Vox, Xbr-1, Xom*). This gradient, with the highest levels of its representatives at the ventral marginal zone, acts antagonistically to the dorsal marginal zone and the dorsal animal zone. By this interaction, intermediate areas (the lateral marginal zones) are established. Furthermore, even during cleavage stages, differences between dorsal and ventral animal blastomeres may be correlated to the unequal distribution and activation of maternal factors (Grunz, 1977; Miller and Moon, 1996; Li *et al.*, 1996).

During gastrulation, the organizer (chordamesoderm) will acquire its final dorsal determination (determination of notochord), which is the prerequisite for its neural inducing activity. However, also during this process genes like *cerberus* are activated in a zone (anterior endomesoderm) formerly not expected to be responsible for the induction of anterior neural structures. The further regionalization of the central nervous system will start during late gastrula up to the neurula stage.

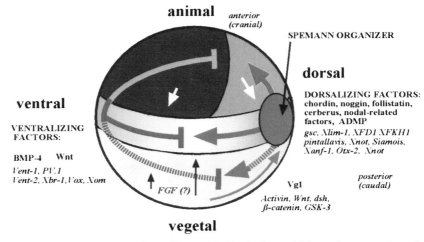

Fig. 7   Model of the formation of the different polarities in the amphibian embryo up to the early gastrula. A gradient of ventralizing signals, with the highest concentrations on the ventral side, acts antagonistically to a gradient of factors localized on the dorsal side (Spemann organizer). The dorsal-side signals also influence the fate of the dorsal vegetal and the marginal zones. During gastrulation, the Spemann organizer (chordamesoderm) induces the dorsal ectoderm to form neural structures. Since in the ventral ectoderm *BMP-4* activity is not suppressed by dorsal signals, it differentiates to epidermis. Substantial evidence exists that factors in the animal hemisphere influence the fate of cells located in the marginal zone and even in the vegetal hemisphere (white arrows). Cortical rotation immediately after insemination causes the formation of the dorsal/ventral polarity by shifting substances (*vgl* and unknown factors) from the vegetal pole to the dorsal side of the early embryo (Nieuwkoop center).

It has been suggested that basic fibroblast growth factor (bFGF) may participate in the anterior–posterior pattern formation (Doniach, 1995). However, the new transgenic *Xenopus* technique has shown that these and other results must be discussed with great care (see also the next section, "Perspectives"). Basic fibroblast growth factor is thought to participate in the formation of mesoderm in the early- and middle-blastula stages (Slack *et al.*, 1987; Ding *et al.*, 1992; Grunz *et al.*, 1987). Furthermore, it has been postulated that bFGF is able to induce neural structures (Kengaku and Okamoto, 1993) or that it could shift the neural pattern from an anterior to a more posterior level (Kengaku and Okamoto, 1995; Doniach, 1995). Under *in vivo* conditions, however, a truncated dominant-negative FGF receptor under the control of simian cytomegalovirus (CMV) promoter drives the expression from the early-gastrula stage, and the embryos develop with normal patterning (Kroll and Amaya, 1996). Another excellent example of the importance of the correct timing of gene expression is illustrated by *Xwnt-8*. While overexpression of this gene product by RNA injection causes dorsalization (Smith and Harland, 1991; Sokol *et al.*, 1991), the more appropriate expression after MBT (midblastula transition) reveals that it is more likely to be involved in the specification of ventral mesoderm (Christian and Moon, 1993) (Figs. 5, 7).

In addition to the puzzling result that a ventralizing gradient is just as important as the dorsal gradient with its source in Spemann organizer, a further gradient can be postulated, acting in direction from the animal pole to the vegetal pole (Fig. 7, white arrows). In contrast to the earlier view that determination factors are present only in the vegetal hemisphere of the egg, there are indications that the animal hemisphere also contains factors essential to the determination of tissues derived from the equatorial zone. When the marginal zone or marginal zone together with the vegetal region were isolated from *Ambystoma mexicanum* early blastula, no differentiation occurred. The differentiation of notochord took place when the vegetal hemisphere was isolated from middle to late blastula or later stages (Tiedemann, 1993; Tiedemann *et al.*, 1996). We could show with *Xenopus* that the isolated vegetal hemisphere of stage-7 blastula will form only somites when the whole animal cap is removed. If this is done at stage 8 or 9, the vegetal part will also form notochord (Grunz, unpublished results). These data indicate that factors are located in the animal half that influence the vegetal hemisphere. It could be shown that animal-to-vegetal, contact-dependent signals pass from the B-tier to the C-tier blastomeres of the 32-cell embryo (Bauer *et al.*, 1996). The nature of these signals acting from the animal to the vegetal hemisphere needs further analysis. One putative candidate may be β-catenin, which is already located in the dorsal animal blastomeres of a four-cell stage (Miller and Moon, 1996; Wylie *et al.*, 1996; Schneider *et al.*, 1996; Li *et al.*, 1996; Larabell *et al.*, 1997).

XIII. Conclusions

A dramatic improvement in our knowledge about the establishment of the embryonic body plan, including the formation of the central nervous system, has taken place during the last decade and just in the last three years. The importance of the Spemann organizer for neural induction has been confirmed by the new molecular techniques. Certain genes coding for homeobox-containing transcription factors (Hollemann *et al.*, 1996) and secreted proteins could be localized predominantly in the organizer area. Very puzzling were the results that other genes and their products, like *Wnt, BMP-4, vent-1*, and *vent-2 (Vox, Xbr-1, Xom)* are expressed in a ventral/dorsal gradient mainly on the ventral and lateral sides of the marginal zone. They act antagonistically to a gradient of factors, with the highest concentration on the dorsal side of the embryo, including the Spemann organizer. Although a large number of very interesting genes has been isolated, probably not all the important ones are yet known to explain the exact mechanisms of neural induction. One example that the probability is high that very interesting genes can still be found is the recent detection of a gene, *cerberus*, that is expressed in the leading edge of the endomesoderm of the Spemann organizer

region. Also, the processes leading to the establishment of the different embryonic gradients before midblastula transition and shortly after the begin of zygotic
activation are not fully understood. It is generally accepted that there are vegetal
→ animal and dorsal ⇔ ventral gradients. In addition, it can be suggested that
factors located in the animal hemisphere exert a substantial influence on the
vegetal part of the embryo (animal → vegetal; Fig. 7, white arrows). The aim for
the future will be not only the isolation and exact analysis of the special and
temporal expression of the different genes and their products, but also their
complex mutual interactions and the detailed study of their regulation (analysis
and function of their promoters and enhancers).

XIV. Future Prospects

The disadvantage of *Xenopus laevis* and other amphibian species is that they
could not be used for genetic studies because of their long generation times and
their tetraploid genome (two similar copies of each chromosome; Kobel and Du
Pascier, 1986). Therefore, other model systems, *Caenorhabditis, Drosophila*,
zebrafish, and mouse, were chosen. The molecular analysis of the processes
leading to a large number of mutants in *Drosophila* and zebrafish have tremen-
dously increased our knowledge about the early steps of embryogenesis
(Nüsslein-Volhard, 1996; Granato and Nüsslein-Volhard, 1996). In the mouse it
was possible to make transgenic animals; in *Xenopus*, similar approaches resulted in the expression of the transgene in only a minority of cells. This mosaic
expression was a severe limitation of the *Xenopus* model system.

Experiments on the gain and loss of function frequently give only limited
information, since the mRNA (wild-type RNA, or dominant-negative mutant
mRNA coding for ligands or receptors) must be injected into early-cleavage
stages, although many genes will function later in development. This fact makes
it difficult to judge the importance of the synthesis and function of maternal and
zygotic expression of the same gene. A big breakthrough therefore is the production of real transgenic *Xenopus* embryos in large quantities by nuclear transplantation of sperm nuclei with introduced transgenes into unfertilized eggs (Kroll
and Amaya, 1996; Amaya and Kroll, 1997). Thus, it will now be possible to
merge the traditional advantages of amphibians (large number and size of eggs,
development outside of the maternal organisms, ease of dissection and microinjection, fairly good knowledge of the anlagenplan in the early-cleavage and
gastrula stages, study of vertebrate-specific mesodermal and neural induction)
with the potential to start the expression of genes at a desired period and spatial
localization during embryogenesis by introducing specific promoters. Kroll and
Amaya themselves have already used the new technique to introduce a transgenic
dominant mutant of FGF (fibroblast growth factor) in *Xenopus* oocyte. In contrast to earlier *in vitro* results (Kengaku and Okamoto, 1995; reviewed by Doniach, 1995), it was shown that FGF is not required *in vivo* for neural induction

and pattern formation of the central nervous system. These data show that over the last 70 years the amphibian embryo (especially the "laboratory frog" *Xenopus laevis*) did not lose its importance for developmental biology. On the contrary, because of all its advantages compared to other invertebrate and vertebrate embryos, the prospects are excellent that it will be used in a growing number of laboratories.

Acknowledgments

Our own studies mentioned in this chapter were supported by the Deutsche Forschungsgemeinschaft and in part by the Forschungspool of the Universität GH Essen.

References

Aberger, F., Schüren, C., Lepperdinger, G., Grunz, H. and Richter, K. (1997). Induction of anterior neural fate by a novel cement gland-specific gene in *Xenopus*. Submitted.

Albers, B. (1987). Competence as the main factor determining the size of the neural plate. *Develop. Growth and Differ.* **29**, 535–545.

Amaya, E., and Kroll, K. L. (1997). A method for generating transgenic frog embryos. *In* "Methods in Molecular Biology: Methods and Protocols" (Paul Sharpe and Ivor Mason, Eds.), in press. Humana Press, Totowa, N.J.

Amaya, E., Stein, P. A., Musci, T. J., and Kirschner, M. W. (1993). FGF signalling in the early specification of mesoderm in *Xenopus*. *Development* **118**, 477–487.

Ancel, P., and Vintemberger, P. (1948). Recherches sur le déterminisme de la symétrie bilaterale dans l'oeuf des amphibiens. *Bull. Biol. France et Belg.* **suppl. 31**, 1–182.

Ariizumi, T., and Asashima, M. (1995). Head and trunk-tail organizing effects of the gastrula ectoderm of *Cynops pyrrhogaster* after treatment with *activin*-A. *Roux's Arch. Dev. Biol.* **204**, 427–435.

Ariizumi, T., Sawamura, K.-I., Uchiyama, H., and Asashima, M. (1991). Dose- and time-dependent mesoderm induction and outgrowth formation by *activin*-A in *Xenopus* laevis. *Int. J. Dev. Biol.* **35**, 407–414.

Asashima, M. (1994). Review: Mesoderm induction during early amphibian development. *Develop. Growth Differ.* **36**, 343–355.

Asashima, M., Nakano, H., Shimada, K., Knoshita, K., Ishii, K., and Shibai, H. U. (1990). Mesodermal induction in early amphibian embryos by *activin* A (erythroid differentiation factor). *Roux's Arch. Dev. Biol.* **198**, 330–335.

Asashima, M., Nakano, H., Uchiyama, H., Sugino, H., Nakamura, T., Eto, Y., Ejima, D., Davids, M., Plessow, S., Cichocka, I., and Kinoshita, K. (1991a). Follistatin inhibits the mesoderm-inducing activity of *activin*-A and the vegetalizing factor from chicken embryo. *Roux's Arch. Dev. Biol.* **200**, 4–7.

Asashima, M., Uchiyama, H., Nakano, H., Eto, Y., Ejima, D., Sugino, H., Davids, M., Plessow, S., Born, J., Hoppe, P., Tiedemann, H., and Tiedemann, H. (1991b). The vegetalizing factor from chicken embryos: Its EDF (*activin*-A)-like activity. *Mech. Development* **34**, 135–141.

Ault, K. T., Dirksen, M. L., and Jamrich, M. (1996). A novel homeobox gene PV.1 mediates induction of ventral mesoderm in *Xenopus* embryos. *Proc. Natl. Acad. Sci. USA.* **93**, 6415–6420.

Bauer, D. V., Best, D. W., Hainski, A. M., and Moody, S. A. (1996). A contact-dependent animal-to-vegetal signal biases neural lineages during *Xenopus* cleavage stages. *Dev. Biol.* **178**, 217–228.

Bautzmann, H., Holtfreter, J., Spemann, H., and Mangold (1932). Versuche zur analyse der induktionsmittel in der embryonalentwicklung. *Naturwissenschaften* **20**, 971–974.

Baxevanis, A. D., and Landsman, D. (1995). The HMG-1 box protein family: Classification and functional relationships. *Nucleic Acids Res.* **23**, 1604–1613.

Becker, U., Tiedemann, H., and Tiedemann, H. (1959). Versuche zur determination von embryonalen amphibiengewebe durch induktionsstoffe in Lösung. *Z. Naturforschg.* **14b**, 608–609.

Behrens, J., von Kries, J. P., Kuhl, M., Bruhn, L., Wedlich, D., Grosschedl, R., and Birchmeier, W. (1996). Functional interaction of β-catenin with the transcription factor LEF-1. *Nature* **382**, 638–642.

Blitz, I. L., and Cho, K. W. Y. (1995). Anterior neurectoderm is progressively induced during gastrulation: The role of the *Xenopus* homeobox gene orthodenticle. *Development* **121**, 993–1004.

Blumberg, B., Wright, C. V. E., De Robertis, E. M., and Cho, K. W. Y. (1991). Organizer-specific homeobox genes in *Xenopus*-laevis embryos. *Science* **253**, 194–196.

Born, J., Hoppe, P., Janeczek, J., Tiedemann, H., and Tiedemann, H. (1986). Covalent coupling of neuralizing factors from *Xenopus* to sepharose beads: No decrease of inducing activity. *Cell Diff.* **19**, 97–101.

Born, J., Janeczek, J., Schwarz, W., Tiedemann, H., and Tiedemann, H. (1989). Activation of masked neural determinants in amphibian eggs and embryos and their release from the inducing tissue. *Cell Diff. Develop.* **27**, 1–7.

Bouwmeester, T., Kim, S. H., Sasai, Y., Lu, B., and De Robertis, E. M. (1996). *cerberus* is a head-inducing secreted factor expressed in the anterior endoderm of Spemann's organizer. *Nature* **382**, 595–601.

Cardellini, P. (1988). Reversal of dorsoventral polarity in *Xenopus* laevis embryos by 180° rotation of the animal micromeres at the eight-cell stage. *Dev. Biol.* **128**, 428–434.

Cho, K. W. Y., Blumberg, B., Steinbeisser, H., and De Robertis, E. M. (1991). Molecular nature of Spemann's organizer—The role of the *Xenopus* homeobox gene goosecoid. *Cell* **67**, 1111–1120.

Christian, J. L., and Moon, R. T. (1993). Interactions between *Xwnt-8* and Spemann organizer signaling pathways generate dorsoventral pattern in the embryonic mesoderm of *Xenopus*. *Genes Develop.* **7**, 13–28.

Chuang, H-H. (1938). Spezifische induktionsleistungen von leber und niere im explantationsversuch. *Biol. Zbl.* **58**, 472–480.

Chuang, H-H. (1939). Induktionsleistungen von frischen und gekochten organteilen (niere, leber) nach ihrer verpflanzung in explantate und verschiedene wirtsregionen von tritonkeimen. *Arch. EntwMech. Org.* **139**, 556–638.

Clement, J. H., Fettes, P., Knöchel, S. Lef, J., and Knöchel,W. (1995). Bone morphogenetic protein 2 (*BMP*-2) in the early development of *Xenopus* laevis. *Mech. Develop.* **52**, 357–370.

Cui, Y. Z., Brown, J. D., Moon, R. T., and Christian, J. L. (1995). *Xwnt-8b*: A maternally expressed *Xenopus* Wnt gene with a potential role in establishing the dorsoventral axis. *Development* **121**, 2177–2186.

Dalcq, A., and Pasteels, J. (1938). Potential morphogénétique, regulation et "axial gradients" de Child. Mireau point des bases physiologiques de la morphogénèse. *Bull. Acad. Med. Belg.* **6**, Ser. 3, 261–308.

Dale, L., Howes, G., Price, B. M. J., and Smith, J. C. (1992). Bone morphogenetic protein-4—A ventralizing factor in early *Xenopus* development. *Development* **115**, 573–585.

Davids, M. (1988). Protein kinases in amphibian ectoderm induced for neural differentiation. *Roux's Arch. Dev. Biol.* **197**, 339–344.

Davids, M., Loppnow, B., Tiedemann, H., and Tiedemann, H. (1987). Neural differentiation of amphibian gastrula ectoderm exposed to phorbol ester. *Roux's Arch. Dev. Biol.* **196**, 137–140.

Dawid, I. B., and Taira, M. (1994). Axis determination in *Xenopus*: Gradients and signals. *Bioessays* **16**, 385–386.

De Robertis, E. M., Morita, E. A., and Cho, K. W. Y. (1991). Gradient fields and homeobox genes. *Development* **112**, 669–678.

Ding, X.-Y., McKeehan, W. L., Xu, J., and Grunz, H. (1992). Spatial and temporal localization of FGF receptors in *Xenopus* laevis. *Roux's Arch. Dev. Biol.* **201**, 334–339.

Dirksen, M. L., and Jamrich, M. (1992). A novel, activin-inducible, blastopore lip-specific gene of *Xenopus*-laevis contains a fork head DNA-binding domain. *Genes Develop.* **6**, 599–608.

Doniach, T. (1992). Induction of anteroposterior neural pattern in *Xenopus* by planar signals. *Development* **Suppl.**, 183–193.

Doniach, T. (1995). Basic FGF as an inducer of anteroposterior neural pattern. *Cell* **83**, 1067–1070.

Doniach, T., Phillips, C. R., and Gerhart, J. C. (1992). Planar induction of anteroposterior pattern in the developing central nervous system of *Xenopus* laevis. *Science* **257**, 542–545.

Drean, G., Leclerc, C., Duprat, A. M., and Moreau, M. (1995). Expression of L-type Ca^{2+} channel during early embryogenesis in *Xenopus* laevis. *Int. J. Dev. Biol.* **39**, 1027–1032.

Fainsod, A. S., Steinbeisser, H., and De Robertis, E. M. (1994). On the function of BmP-4 in patterning the marginal zone of the *Xenopus* embryo. *EMBO J.* **13**, 5015–5025.

Frohnhöfer, H. G., and Nüsslein-Volhard, C. (1986). Organization of anterior pattern in the *Drosophila* embryo by the maternal gene bicoid. *Nature* **324**, 120–125.

Fujisue, M., Kobayakawa, Y., and Yamana, K. (1993). Occurrence of dorsal axis–inducing activity around the vegetal pole of an uncleaved *Xenopus* egg and displacement to the equatorial region by cortical rotation. *Development* **118**, 163–170.

Fukui, A., and Asashima, M. (1994). Control of cell differentiation and morphogenesis in Amphibian development. *Int. J. Dev. Biol.* **38**, 257–266.

Funayama, N., Fagotto, F., Mccrea, P., and Gumbiner, B. M. (1995). Embryonic axis induction by the armadillo repeat domain of beta-catenin: Evidence for intracellular signaling. *J. Cell Biol.* **128**, 959–968.

Gallagher, B. C., Hainski, A. M., and Moody, S. A. (1991). Autonomous differentiation of dorsal axial structures from an animal cap cleavage stage blastomere in *Xenopus*. *Development* **112**, 1103–1114.

Gawantka, V., Delius, H., Hirschfeld, K., Blumenstock, C., and Niehrs, C. (1995). Antagonizing the Spemann organizer: Role of the homeobox gene *Xvent-1*. *EMBO J.* **14**, 6268–6279.

Gebhardt, D. O. E., and Nieuwkoop, P. D. (1964). The influence of lithium on the competence of the ectoderm of Ambystoma mexicanum. *J. Embryol. Exp. Morph.* **12**, 317–331.

Gerhart, J., Danilchik, M., Doniach, T., Roberts, S., Rowning, B., and Stewart, R. (1989). Cortical rotation of the *Xenopus* egg: Consequences for the anteroposterior pattern of embryonic dorsal development. *Development* **107 Suppl.**, 37–51.

Gilbert, S. F., and Saxén, L. (1993). Spemann organizer—Models and molecules. *Mech. Develop.* **41**, 73–89.

Godsave, S. F., and Slack, J. M. W. (1989). Clonal analysis of mesoderm induction in *Xenopus* laevis. *Dev. Biol.* **134**, 486–490.

Goetschy, J. F., Letourneur, O., Cerletti, N., and Horisberger, M. A. (1996). The unglycosylated extracellular domain of type-II receptor for transforming growth factor-beta—A novel assay for characterizing ligand affinity and specificity. *Eur. J. Biochem.* **241**, 355–362.

Gont, L. K., Fainsod, A., Kim, S. H., and De Robertis, E. M. (1996). Overexpression of the homeobox gene Xnot-2 leads to notochord formation in *Xenopus*. *Dev. Biol.* **174**, 174–178.

Graff, J. M., Thies, R. S., Song, J. J., Celeste, A. J., and Melton, D. A. (1994). Studies with a *Xenopus* BMP receptor suggest that ventral mesoderm–inducing signals override dorsal signals in vivo. *Cell* **79**, 169–179.

Granato, M., and Nüsslein-Volhard, C. (1996). Fishing for genes controlling development. *Curr. Opin. Gen. Develop.* **6**, 461–468.

Green, J. B. A., and Smith, J. C. (1991). Growth factors as morphogens—Do gradients and thresholds establish body plan? *Trends Genetics* **7**, 245–250.

Green, J. B. A., Smith, J. C., and Gerhart, J. C. (1994). Slow emergence of a imultithreshold response to *activin* requires cell-contact-dependent sharpening but not prepattern. *Development* **120**, 2271–2278.

Grunz, H. (1968). Experimentelle untersuchungen über die kompetenzverhältnisse früher entwicklungsstadien des amphibien-ektoderms. *Roux's Arch. Dev. Biol.* **160**, 344–374.

Grunz, H. (1969). Hemmung der reaggregation dissoziierter amphibienzellen durch inhibitoren der RNS und Proteinsynthese. *Roux's Arch. Dev. Biol.* **163**, 184–196.

Grunz, H. (1977). The differentiation of the four animal and the four vegetal blastomeres of the eight-cell stage of *Triturus alpestris*. *Roux's Arch. Dev. Biol.* **181**, 267–277.

Grunz, H. (1979). Change of the differentiation pattern of amphibian ectoderm after the increase of the initial cell mass. *Roux's Arch. Dev. Biol.* **187**, 49–57.

Grunz, H. (1983). Change in the differentiation pattern of *Xenopus* laevis ectoderm by variation of the incubation time and concentration of vegetalizing factor. *Roux's Arch. Dev. Biol.* **192**, 130–137.

Grunz, H. (1990). Homoiogenetic neural inducing activity of the presumptive neural plate of *Xenopus*-laevis. *Develop. Growth Differen.* **32**, 583–589.

Grunz, H. (1992). *Suramin* changes the fate of Spemann's organizer and prevents neural induction in *Xenopus*-laevis. *Mech. Develop.* **38**, 133–142.

Grunz, H. (1993a). The dorsalization of Spemann's organizer takes place during gastrulation in *Xenopus* laevis embryos. *Develop. Growth Differ.* **35**, 21–28.

Grunz, H. (1993b). Factors controlling the determination of the body plan in the amphibian embryo. *Acta Biologiae Experimentalis Sinica* **26**, 317–341.

Grunz, H. (1994). The four animal blastomeres of the eight-cell stage of *Xenopus* laevis are intrinsically capable of differentiating into dorsal mesodermal derivatives. *Int. J. Devel. Biol.* **38**, 69–76.

Grunz, H. (1996). Factors responsible for the establishment of the body plan in the amphibian embryo. *Int. J. Devel. Biol.* **40**, 279–289.

Grunz, H., and Staubach, J. (1979). Cell contacts between chorda-mesoderm and the overlaying neuroectoderm (presumptive central nervous system) during the period of primary embryonic induction in amphibians. *Cell Diff.* **14**, 59–65.

Grunz, H., and Tacke, L. (1986). The inducing capacity of the presumptive endoderm of *Xenopus* laevis studied by transfilter experiments. *Roux's Arch. Dev. Biol.* **195**, 467–473.

Grunz, H., and Tacke, L. (1989). Neural differentiation of *Xenopus* laevis ectoderm takes place after disaggregation and delayed reaggregation without inducer. *Cell Differen. Develop.* **28**, 211–218.

Grunz, H., and Tacke, L. (1990). Extracellular matrix components prevent neural differentiation of disaggregated *Xenopus* ectoderm cells. *Cell Differen. Develop.* **32**, 117–124.

Grunz, H., and Tiedemann, H. (1977). Influence of cyclic nucleotides on amphibian ectoderm. *Roux's Arch. Dev. Biol.* **181**, 261–265.

Grunz, H., Born, J., Tiedemann, H., and Tiedemann, H. (1986). The activation of a neuralizing factor in the neural plate is correlated with its homoiogenetic-inducing activity. *Roux's Arch. Dev. Biol.* **195**, 464–466.

Grunz, H., McKeehan, W. L., Knöchel, W., Born, J., Tiedemann, H., and Tiedemann, H. (1987). Induction of mesodermal tissue by acidic and basic heparin binding growth factors. *Cell Diff.* **22**, 183–190.

Grunz, H., Schüren, C., and Richter, K. (1995). The role of vertical and planar signals during the early steps of neural induction. *Int. J. Dev. Biol.* **39**, 539–543.

Gualandris, L., and Duprat, A. M. (1981). A rapid experimental method to study primary induction. *Differentiation* **20**, 270–273.

Gumbiner, B. M. (1995). Signal-transduction by β-catenin. *Curr. Opin. Cell. Biol.* **7**, 634–640.

Gurdon, J. B. (1988). A community effect in animal development. *Nature* **336**, 772–774.

Gurdon, J. B. (1992). The generation of diversity and pattern in animal development. *Cell* **68**, 185–199.

Gurdon, J. B., Harger, P., Mitchell, A., and Lemaire, P. (1994). Activin signalling and response to a morphogen gradient. *Nature* **371**, 487–492.

Gurdon, J. B., Mitchell, A., and Ryan, K. (1996). An experimental system for analyzing response to a morphogen gradient. *Proc. Nat. Acad. Sci. USA.* **93**, 9334–9338.

Haffter, P., and Nüsslein-Volhard, C. (1996). Large-scale genetics in a small vertebrate, the zebrafish. *Int. J. Dev. Biol.* **40**, 221–227.

Hamburger, V. (1988). "The Heritage of Experimental Embryology: Hans Spemann and the Organizer." Oxford University Press, New York.

Harger, P. L., and Gurdon, J. B. (1996). Mesoderm induction and morphogen gradients. *Sem. Cell Dev. Biol.* **7**, 87–93.

He, X., Saint-Jeannet, J. P., Woodgett, J. R., Varmus, H. E., and Dawid, I. B. (1995). Glycogen synthase kinase-3 and dorsoventral patterning in *Xenopus* embryos. *Nature* **374**, 617–622.

Hemmati-Brivanlou, A., and Melton, D. A. (1992). A truncated activin receptor inhibits mesoderm induction and formation of axial structures in *Xenopus* embryos. *Nature* **359**, 609–614.

Hemmati-Brivanlou, A., and Melton, D. A. (1994). Inhibition of activin receptor signaling promotes neuralization in *Xenopus*. *Cell* **77**, 273–281.

Hemmati-Brivanlou, A., Kelly, O. G., and Melton, D. A. (1994). Follistatin, an antagonist of activin, is expressed in the Spemann organizer and displays direct neuralizing activity. *Cell* **77**, 283–295.

Hogan, B. L. M. (1996). Bone morphogenetic proteins in development. *Curr. Opin. Genet. Develop.* **6**, 432–438.

Hollemann, T., Bellefroid, E., Stick, R., and Pieler, T. (1996). Zinc finger proteins in early *Xenopus* development. *Int. J. Dev. Biol.* **40**, 291–295.

Hollemann, T., Pieler, T. and Grunz, H. (1997). Vertical versus planar signals during early neural induction in *Xenopus* and *Triturus alpestris*. In preparation.

Holley, S. A., Jackson, P. D., Sasai, Y., Lu, B., De Robertis, E. M., Hoffmann, F. M., and Ferguson, E. L. (1995). A conserved system for dorsal–ventral patterning in insects and vertebrates involving sog and *chordin*. *Nature* **376**, 249–253.

Holtfreter, J. (1933a). Nachweis der induktionsfähigkeit abgetöteter keimteile. isolations—und transplantations—versuche. *Arch. Entw. Mech. Org.* **128**, 584–633.

Holtfreter, J. (1933b). Die totale exogastrulation, eine selbstablösung des ektoderms von entomesoderm. Entwicklung und funktionelles verhalten nervenloser organe. *Arch. Entw. Mech. Org.* **129**, 669–793.

Holtfreter, J. (1934). Der einfluss thermischer, mechanischer und chemischer eingriffe auf die induzierfähigkeit von triton-keimteilen. *Roux's Arch. Dev. Biol.* **132**, 225–306.

Holtfreter, J. (1947). Neural induction in explants which have passed through a sublethal cytolysis. *J. Exp. Zool.* **106**, 197–222.

Hoppler, S., Brown, J. D., and Moon, R. T. (1996). Expression of a dominant-negative *Wnt* blocks induction of MyoD in *Xenopus* embryos. *Genes Develop.* **10**, 2805–2817.

Hopwood, N. D., and Gurdon, J. B. (1990). Activation of muscle genes without myogenesis by ectopic expression of MyoD in frog embryo cells. *Nature* **347**, 197–200.

Houliston, E. (1994). Microtubule translocation and polymerization during cortical rotation in *Xenopus* eggs. *Development* **120**, 1213–1220.

Houliston, E., and Elinson, R. P. (1991). Evidence for the involvement of microtubules, ER, and kinesin in the cortical rotation of fertilized frog eggs. *J. Cell Biol.* **114**, 1017–1028.

Huber, O., Bierkamp, C., and Kemler, R. (1996a). Cadherins and catenins in development. *Curr. Opin. Cell Biol.* **8**, 685–691.

Huber, O., Korn, R., Mclaughlin, J., Ohsugi, M., Herrmann, B. G., and Kemler, R. (1996b). Nuclear localization of beta-catenin by interaction with transcription factor LEF-1. *Mech. Develop.* **59,** 3–10.

Itoh, K., Tang, T. L., Neel, B. G., and Sokol, S. Y. (1995). Specific modulation of ectodermal cell fates in *Xenopus* embryos by glycogen synthase kinase. *Development* **121,** 3979–3988.

Janeczek, J., John, M., Born, M., Tiedemann, H., and Tiedemann, H. (1984). Inducing activity of subcellular fractions from amphibian embryos. *Roux's Arch. Dev. Biol.* **193,** 1–12.

Janeczek, J., Born, J., Hoppe, P., and Tiedemann, H. (1992). Partial characterization of neural-inducing factors from *Xenopus*-gastrulae—Evidence for a larger protein complex containing the factor. *Roux's Arch. Dev. Biol.* **201,** 30–35.

John, M., Janeczek, J., Born, J., Hoppe, P., Tiedemann, H., and Tiedemann, H. (1983). Neural induction in amphibians. Transmission of a neuralizing factor. *Roux's Arch. Dev. Biol.* **192,** 45–47.

John, M., Born, J., Tiedemann, H., and Tiedemann, H. (1984). Activation of a neuralizing factor in amphibian ectoderm. *Roux's Arch. Dev. Biol.* **193,** 13–18.

Johnston, D. S., and Nüsslein-Volhard, C. (1992). The origin of pattern and polarity in the *Drosophila* embryo. *Cell* **68,** 201–219.

Jones, C. M., Lyons, K. M., Lapan, P. M., Wright, C. V. E., and Hogan, B. L. M. (1992). DVR-4 (bone morphogenetic protein-4) as a posterior-ventralizing factor in *Xenopus* mesoderm induction. *Development* **115,** 639–647.

Jones, C. M., Kuehn, M. R., Hogan, B. L. M., Smith, J. C., and Wright, C. V. E. (1995). Nodal-related signals induce axial mesoderm and dorsalize mesoderm during gastrulation. *Development* **121,** 3651–3662.

Jones, C. M., Dale, L., Hogan, B. L. M., Wright, C. V. E., and Smith, J. C. (1996). Bone morphogenetic protein 4 (*BMP-4*) exerts its ventralizing effects on embryonic mesoderm during gastrula stages in *Xenopus* development. *Development* **122,** 1545–1544.

Kageura, H., and Yamana, K. (1983). Pattern regulation in isolated halves and blastomeres of early *Xenopus laevis*. *J. Embryol. Exp. Morph.* **74,** 221–234.

Kageura, H., and Yamana, K. (1984). Pattern regulation in defect embryos of *Xenopus* laevis. *Dev. Biol.* **101,** 410-415.

Kao, K. R., and Elinson, R. P. (1988). The entire mesodermal mantle behaves as Spemann's organizer in dorsoanterior enhanced *Xenopus* laevis embryos. *Dev. Biol.* **127,** 64–77.

Kao, K. R., Masui, Y., and Elinson, R. P. (1986). Lithium-induced respecification of pattern in *Xenopus* laevis embryos. *Nature* **322,** 371–373.

Kato, K., and Gurdon, J. B. (1994). An inhibitory effect of *Xenopus* gastrula ectoderm on muscle cell differentiation and its role for dorsoventral patterning of mesoderm. *Dev. Biol.* **163,** 222–229.

Kemler, R. (1993). From cadherins to catenins: Cytoplasmic protein interactions and regulation of cell adhesion. *TIG* **9,** 317–321.

Kengaku, M., and Okamoto, H. (1993). Basic fibroblast growth factor induces differentiation of neural tube and neural crest lineages of cultured ectoderm cells from *Xenopus* gastrula. *Development* **119,** 1067–1078.

Kengaku, M., and Okamoto, H. (1995). bFGF as a possible morphogen for the anteroposterior axis of the central nervous system in *Xenopus*. *Development* **121,** 3121–3130.

Kessler, D. S., and Melton, D. A. (1995). Induction of dorsal mesoderm by soluble, mature *vg1* protein. *Development* **121,** 2155–2164.

King, M. L. (1996). Molecular basis for cytoplasmic localization. *Develop. Gen.* **19,** 183–189.

Klein, P. S., and Melton, D. A. (1996). A molecular mechanism for the effect of lithium on development. *Proc. Natl. Acad. Sci. USA.* **93,** 8455–8459.

Kloc, M., and Etkin, L. D. (1994). Delocalization of *vg1* mRNA from the vegetal cortex in *Xenopus* oocytes after destruction of Xlsirt RNA. *Science* **265,** 1101–1103.

Kloc, M., and Etkin, L. D. (1995). Two distinct pathways for the localization of RNAs at the vegetal cortex in *Xenopus* oocytes. *Development* **121,** 287–297.

Knöchel, S., Lef, J., Clement, J., Klocke, B., Hille, S., Köster, M., and Knöchel, W. (1992). *activin*-A-induced expression of a fork head related gene in posterior chordamesoderm (notochord) of *Xenopus* laevis embryos. *Mech. Devel.* **38**, 157–165.

Knöchel, W., Born, J., Hoppe, P., Loppnow-Blinde, B., Tiedemann, H., Tiedemann, H., McKeehan, W., and Grunz, H. (1987). Mesoderm inducing factors: Their possible relationship to heparin-binding growth factors and transforming growth factor-β. *Naturwissenschaften* **74**, 604–606.

Kobel, H. R., and Du Pascier, L. (1986). Genetics of polyploid *Xenopus. Trends Genet.* **2**, 310–315.

Köster, M., Plessow, S., Clement, J. H., Lorenz, A., Tiedemann, H., and Knöchel, W. (1991). Bone morphogenetic protein 4 (*BMP-4*), a member of the TGF-beta family, in early embryos of *Xenopus*-laevis—Analysis of mesoderm inducing activity. *Mech. Develop.* **33**, 191–200.

Kroll, K. L., and Amaya, E. (1996). Transgenic *Xenopus* embryos from sperm nuclear transplantations reveal FGF signaling requirements during gastrulation. *Development* **122**, 3173–3183.

Kühl, M., and Wedlich, D. (1996). *Xenopus* cadherins: Sorting out types and functions in embryogenesis. *Dev. Dynam.* **207**, 121–134.

Kuusi, T. (1951a). Über die chemische natur der induktionsstoffe mit besonderer berücksichtigung der rolle der proteine und der nuklein-säuren. Diss., Helsinki. *Ann. Soc. Zool.-Bot. Fenn. Vanarno*, **14**(4), 1–98.

Kuusi, T. (1951b). Über die chemische natur der induktionsstoffe im implantatversuch bei triton. *Experientia* **7**, 299–300.

Ladher, R., Mohun, T. J., Smith, J. C., and Snape, A. M. (1996). *Xom*: A *Xenopus* homeobox gene that mediates the early effects of *BMP-4. Development* **122**, 2385–2394.

Larabell, C. A., Torres, M., Rowning, B. A., Yost, C., Miller, J. R., Wu, M., Kimelman, D., and Moan, R. T. (1997). Establishment of the dorso–ventral axis in *Xenopus* embryos is presaged by early asymmetries in beta-catenin that are modulated by the Wnt signaling pathway. *J. Cell. Biol.* **136**, 1123–1136.

Laudet, V., Stehelin, D., and Clevers, H. (1993). Ancestry and diversity of the HMG box superfamily. *Nucleic Acids Res.* **21**, 2493–2501.

Lemaire, P., Garrett, N., and Gurdon, J. B. (1995). Expression cloning of Siamois, a *Xenopus* homeobox gene expressed in dorsal-vegetal cells of blastulae is able to induce a complete secondary axis. *Cell* **81**, 85–94.

Li, S. H., Mao, Z. R., Yan, S. Y., and Grunz, H. (1996). Isolated dorsal animal blastomeres of *Xenopus* laevis are capable to form mesodermal derivatives, while the ventral animal blastomeres differentiate into ciliated epidermis only. *Zool. Sci.* **13**, 125–131.

London, C., Akers, R., and Phillips, C. (1988). Expression of Epi 1, an epidermis-specific marker in *Xenopus* laevis embryos, is specified prior to gastrulation. *Dev. Biol.* **129**, 380–389.

Lopashov, G. V., Selter, H., Montenarh, M., Knöchel, W., Grunz, H., Tiedemann, H., and Tiedemann, H. (1992). Neural inducing factors in neuroblastoma and retinoblastoma cell lines—Extraction with acid ethanol. *Naturwissenschaften* **79**, 365–367.

Lustig, K. D., Kroll, K., Sun, E., Ramos, R., Elmendorf, H., and Kirschner, M. W. (1996). *Xenopus* nodal-related gene that acts in synergy with *noggin* to induce complete secondary axis and notochord formation. *Development* **122**, 3275–3282.

Maéno, M., Ong, R. C., and Kung, H. F. (1992). Positive and negative regulation of the differentiation of ventral mesoderm for erythrocytes in *Xenopus*-laevis. *Develop. Growth Diff.* **34**, 567–577.

Maéno, M., Ong, R. C., Suzuki, A., Ueno, N., and Kung, H. F. (1994a). A truncated bone morphogenetic protein 4 receptor alters the fate of ventral mesoderm to dorsal mesoderm: Roles of animal pole tissue in the development of ventral mesoderm. *Proc. Acad. Sci. USA* **91**, 10260–10264.

Maéno, M., Ong, R. C., Xue, Y., Nishimatsu, S., Ueno, N., and Kung, H. F. (1994b). Regulation of primary erythropoiesis in the ventral mesoderm of *Xenopus* gastrula embryo—Evidence for the expression of a stimulatory factor(s) in animal pole tissue. *Dev. Biol.* **161**, 522–529.

Mangold, O. (1933). Über die induktionsfähigkeit der verschiedenen bezirke der neurula von urodelen. *Naturwissenschaften* **21**, 761–766.

Mangold, O., and Spemann, H. (1927). Über die induktion von medullarplatte durch medullarplatte beim jüngeren keim, ein beispiel homoiogenetischer oder assimilatorischer induktion. *Roux's Arch. Entw. Mech. Org.* **109**, 599–638.

Massagué, J., and Weis-Garcia, F. (1996). Serine/threonine kinase receptors: Mediators of transforming growth factor beta family signals. *Cancer Surv.* **27**, 41–64.

Masui, Y. (1961). Mesodermal and endodermal differentiation of the presumptive ectoderm of *Triturus* gastrula through influence of lithium ion. *Experientia (Basel)* **17**, 458–459.

Mikhailov, A. T., and Gorgolyuk, N. A. (1987). Biochemistry of embryonic induction: Identification and characterization of morphogenetic factors. *Sov. Sci. Rev. F. Physiol. Gen. Biol.* **1**, 267–306.

Mikhailov, A. T., and Gorgolyuk, N. A. (1989). Embryonic brain derived neuralizing factor. *Cell Diff. Dev.* **27 (Suppl.)**, 70–80.

Mikhailov, A. T., Gorgolyuk, N. A., Tacke, L., Mykhoyan, M. M., and Grunz, H. (1995). Partially purified factor from embryonic chick brain can provoke neuralization of Rana temporaria and Trirurus *alpestris* but not *Xenopus* laevis early gastrual ectoderm. *Int. J. Dev. Biol.* **39**, 317–325.

Miller, J. R., and Moon, R. T. (1996). Signal transduction through beta-catenin and specification of cell fate during embryogenesis. *Genes Develop.* **10**, 2527–2539.

Minuth, M., and Grunz, H. (1980). The formation of mesodermal derivates after induction with vegetalizing factor depends on secondary cell interactions. *Cell Diff.* **3**, 229–238.

Molenaar, M., Van de Wetering, M., Oosterwegel, M., Peterson-Maduro, J., Godsave, S., Korinek, V., Roose, J., Destree, O., and Clevers, H. (1996). Xtcf-3 transcription factor mediates β-catenin-induced axis formation in *Xenopus* embryos. *Cell* **86**, 391–399.

Moos, M., Wang, S. W., and Krinks, M. (1995). Anti-dorsalizing morphogenetic protein is a novel TGF-beta homolog expressed in the Spemann organizer. *Development* **121**, 4293–4301.

Moreau, M., Leclerc, C., Gualandrisparisot, L., and Duprat, A. M. (1994). Increased internal Ca^{2+} mediates neural induction in the amphibian embryo. *Proc. Nat. Acad. Sci. USA* **91**, 12639–12643.

Nakamura, O., and Toivonen, S., Eds. (1978). "Organizer. A milestone of a half-century from Spemann." Elsevier/North-Holland Biomedical Press, New York.

Nakamura, T., Takio, K., Eto, Y., Hiroshiro, S., Titani, K., and Sugino, H. (1990). Activin-binding protein from rat ovary is *follistatin*. *Science* **247**, 836–838.

Nascone, N., and Mercola, M. (1995). An inductive role for the endoderm in *Xenopus* cardiogenesis. *Development* **121**, 515–523.

Needham, J., Waddington, C. H., and Needham, D. M. (1934). Physico-chemical experiments on the amphibian organizer. *Proc. R. Soc. Lond., Ser. B.* **114**, 393–422.

Nieuwkoop, P. D. (1969). The formation of the mesoderm in urodelean amphibians. I. Induction by the endoderm. *Roux's Arch. Dev. Biol.* **162**, 341–373.

Nieuwkoop, P. D., and Faber, J. (1956). Normal table of *Xenopus* laevis (Daudin). North Holland, Amsterdam.

Nieuwkoop, P. D., and Koster, K. (1995). Vertical versus planar induction in amphibian early development. *Develop. Growth Diff.* **37**, 653–668.

Nieuwkoop, P. D., Boterenbrood, E. C., Kremer, A., Bloemsma, F. F. S. N., Hoessels, E. L. M. J., Meyer, G., and Verheyen, F. J. (1952). Activation and organization of the central nervous system in amphibians. I. Induction and activation. II. Differentiation and organization. III. Synthesis of a new working hypothesis. *J. Exp. Zool.* **120**, 1–108.

Nieuwkoop, P. D., Johnen, A., and Albers, B. (1985). "The Epigenetic Nature of Early Chordate Development. Inductive Interaction and Competence." Cambridge University Press, New York.

Niu, M. C., and Twitty, V. C. (1953). The differentiation of gastrula ectoderm in medium conditioned by axial mesoderm. *Proc. Nat. Acad. Sci. USA* **39**, 985–989.

Nüsslein-Volhard, C. (1994). Of flies and fishes. *Science* **266**, 572–574.

Nüsslein-Volhard, C. (1996). Gradients that organize embryo development. *Scientific American* **275**, 54.

Nüsslein-Volhard, C., Frohnhöfer, H. G., and Lehmann, R. (1987). Determination of antero-posterior polarity in *Drosophila*. *Science* **238**, 1675–1681.

Okada, T. S. (1954). Experimental studies on the differentiation of endodermal organs in amphibians. II. Differentiation potencies of the presumptive endoderm in the presence of the mesodermal tissue. *Mem. Coll. Sci. Univ. Kyoto Ser. B.* **21**, 7–14.

Onichtchouk, D., Gawantka, V., Dosch, R., Delius, H., Hirschfeld, K., Blumenstock, C., and Niehrs, C. (1996). The *Xvent-2* homeobox gene is part of the *BMP-4* signalling pathway controlling dorsoventral patterning of *Xenopus* mesoderm. *Development* **122**, 3045–3053.

Oschwald, R., Clement, J. H., Knöchel, W., and Grunz, H. (1993). *suramin* prevents transcription of dorsal marker genes in *Xenopus-laevis* embryos, isolated dorsal blastopore lips and activin-A induced animal caps. *Mech. Dev.* **43**, 121–133.

Otte, A. P., Koster, C. H., Snoek, G. T., and Durston, A. J. (1988). Protein kinase C mediates neural induction in *Xenopus* laevis. *Nature* **334**, 618–620.

Otte, A. P., RunVan, P., Heideveld, M., vanDriel, R., and Durston, A. J. (1989). Neural induction is mediated by cross-talk between the protein kinase C and cyclic AMP pathways. *Cell* **58**, 641–648.

Otte, A. P., Kramer, I. M., and Durston, A. J. (1991). Protein kinase C and regulation of the local competence of *Xenopus* Ectoderm. *Science* **251**, 570–573.

Pannese, M., Polo, C., Andreazzoli, M., Vignali, R., Kablar, B., Barsacchi, G., and Boncinelli, E. (1995). The *Xenopus* homologue of Otx2 is a maternal homeobox gene that demarcates and specifies anterior body regions. *Development* **121**, 707–720.

Papalopulu, N., and Kintner, C. (1996). A *Xenopus* gene, Xbr-1, defines a novel class of homeobox genes and is expressed in the dorsal ciliary margin of the eye. *Dev. Biol.* **174**, 104–114.

Penzel, R., Oschwald, R., Chen, R., Tacke, L., and Grunz, H. (1997). Characterization and early embryonic expression of a neural specific transmission factor xSox3 in *Xenopus*. Submitted.

Piccolo, S., Sasai, Y., Lu, B., and De Robertis, E. (1996). Dorsoventral patterning in *Xenopus*: Inhibition of ventral signals by direkt binding of *chordin* to *BMP-4*. *Cell* **86**, 1–20.

Pierandrei-Amaldi, P., and Amaldi, F. (1994). Aspects of regulation of ribosomal protein synthesis in *Xenopus* laevis—Review. *Genetica* **94**, 181–193.

Reilly, K. M., and Melton, D. A. (1996). Short-range signaling by candidate morphogens of the TGF beta family and evidence for a relay mechanism of induction. *Cell* **86**, 743–754.

Roux, W. (1885) Beiträge zur entwicklungsmechanik des Embryo, III. Über die bestimmung der hauptrichtungen des froschembryo im ei und über die erste teilung des froscheies. *Breslau. ärztl. Z., Ges. Abh.* II, 277–343.

Roux, W. (1887) Beiträge zur entwicklungsmechanik des embryo, IV. Die bestimmung der medianebene des froschembryo durch die copulationsrichtung des ei-kernes und des spermakernes. *Arch. mikrosk. Anat.* **29**, 344–418.

Ruiz i Altaba, A. (1992). Planar and vertical signals in the induction and patterning of the *Xenopus* nervous system. *Development* **115**, 67–80.

Ruiz i Altaba, A., and Jessell, T. M. (1992). Pintallavis, a gene expressed in the organizer and midline cells of frog embryos—Involvement in the development of the neural axis. *Development* **116**, 81–93.

Saha, M. S., and Grainger, R. M. (1992). A labile period in the determination of the anterior–posterior axis during early neural development in *Xenopus*. *Neuron* **8**, 1003–1014.

Sargent, T. D., Jamrich, M., and Dawid, I. B. (1986). Cell interactions and the control of gene activity during early development of *Xenopus* laevis. *Dev. Biol.* **114**, 238–246.

Sasai, Y., Lu, B., Steinbeisser, H., Geissert, D., Gont, L. K., and De Robertis, E. M. (1994). *Xenopus chordin*: A novel dorsalizing factor activated by organizer-specific homeobox genes. *Cell* **79**, 779–790.

Sasai, Y., Lu, B., Steinbeisser, H., and De Robertis, E. M. (1995). Regulation of neural induction by the Chd and Bmp-4 antagonistic patterning signals in *Xenopus. Nature* **376,** 333–336.

Sater, A. K., and Jacobson, A. G. (1990). The role of the dorsal lip in the induction of heart mesoderm in *Xenopus* laevis. *Development* **108,** 461–470.

Sato, S. M., and Sargent, T. D. (1991). Localized and inducible expression of *Xenopus*-posterior (Xpo), a novel gene active in early frog embryos, encoding a protein with a CCHC finger domain. *Development* **112,** 747–753.

Saxén, L. (1961). Transfilter neural induction of amphibian ectoderm. *Develop. Biol.* **3,** 140–152.

Saxén, L. (1989). Neural induction. *Int. J. Dev. Biol.* **33,** 21–48.

Saxén, L., and Toivonen, S. (1961). The two-gradient hypothesis in primary induction. The combined effect of two types of inductors mixed in different ratios. *J. Embryol. Exp. Morph.* **9,** 514–533.

Saxén, L., and Toivonen, S. (1962). Primary embryonic induction. *In* (D. R. Newth, H. Abercrombie, L. Brent, and J. Mynard Smith, Eds.), Logos Press/Academic Press, London.

Saxén, L., Toivonen, S., and Vainio, T. (1964). Initial stimulus and subsequent interactions in embryonic induction. *J. Embryol. Exp. Morph.* **12,** 333–338.

Scharf, S. R., and Gerhart, J. C. (1980). Determination of the dorsal–ventral axis in eggs of *Xenopus* laevis: Complete rescue of UV-impaired eggs by oblique orientation before first cleavage. *Dev. Biol.* **79,** 181–198.

Schmidt, J. E., Suzuki, A., Ueno, N., and Kimelman, D. (1995). Localized *BMP-4* mediates dorsal–ventral patterning in the early *Xenopus* embryo. *Dev. Biol.* **169,** 37–50.

Schmidt, J. E., von Dassow, G., and Kimelman, D. (1996). Regulation of dorsal–ventral patterning: The ventralizing effects of the novel *Xenopus* homeobox gene. *Vox. Development* **122,** 1711–1721.

Schneider, S., Steinbeisser, H., Warga, R. M., and Hausen, P. (1996). Beta-catenin translocation into nuclei demarcates the dorsalizing centers in frog and fish embryos. *Mech. Develop.* **57,** 191–198.

Sive, H. L. (1993). The frog prince-ss—A molecular formula for dorsoventral patterning in *Xenopus. Genes Develop.* **7,** 1–12.

Sive, H. L., and Bradley, L. (1996). A sticky problem: The *Xenopus* cement gland as a paradigm for anteroposterior patterning. *Dev. Dynam.* **205,** 265–280.

Slack, J. M. W., Darlington, B. G., Heath, J. K., and Godsave, S. F. (1987). Mesoderm induction in early *Xenopus* embryos by heparin-binding growth factors. *Nature* **326,** 197–200.

Smith, J. C. (1987). A mesoderm-inducing factor is produced by a *Xenopus* cell line. *Devel.* **99,** 3–14.

Smith, J. C., Price, B. M. J., Van Nimmen, K., and Huylebroeck, D. (1990). Identification of a potent *Xenopus* mesoderm-inducing factor as a homologue of *activin*-A. *Nature* **345,** 729–731.

Smith, W. C., and Harland, R. M. (1991). Injected Xwnt-8 RNA acts early in *Xenopus* embryos to promote formation of a vegetal dorsalizing center. *Cell* **67,** 753–765.

Smith, W. C., and Harland, R. M. (1992). Expression cloning of *noggin,* a new dorsalizing factor localized to the Spemann organizer in *Xenopus* embryos. *Cell* **70,** 829–840.

Smith, W. C., Mckendry, R., Ribisi, S., and Harland, R. M. (1995). A nodal-related gene defines a physical and functional domain within the Spemann organizer. *Cell* **82,** 37–46.

Sokol, S., and Melton, D. A. (1991). Pre-existent pattern in *Xenopus* animal pole cells revealed by induction with activin. *Nature* **351,** 409–411.

Sokol, S., Christian, L., Moon, R. T., and Melton, D. A. (1991). Injected *Wnt* RNA induces a complete body axis in *Xenopus* embryos. *Cell* **67,** 741–752.

Spemann, H., and Mangold, H. (1924). Über induktion von embryonalanlagen durch implantation artfremder organisatoren. *Roux's Arch. Dev. Biol.* **100,** 599–638.

Steinbeisser, H., Fainsod, A., Niehrs, C., Sasai, Y., and Derobertis, E. M. (1995). The role of gsc and *BMP-4* in dorsal–ventral patterning of the marginal zone in Xenopus: A loss-of-function study using antisense RNA. *EMBO J.* **14,** 5230–5243.

Stern, C. D., and Ireland, G. W. (1993). HGF-SF—a neural inducing molecule in vertebrate embryos, hepatocyte growth factor—scatter factor (Hgf-Sf), and the C-Met receptor. *Experientia Supplementa* **65,** 369–380.

Suzuki, A., Thies, R. S., Yamaji, N., Song, J. J., Wozney, J. M., Murakami, K., and Ueno, N. (1994). A truncated bone morphogenetic protein receptor affects dorsal–ventral patterning in early *Xenopus* embryo. *Proc. Acad. Sci. USA* **91,** 10255–10259.

Tacke, L., and Grunz, H. (1988). Close juxtaposition between inducing chordamesoderm and reacting neuroectoderm is a prerequisite for neural induction in *Xenopus* laevis. *Cell. Diff.* **24,** 33–44.

Taira, M., Jamrich, M., Good, P. J., and Dawid, I. B. (1992). The LIM domain-containing homeobox gene Xlim-1 is expressed specifically in the organizer region of *Xenopus* gastrula embryos. *Genes Develop.* **6,** 356–366.

Takeichi, M. (1995). Morphogenetic roles of classical cadherins. *Curr. Opin. Cell Biol.* **7,** 619–627.

Thomsen, G. H., and Melton, D. A. (1993). Processed *vg1* protein is an axial mesoderm inducer in *Xenopus. Cell* **74,** 433–441.

Tiedemann, H. (1993). Mesoderm differentiation in early amphibian embryos depends on the animal cap. *Roux's Arch. Dev. Biol.* **203,** 28–33.

Tiedemann, H., and Tiedemann, H. (1956). Versuche zur chemischen kennzeichnung von embryonalen induktionsstoffen. *Z. Physiol. Chem.* **306,** 7–32.

Tiedemann, H., and Tiedemann, H. (1964). Das induktionsvermögen geeigneter induktionsfaktoren im kombinationsversuch. *Rev. Suisse Zool.* **71,** 117–137.

Tiedemann, H., Grunz, H., McKeehan, W. L., Knöchel, W., Born, J., Hoppe, P., Loppnow-Blinde, B., Tiedemann, H., and Volk, R. (1988). Mesoderm induction by heparin-binding growth factors and transforming growth factor-β. *J. Cell Biochem.* **Suppl. 12A,** 159.

Tiedemann, H., Tiedemann, H., Grunz, H., and Knöchel, W. (1995). Molecular mechanisms of tissue determination and pattern formation in amphibian embryos. *Naturwissenschaften* **82,** 123–134.

Tiedemann, H., Asashima, M., Born, J., Grunz, H., Knöchel, W., and Tiedemann, H. (1996). Determination, induction and pattern formation in early amphibian embryo. *Develop. Growth Diff.* **38,** 575.

Toivonen, S. (1938). Spezifische induktionsleistungen von abnormen induktoren im implantationsversuch. *Ann. Zool. Soc. Zool. Bot. Fenn. Vanamo* **6,** 1–8.

Toivonen, S., and Saxén, L. (1955). The simultaneous inducing action of liver and bone-marrow of the guinea-pig in implantation and explanation experiments with embryos of *Triturus. Exp. Cell Res (Suppl.)* **3,** 346–357.

Toivonen, S., and Wartiovaara, J. (1976). Mechanisms of cell interaction during primary embryonic induction studied in transfilter experiments. *Differentiation* **5,** 61–66.

Tonissen, K. F., Drysdale, T. A., Lints, T. J., Harvey, R. P., and Krieg, P. A. (1994). Xnkx-2.5, a *Xenopus* gene related to Nkx-2.5 and Tinman—Evidence for a conserved role in cardiac development. *Dev. Biol.* **162,** 325–328.

Townes, P. L., and Holtfreter, J. (1955). Directed movements and selective adhesion of embryonic amphibian cells. *J. Exp. Zool.* **128,** 53–120.

Ueno, N., Ling, N., Ying, S., Esch, F., Shimasaki, S., and Guillemin, R. (1987). Isolation and partial characterization of *follistatin*: A single-chain Mr 35,000 monomeric protein that inhibits the release of follicle-stimulating hormone. *Proc. Natl. Acad. Sci. USA* **84,** 8282–8286.

Vincent, J. P., and Gerhart, J. C. (1987). Subcortical rotation in *Xenopus* eggs: An early step in embryonic axis specification. *Dev. Biol.* **123,** 526–539.

Von Dassow, G., Schmidt, J. E., and Kimelman, D. (1993). Induction of the *Xenopus* organizer—

Expression and regulation of Xnot, a novel FGF and activin-regulated homeo box gene. *Genes Develop.* **7,** 355–366.

Vriz, S., Joly, C., Boulekbacke, H., and Condamine, H. (1996). Zygotic expression of the zebra-fish Sox-19, an HMG box-containing gene, suggests an involvement in central nervous system development. *Molec. Brain Res.* **40,** 221–228.

Weeks, D. L., and Melton, D. A. (1987). A maternal mRNA localized to the animal pole of *Xenopus* eggs encodes a subunit of mitochondrial ATPase. *Proc. Nat. Acad. Sci. USA* **84,** 2798–2802.

Wilson, P. A., and Hemmati-Brivanlous, A. (1995). Induction of epidermis and inhibition of neu-ral fate by *BMP-4. Nature* **376,** 331–333.

Wilson, P. A., and Melton, D. A. (1994). Mesodermal patterning by an inducer gradient depends on secondary cell–cell communication. *Curr. Biol.* **4,** 676–686.

Wright, E. M., Snopek, B., and Koopman, P. (1993). Seven new members of the Sox gene family expressed during mouse development. *Nucleic Acids Res.* **21,** 744.

Wylie, C., Kofron, M., Payne, C., Anderson, R., Hosobuchi, M., Joseph, E., and Heasman, J. (1996). Maternal beta-catenin establishes a "dorsal signal" in early *Xenopus* embryos. *Development* **122,** 2987–2996.

Xu, R. H., Kim, J. B., Taira, M., Zhan, S. I., Sredni, D., and Kung, H. F. (1995). A dominant negative bone morphogenetic protein 4 receptor causes neuralization in *Xenopus* ectoderm. *Biochem. Biophys. Res. Comm.* **212,** 212–219.

Yamada, T. (1947). An extension of the double potential theory in morphogenesis. *Zool. Mag.* **57,** 124–126.

Yamada, T. (1950). Regional differentiation of the isolated ectoderm of the *Triturus* gastrula in-duced through a protein extract. *Embryologia* **1,** 1–20.

Yamada, T. (1958). Induction of specific differentiation by samples of proteins and nucleoproteins in the isolated ectoderm of *Triturus*-gastrulae. *Experientia* **14,** 81–87.

Yang-Snyder, J., Miller, J. R., Brown, J. D., Lai, C. J., and Moon, T. T. (1996). A *frizzled* homo-log in a vertebrate *Wnt* signaling pathway. *Curr. Biol.* **6,** 1302–1306.

Yost, C., Torres, M., Miller, R. R., Huang, E., Kimelman, D., and Moon, R. T. (1996). The axis-inducing activity, stability, and subcellular distribution of beta-catenin is regulated in *Xenopus* embryos by glycogen synthase kinase 3. *Genes Develop.* **10,** 1443–1454.

Zaraisky, A. G., Ecochard, V., Kazanskaya, O. V., Lukyanov, S. A., Fesenko, I. V., and Duprat, A. M. (1995). The homeobox-containing gene XANF-1 may control development of the Spemann organizer. *Development* **121,** 3839–3847.

Zhou, Y., and King, M. L. (1996). RNA transport to the vegetal cortex of *Xenopus* oocytes. *Dev. Biol.* **179,** 173–183.

7

Paradigms to Study Signal Transduction Pathways in *Drosophila*

Lee Engstrom
Muncie Center for Medical Education
Indiana University School of Medicine
Ball State University
Muncie, Indiana 47306

Elizabeth Noll and Norbert Perrimon
Department of Genetics
Howard Hughes Medical Institute
Harvard Medical School
Boston, Mass. 02115

I. Introduction

Ever since early experimental embryologists demonstrated that organisms are formed by epigenetic mechanisms, not by growth of preformed animalcules, scientists have attempted to understand the complex processes transforming a

Current Topics in Developmental Biology, Vol. 35
229

single cell into the newborn offspring. During the past 20 years many of the mechanisms involved in this transformation have been revealed, beginning with the pioneering work of E. B. Lewis, C. Nusslein-Volhard, E. Wieschaus (Lewis, 1978; Nusslein-Volhard and Wieschaus, 1980), and other *Drosophila* and *Caenorhabditis* developmental geneticists and molecular biologists. It now appears that the understanding of these mechanisms, gleaned mostly from flies and worms, can, in many cases, be applied to most animals. The molecules, gradients, interactions, and movements involved in embryonic development flow in a continuum of increasing complexity of structure and function.

The genetic approach to gain insight into these developmental processes has proven very fruitful. The identification of molecular mechanisms responsible for embryonic pattern formation has led to a fairly extensive understanding of the steps involved in the creation of the larval body plan. The initial information for the assignment of positional information in the embryo is in the form of gradients of activities set up during oogenesis and early embryogenesis. Overlapping gradients organize groups of cells into subdivisions along the anterior–posterior and dorsal–ventral axes (see reviews by St. Johnston and Nusslein-Volhard, 1992; Struhl, 1989). Each subdivision then acquires a determined state, or genetic address, which consists of a combination of various genes being turned on or off. This combination of genetic activities tells the cells where they are and how to interact appropriately with cells in neighboring subdivisions. These interactions then initiate new gradients, which produce further subdivisions, and these hierarchical events ultimately define what each cell within each subdivision will become and how it will function.

In this review we have focused our discussion on the mechanisms by which groups of cells alter the developmental behavior of other cells during oogenesis and embryogenesis. First, we describe the processes of oogenesis and embryogenesis in *Drosophila melanogaster*, and second, we describe how various genetic analyses have led to the characterizations of genes involved in specific cell–cell interaction processes. Third, we review some of the molecular pathways used to build the egg and to pattern the embryo. These evolutionarily conserved pathways provide unique paradigms to apply genetic tools to identify novel components and characterize their functions. It now appears that many of the signaling pathways are used repeatedly during development and have been conserved during evolution. These pathways identify common routes for extracellular information to be integrated, processed and used by cells. Perhaps not surprisingly, given their instructive roles in development, abnormal expression of many of the components of these signaling pathways lead to oncogenesis in vertebrates.

II. *Drosophila* Oogenesis

Oogenesis has been comprehensively reviewed recently by both Spradling (1993) and Ray and Schupbach (1996). Briefly, *Drosophila* females possess two polytrophic, meroistic ovaries (Fig. 1), each consisting of 15 to 20 ovarioles producing a progression of eggs in an assembly line–like manner. Anteriorly, each ovariole has an elongated germarium containing germline and somatic stem cells. The germline-derived stem cells divide asymmetrically to form another stem cell and a cystoblast. The cystoblast in turn undergoes four successive divisions, forming 16 cystocytes connected by cytoplasmic bridges (ring canals) due to incomplete cytokinesis. The somatically derived stem cells divide to form a single layer surrounding the cystocytes and are referred to as *follicle cells*. Several interactive signals are exchanged between germline cells and follicle cells during the movement of cysts through the germarium. These signals control various processes, such as proliferation of follicle cells, their differentiation into epithelia surrounding the cyst versus stalk cells, nurse cell versus oocyte differentiation, oocyte position, and other facets of egg chamber formation. This egg chamber, or follicle, buds from the posterior end of the germarium and progresses posteriorly in the vitellarium, undergoing further development as it proceeds. All chambers are composed of three cell types: 15 germline-derived nurse cells interconnected via ring canals to a single, germline-derived, posteriorly located oocyte and many proliferating somatic follicle cells.

The development of egg chambers in the vitellarium has been divided morphologically into 14 continuous stages by King (1970). The first seven stages are previtellogenic and characterized by growth of the nurse cells and their polyploidization. The oocyte also grows during stages 1–7, but forms a germinal vesicle with disappearance of synaptonemal complexes, condensation of its chromatin into a karyosome, and formation of an "endobody" (Mahowald and Kambysellis, 1980). The follicle cells proliferate as a simple epithelium until there are about 1,000 cells at stage 5. Then they stop dividing and undergo a period of polyploidization.

The oocyte begins accumulating yolk at stage 8, continuing through stage 10. At stage 9 most follicle cells migrate posteriorly over the oocyte, leaving an almost squamous layer of follicle cells over the anterior nurse cells. The most posterior follicle cells assume a somewhat more columnar shape. Also at this stage, a small group of 6–10 anterior follicle cells carry out a dramatic migration. While maintaining contact with one another, they migrate between the nurse cells and come to rest on the anterior end of the oocyte. As we will discuss later, the different types of follicle cells play important roles in signaling positional information to the egg and embryo.

Although some cytoplasmic products from the nurse cells preferentially flow to the oocyte throughout previtellogenic stages, starting at stage 10 major degeneration of the nurse cells occurs, with concomitant "dumping" of their cytoplasmic contents into the oocyte. By stage 12, the nurse cells are reduced to a cluster of nuclear and other debris at the anterior end of the oocyte. During the same period, the follicle cell spreading continues until the oocyte is completely covered as the oocyte expands anteriorly. The nurse cell remnants eventually disintegrate and are resorbed.

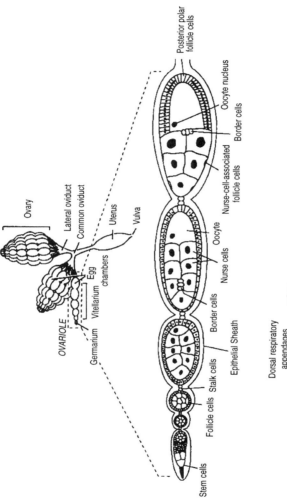

Fig. 1 *Drosophila* oogenesis. See text for description. Reproduced with permission from Duffy and Perrimon (1994).

Starting at stage 9, follicle cells begin secreting material that will form the vitelline membrane. As the follicle cells progressively cover the expanding oocyte, the vitelline membrane is completed and the layers of the eggshell are laid down over the vitelline membrane. The eggshell is formed by groups of follicle cells with specific movements, positions, surface features, and temporally controlled secretions. Specialized eggshell regions are produced in very tightly controlled spatial and temporal arrays. These regions include the micropyle, dorsal appendages, the operculum in the anterior, dorsal, and ventral portions of the main eggshell body and the posterior cap. The control of the activities of the follicle cells producing these regions is not understood, although most mutations that disrupt the dorsal–ventral and anterior–posterior axes of the embryo also disrupt the eggshell pattern. Thus, cell–cell signaling pathways are likely to control the activities of the follicle cell groups.

III. *Drosophila* Embryogenesis

Eggs become rigid, are fertilized, and complete meiosis as they pass through the uterus. Embryonic development begins immediately and is divided into 17 morphological stages (Fig. 2; see also Campos-Ortega and Hartenstein, 1985; Bate and Martinez-Arias, 1993). Briefly, the first three stages are occupied by nine rapid, quite synchronous nuclear cleavage divisions, migration of most of the nuclei to the peripheral cytoplasm, and formation of the polar buds at the posterior end. Stage 4 represents the syncytial blastoderm stage, with four more nuclear cleavage cycles, and cytokinesis of the pole cells at the posterior tip. The approximately 6,000 nuclei of the syncytio-blastoderm become cellularized during stage 5, followed by gastrulation movements and morphogenesis at stage 6. During these early stages the cascade of signaling pathways have set up distinct anterior–posterior patterns of positional information, so terminal regions (acron and telson), head, abdomen, and thorax can be distinguished by the presence of specifically expressed "address molecules." Dorsal and ventral differences can also be visualized at this time. Gastrulation and other morphogenetic movements continue during the remaining stages. These include: (1) the mesoderm forms via ventral furrow invagination, posterior midgut formation by the posterior plate (carrying the pole cells) dorsally, and cephalic furrow formation (Fig. 2, stage 6); (2) the posterior midgut invaginates, internalizing the pole cells, and the anterior midgut invagination appears (Fig. 2, stage 7); (3) the germ band extends around the posterior end along the dorsal side to about 65% of egg length, and the cephalic furrow mostly disappears (Fig. 2, stage 8); (4) germ-band extension continues slowly, the stomodeal invagination becomes visible, and segmentation may be visualized by the clusters of neuroblasts along the ventral ectoderm (Fig. 2, stage 9); (5) the germ band reaches maximum extension at 75% egg length, the stomodeal invagination forms and deepens, and segments become visible as

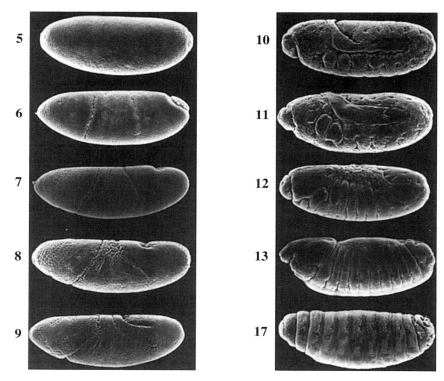

Fig. 2 *Drosophila* embryogenesis. See text for descriptions. These scanning electron micrographs were a gift of Drs. A. P. Mahowald and R. Turner. The various embryonic stages are indicated.

shallow indentations of the ectoderm (Fig. 2, stage 10); (6) the parasegmental furrows become clearly visible, gnathal segments become visible, and a space appears between the germ band and the vitelline membrane at the posterior tip, signaling the beginning of germ-band shortening (Fig. 2, stage 11); (7) the germ band continues to shorten, from about 75% egg length to about 20% egg length, with an increase in the segmentation (Fig. 2, stage 12); (8) germ-band retraction is completed, the clypeolabrum thins and begins retraction, and the labium moves to the ventral midline (Fig. 2, stage 13); (9) head involution and dorsal closure continue, (10) dorsal closure and segmentation is completed, and three constrictions of the gut appear; (11) a dorsal ridge grows anteriorly to cover the clypeolabrum forming the dorsal sac, and the ventral nerve cord retracts forward to about 40% of egg length; (12) elongation of the gut into a tube, nerve cord retraction, and muscle movements begin and continue until hatching (stage 17). From egg deposition to the hatching of first-instar larva, development occupies about 22 hours at 25°C.

IV. Genetic Approaches to the Study of Oogenesis and Embryogenesis

Over the past 20 years, large collections of mutations have been generated by mutagenizing flies and then exhaustively screening them for their effects on embryonic patterning. Three different screens that are described shortly in some detail have proven fruitful in identifying genes involved in patterning events: (1) screens for mutations that cause embryonic lethality, (2) screens for mutations associated with female sterility, and (3) germline clone mosaic screens of zygotic lethal mutations.

One strategy for identifying embryonic lethal mutations on the second chromosome is shown in Fig. 3. From these types of genetic screens, some of the genetic controls of embryonic patterning based on observations of larval cuticles produced by these embryonic lethal mutations were found (Nusslein-Volhard and Wieschaus, 1980; Jurgens *et al.*, 1984; Nusslein-Volhard *et al.*, 1984; Wieschaus *et al.*, 1984). Nearly all of those mutations affecting the anterior–posterior pattern could be classified into four types of segmental disruption: Gap mutations cause a deletion of one to several segments along the anterior–posterior axis; pair-rule mutations cause a deletion of alternating segments or parts of segments; segment-polarity mutations cause deletions of portions of all segments; and homeotic mutations (Lewis, 1978) cause a segment or segments to lose its/their identity and to be transformed into other segment(s).

Other mutagenesis screens have been designed and carried out to identify genes whose disruption produces female sterility (Gans *et al.*, 1975; Mohler, 1977; Komitopoulou *et al.*, 1983; Perrimon *et al.*, 1986; Nusslein-Volhard *et al.*, 1987; Schupbach and Wieschaus, 1986, 1989; Erderlyi and Szabad, 1989; Szabad *et al.*, 1989). As an example, a scheme for screening mutations for female sterility on the second chromosome is shown in Fig. 3. Screens for female-sterile mutations have identified mutations that disrupt some aspect of oogenesis: The mutation blocks egg production at some point so that no viable eggs are produced. In addition, they have isolated maternal-effect mutations where the mutation disrupts the production of an informational signal needed by the embryo after fertilization, or it interferes with the correct placement of such signals. There are two general types of maternal-effect mutations: strict maternal-effect mutations, whose embryonic phenotype is unaffected by the zygotic genotype; maternal-effect-rescuable mutations, whose embyronic phenotype is less severe or wild type if the egg is fertilized by a nonmutant sperm, so the embryo is heterozygous for the normal and mutant alleles at that locus.

The female-sterile screens have identified a small number of mutations that identify maternal functions involved in organizing patterns of the early embryo. These maternal systems have been classified into four phenotypic classes (Nusslein-Volhard *et al.*, 1987; St. Johnston and Nusslein-Volhard, 1992): The anterior group specifies the head and thorax as well as the anterior identity of

Treat males with a mutagen like ethylmethane sulfonate

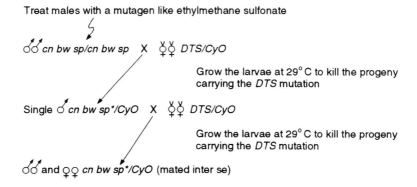

Fig. 3 Screen for embryonic-lethal and female-sterile mutations. In this mutagenesis scheme, lines heterozygous for a mutagenized chromosome are established and examined for embryonic lethality. If homozygous animals are recovered, females are tested for fertility. The mutagenized chromosome (*cn bw sp*) carries three visible markers: *cinnabar (cn), brown (bw)*, and *speck (sp)*. *DTS* is a dominant temperature-sensitive lethal mutation, and *CyO* is a second chromosome balancer. Mutagenized chromosome indicated by *.

the acron; the posterior group specifies the abdomen and the germline cells; the terminal group specifies the nonsegmented anterior acron and posterior telson; and the dorsoventral group specifies the pattern along that axis.

A survey of the female-sterile mutations has revealed that only a few (perhaps not more than 40) genes in *Drosophila* are used strictly during oogenesis (Perrimon *et al.*, 1986; Nusslein-Volhard *et al.*, 1987; Schupbach and Wieschaus, 1989). Perrimon *et al.* (1986) found that when only a single mutant allele, obtained from such screens, exhibits female sterility, this allele may not be an amorphic (i.e., null activity), and that amorphic alleles may mutate to zygotic lethality, reflecting their pleiotropic effects. This observation suggested that many genes that play a critical role in early developmental events may have been missed from the female-sterile screens. This prompted the analysis of the maternal effect of zygotic lethal mutations in germline mosaics (Perrimon *et al.*, 1984a,b, 1989, 1996). These studies were possible because of the development of the "dominant female sterile" (DFS) technique (Perrimon and Gans, 1983) and its subsequently improved version, the FLP-DFS technique (Chou and Perrimon, 1992, 1996; Fig. 4), which allows analysis of a large number of mutations in germline mosaics. The FLP-DFS technique takes advantage of FLP-recombinase to generate chromosomal, site-specific recombination (Golic, 1991) as well as a

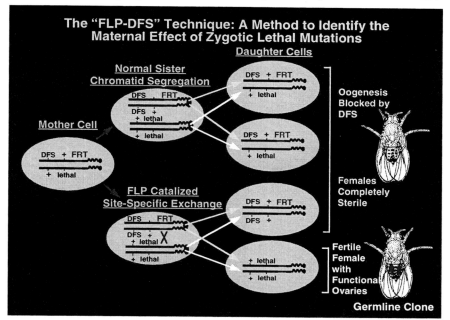

Fig. 4 The "FLP-DFS" technique. A chromosomal exchange that occurs in the euchromatin of a fly of genotype *DFS + FRT/ + lethal FRT; FLP/ +* is shown. The *FRT* insertion is located proximally to both *DFS* and *lethal*. *hsp70-FLP* from another chromosome site can provide recombinase activity following heat shock induction to catalyze site-specific chromosomal exchange at the position of the *FRT* sequences. *FLP*-catalyzed recombination results in the recovery of almost 100% of females with *lethal/lethal* germline clones (lowest branch). Adapted from Chou and Perrimon (1996). *Nomenclature*: Atrophic ovaries are shown as yellow empty ovals, and ovaries with developed ovarioles as red filled ovals. *FLP*-recombinase target sequences (*FRT*) are depicted as red boxes. Dominant female sterile (*DFS*), recessive zygotic lethal mutation (*lethal*).

dominant-female-sterile (DFS) mutation to easily detect female germline mosaics (Perrimon and Gans, 1983). An example of a screen to detect the maternal-effect phenotypes of zygotic lethal mutations on the second chromosome is shown in Fig. 5.

In examining X-linked lethal loci (Perrimon *et al.*, 1984a, 1989), a number of late zygotic lethal loci were identified that produce specific maternal-effect phenotypes. Molecular analyses of some of these genes revealed that some of them encode elements of the signaling pathways that interpret maternal or zygotic cues for pattern formation. For example, the *D-raf, corkscrew*, and *D-sor1* genes, which mutate to late zygotic lethality, encode components of the Torso receptor tyrosine kinase signaling pathway (Duffy and Perrimon, 1994). Similarly, *dishevelled, porcupine*, and *zeste-white 3* encode molecules involved in Wingless signaling. (These examples will be discussed more extensively later.)

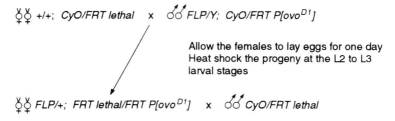

Allow the females to lay eggs for one day
Heat shock the progeny at the L2 to L3
larval stages

♀♀ *FLP/+; FRT lethal/FRT P[ovo^D1]*   x   ♂♂ *CyO/FRT lethal*

**Examine the germline clone phenotypes for oogenesis defects or maternal
effect phenotypes (strict or paternally rescuable).**

Fig. 5   Screen to analyze the maternal effect of zygotic lethal mutations. FRT is the target for the
FLP recombinase under control of a heat-shock promoter. *P[ovo]^D1* is the dominant female-sterile
mutation onto the right arm of chromosome 2 (see Chou et al., 1993; Chou and Perrimon, 1996). *CyO*
is a balancer (Lindsley and Zimm, 1992).

The results obtained by looking at lethal loci on the X chromosome, which
represents about 20% of the total *Drosophila* genome, prompted a search for
lethal loci with maternal effects on the autosomes. Large screens to identify
zygotic lethal mutations with specific phenotypes have been undertaken, and
many new autosomal loci involved in various aspects of oogenesis and embryogenesis have already been isolated (Perrimon *et al.*, 1996, and unpublished).

V. Paradigms to Dissect Components of Receptor Tyrosine Kinase Signaling Pathways

A. Determination of Embryonic Cell Fates at the Termini of the Embryo

The terminal, or Torso (Tor), pathway involves intercellular communication via a
receptor tyrosine kinase (RTK) to specify positional fates (Fig. 6; for review see

———————————————————————————————————→

Fig. 6   The terminal pathway: A paradigm to study an RTK pathway in *Drosophila*. At the termini
of the egg, localized Tor ligand in the perivitelline fluid activates the Tor RTK, which triggers a
phosphorylation cascade to determine the acron and telson regions. In terminal mutants, the signaling
pathway triggered by activated Tor does not operate, and the embryos develop with deleted terminal
structures. Anteriorly, the missing cuticular structures encompass part of the head skeleton. Posteriorly, all structures posterior to the seventh abdominal segment are deleted. Constitutive activation of
Tor leads to an expansion of terminal regions toward the center of the embryo. These embryos
develop enlarged head and tail regions as well as repressed abdominal segmentation. The expression
of *tll* in wild-type, *tor* loss-of-function, and *tor* gain-of-function allele embryos is shown. The
expression of *tll* mRNA is visualized by *in situ* hybridization to cellular blastoderm embryos. All
embryos are oriented with the anterior to the left and dorsal up. See text for details of the Tor pathway
shown on the rightmost part of the figure.

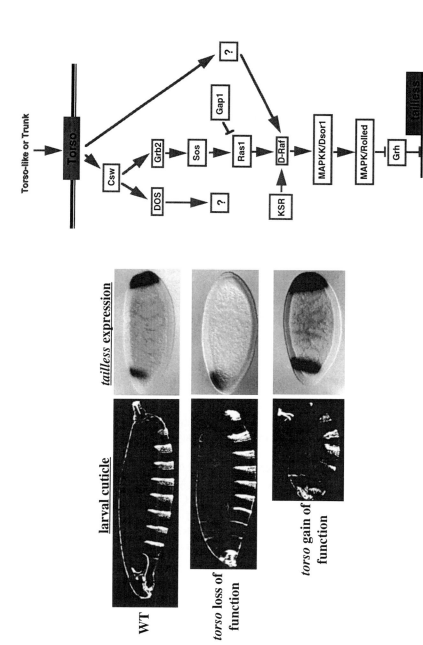

Duffy and Perrimon, 1994). *tor* encodes an RTK, which is transcribed during oogenesis and expressed uniformly at the preblastoderm stage (Casanova and Struhl, 1989). It is thought that Tor becomes activated only at the embryonic termini because its ligand, encoded by either *torso-like* or *trunk*, is localized in the perivitelline space at the egg poles (Casanova and Struhl, 1993; Sprenger and Nusslein-Volhard, 1993). Following its activation, Tor RTK molecules direct the formation of specialized anterior and posterior terminal structures, termed the *acron* and *telson*, respectively (reviews by St. Johnston and Nusslein-Volhard, 1992; Duffy and Perrimon, 1994). Loss-of-function mutations in components of the terminal systems result in a reduction or an elimination of terminal structures; i.e., the anterior head skeleton is collapsed and all structures posterior to abdominal segment 7 are deleted. Conversely, gain-of-function mutations in the terminal pathway lead to an expansion of the terminal regions, at the expense of the central segmented regions.

Two of the primary targets of the Tor RTK are the zygotic genes *tailless (tll)* and *huckebein (hkb)*, which both encode transcription factors. *tll* is a member of the nuclear receptor superfamily (Pignoni *et al.*, 1990), and *hkb* encodes an Sp1/egr-like zinc finger protein (Bronner *et al.*, 1994). Both genes are expressed in two distinct domains at the embryonic termini (Pignoni *et al.*, 1990; Weigel *et al.*, 1990). Consistent with the model that *tll* and *hkb* encode zygotic effectors of Tor, they are not expressed posteriorly in loss-of-function *tor* alleles but are ectopically expressed in gain-of-function alleles. Furthermore, the phenotype of embryos derived from Tor gain-of-function mutant mothers (embryos that show enlarged terminal domains) is partially suppressed by the removal of zygotic *tll* function, indicating that *tll* is a downstream target of Tor signaling (Klingler *et al.*, 1988).

Through genetic analyses of the Tor signaling pathway, a number of genes have been identified that are required for signal transmission (Fig. 6). Signaling from the activated Tor receptors involves an evolutionarily conserved cassette of signaling molecules that include Drk (a Grb2/Sem-5 homologue), Sos (Son of sevenless), Ras1, Gap1, D-Raf, Ksr, Dsor1 (MEK), Rolled (MAPK), and the SH2-domain-containing tyrosine phosphatase Corkscrew (Csw) (review by Duffy and Perrimon, 1994; Hou *et al.*, 1995; Therrien *et al.*, 1995). Activation of the Sos guanine-nucleotide releasing factors (GRFs) expedite the GDP/GTP exchange on p21^{ras}/Ras GTPase. Ras in turn activates the D-raf Ser/Thr kinase, which itself activates the Dsor1 Thr/Tyr kinase. This kinase cascade ends with activation of a mitogen-activated protein kinase (MAPK) encoded by the *rolled* gene. The current model is that Rolled regulates transcription in part by blocking the activity of the transcription factor Grainy head (Grh) that acts as a transcriptional repressor (Liaw *et al.*, 1995). This system is very similar to the signaling pathway activated by mammalian RTKs (see review by Perrimon, 1994a) and thus provides a genetic paradigm to dissect the subtleties of RTK signaling and identify novel components of this pathway.

B. Establishment of Dorsal Follicle Cell Fates During Oogenesis

Studies of the establishment of dorsal–ventral (D/V) cell fates during oogenesis have provided an additional paradigm to identify components involved in RTK signaling. During midoogenesis, the oocyte nucleus and *gurken* (*grk*) mRNA and protein are localized to the anterior dorsalmost region of the oocyte (see reviews by Stein, 1995; Ray and Schupbach, 1996; Fig. 7A). The Grk protein is related to transforming growth factor-α (TGFα) and is presumably the ligand for DER, the *Drosophila* epithelial growth factor receptor, originally isolated as the *torpedo* gene, which is expressed in follicle cells. Localized activation of DER by Grk in dorsal follicle cells regulates, as previously described for Tor, the activity of the Ras1, D-Raf, Dsor1, and Rolled signaling cascade (Brand and Perrimon, 1994).

Several maternally expressed genes are involved in localizing the mRNA from the *grk* gene to the cortical vicinity of the anterior-dorsal oocyte nucleus (Fig. 7B—see color insert). Two of these, *fs(1)K10* and *squid*, resemble nuclear proteins found in other species and are confined to the nucleus after it arrives at its anterior-dorsal location (see review by Ray and Schupbach, 1996). The significance of the nuclear location of these proteins and their interaction with *grk* mRNAs has yet to be elucidated. Females homozygous for mutations of either of these genes lay dorsalized eggs, and fertilized eggs develop into dorsalized embryos. Three other genes are required for this RNA localization: *cappuccino* (*capu*), *orb*, and *spire*. Lantz *et al.* (1994) and Christerson and McKearin (1994) have shown that the Orb protein contains a motif found in members of a family of single-stranded nucleic acid-binding proteins. *capu* encodes a protein that resembles vertebrate formins (Emmons *et al.*, 1995), but its biochemical function has yet to be characterized, and *spire* has also not yet been molecularly characterized.

Manseau and Schupbach (1989) have also shown that *capu* and *spire* both effect anterior–posterior as well as D/V polarity, perhaps because they both interfere with normal mRNA localization by disrupting cytoskeletal elements in the oocyte cytoplasm. Females mutant for these two genes lay dorsalized eggs, but the phenotype is quite variable, again suggesting more universal effects on egg organization (Manseau and Schupbach, 1989). Females mutant for *orb* lay lateralized and/or ventralized eggs, and *grk* mRNAs are found throughout the anterior cytoplasm (Roth and Schupbach, 1994). In addition, the localization of Oskar mRNAs in the posterior determination pathway is disrupted in *orb* embryos. Similar effects are produced in *homeless (hls)* embryos (Gillespie and Berg, 1995). That both Orb and Hls proteins have homologies to known RNA-binding proteins may suggest that they are involved in more generalized RNA interactions.

Thus, studies on the establishment of D/V polarity of the eggshell not only have provided a model system to analyze RTK signaling but also allowed studies on the mechanism underlying subcellular localization of mRNAs in a cell.

A   **Dorsal**

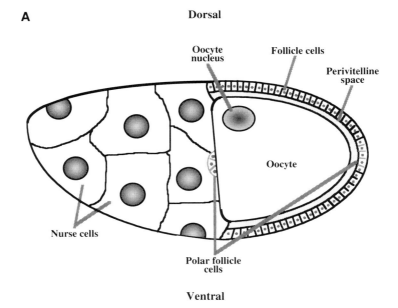

Ventral

Fig. 7 Dorsoventral patterning during embryogenesis: A paradigm to study RTK and NFκb/Iκb signaling pathways in *Drosophila*. **A.** Wild-type stage-10 egg chamber. A localized source of Grk protein activates DER in dorsal/anterior follicle cells. **B.** Genes involved in *grk* mRNA localization are indicated. **C.** Establishment of dorso/ventral embryonic polarity. (See also color insert) See text for details and review by Stein.

VI. Dorsal–Ventral Patterning of the *Drosophila* Embryo: A Paradigm to Dissect the NFκB/IκB Signaling Pathway

The establishment of embryonic D/V polarity provides a unique paradigm to study the mechanism of activation and regulation of the NFκb/Iκb signaling pathway. Twelve maternally expressed genes have been characterized that provide a comprehensive view of the steps involved in this process (see recent reviews by Stein, 1995; Ray and Schupbach, 1996; see Fig. 7C—see color insert). Briefly, during oogenesis the activities of three genes, *windbeutel* (*wind*), *pipe* (*pip*), and *nudel* (*nud*), are required in the follicle cells, to initiate proper dorsoventral embryonic patterning. The *wind* and *pip* genes have not been fully characterized but perhaps are involved in establishing the appropriate ventral distribution of Nud protein in the vitelline membrane. The asymmetric localization and activity of Nud initiates and/or spatially organizes a serine protease cascade involving the sequential action of the Gastrulation-defective, Snake, and Easter proteins, which locally activate the Toll ligand, Spatzle.

B

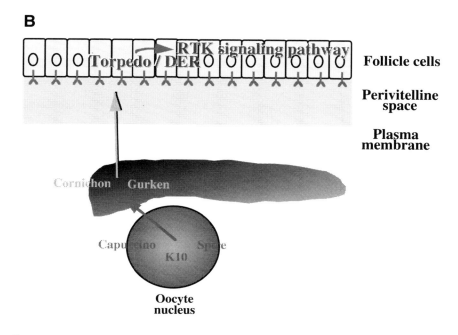

Follicle cells

Perivitelline space

Plasma membrane

c

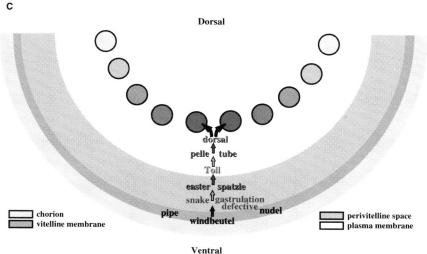

Fig. 7 Continued

Transduction of the signal from the Toll receptor requires the activity of two genes, the serine/threonine kinase *pelle* and the novel protein *tube*. In the absence of signaling, Dorsal (Dl, a fly homologue of NFκB/c-rel) is associated with Cactus (Cac, a homologue of IκB). Release of Dl from Cac allows Dl to enter the nucleus, where it regulates the domain of expression of a number of downstream target genes, such as *twist, snail*, and *zerknult* (see later Fig. 10B). The expression of the zygotic genes then initiates the embryonic pattern along the D/V axis, with the subsequent determination, differentiation, and movements that determine the establishment of D/V cell fates.

VII. Segment Polarity Genes: Paradigms to Dissect the Signaling Pathways Activated by Wnt and Hedgehog Proteins

Each segment of the larva is composed of alternating metameric units consisting of naked cuticle and cuticle covered with some hairs called denticles (Fig. 8A). The "segment polarity" genes specify cellular identities within the embryonic epidermis and in "segment polarity" mutant embryos, cell fates within the segmental units are not maintained properly, leading to various cuticular defects (Nusslein-Volhard and Wieschaus, 1980; see reviews by Siegfried and Perrimon, 1994; Klingensmith and Nusse, 1994; Perrimon, 1994b). Some of the segment-polarity genes, such as *wingless (wg), engrailed (en)*, and *hedgehog (hh)* are expressed in stripes within the embryonic epidermis, and their initial expression is controlled by the localized activities of pair-rule genes that were expressed earlier in development.

During embryonic segmentation, the homeobox transcription factor En as well as the secreted protein Hh are expressed in a stripe of cells adjacent, and posterior, to *wg*-expressing cells. The maintenance of the expression of these genes is critical to patterning, since in the absence of their function, segmentation does not occur. In the epidermis, Wg is required to maintain the expression of the transcription factor En in adjacent cells; En in turn controls the expression of *hh*. In mutants lacking *wg* activity, *en* and *hh* expression initiates normally but fades with time. Similarly, the activities of En and Hh are required for maintenance of *wg* expression in adjacent cells (see review by Perrimon, 1994b). The Wg protein is secreted, and in the embryo has been detected as far as two to three cells away from the cells where it is expressed, including the adjacent En-expressing cells (van den Heuvel *et al.*, 1989; Gonzales *et al.*, 1991). Wg is thought to have a second, distinct autoregulatory role, as well, that might reflect an autocrine *wg* activity that is independent of signaling via *en* (Hooper, 1994; Yoffe *et al.*, 1995, Manoukian *et al.*, 1995).

A

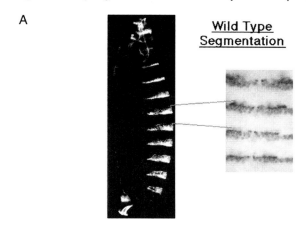

B

Phenotypes of Mutants in the Wingless Signaling Pathway

Loss of Wingless Signaling   **Constitutive Wingless Signaling**

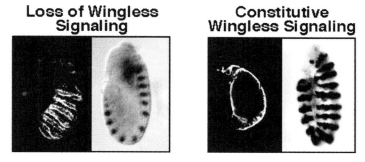

Fig. 8 Patterning of the embryonic segment: A paradigm to study Wnt and Hedgehog signaling pathways. **A.** The wild-type cuticular pattern of a first-intar larva is shown. In this darkfield micrograph, the areas covered with denticle belts and naked cuticle are clearly visible. The segmental expression of *wingless* mRNA and Engrailed protein is shown. (See volume cover for color representation.) **B.** Null *wg* mutations result in embryos that fail to secrete the smooth, or "naked," cuticle; thus the entire ventral surface of the cuticle is covered with denticles. In a *wg* mutant embryo, En expression in the epidermis is lost. Note that there is residual En expression in the central nervous system, however. Ectopic expression of Wg, or constitutive activation of the Wg pathway, leads to the opposite phenotype, a naked cuticle and expansion of the En expression domain. Embryos mutant for *hh*, which are not shown in this figure, exhibit a cuticle mutant phenotype very similar to loss of *wg* functions. **C.** The Wingless signaling pathway. See text for details and abbreviations. Adapted from Perrimon (1996). **D.** The Hedgehog signaling pathway. See text for details and abbreviations. Adapted from Perrimon (1996).

C

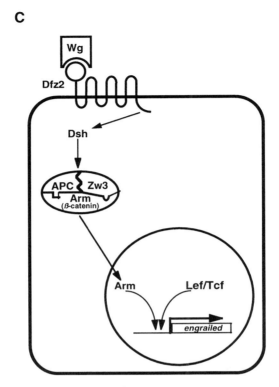

Fig. 8 Continued

A. The Wingless Signaling Pathway

Analysis of mutants that have cuticle phenotypes similar to *wg* and *hh* mutants, as well as mutants that have phenotypes mimicking overstimulation of these pathways, has resulted in the identification of molecules involved in *wg* signaling (see reviews by Perrimon, 1994b, 1995, 1996; see Fig. 8B). Genetic interaction studies have identified four genes that act in the *wg* signaling pathway: *porcupine* (*porc*), *dishevelled* (*dsh*), *zeste-white 3* (*zw3*, also known as *shaggy*), and *armadillo* (*arm*) (Siegfried *et al.*, 1994, Noordermeer *et al.*, 1994, Peifer *et al.*, 1994). Recently, a homologue of the *Drosophila frizzled* gene (Dfz2) that belongs to the Frizzled (Fz) family of serpentine proteins (Wang *et al.*, 1996) has been cloned and shown to be able to act as a receptor for Wg in tissue culture cells (Bhanot *et al.*, 1996).

The current model is that Porc is involved in the processing of Wg protein

D

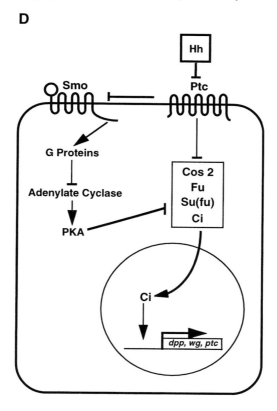

Fig. 8 Continued

(Kadowaki *et al.*, 1996). Wg binds to the Wg receptor, Dfz2 presumably, in the adjacent cells and initiates a signal transduction cascade (Fig. 8C). A candidate for a target of Dfz2 is Dsh, a novel cytoplasmic phosphoprotein absolutely essential for transduction of the Wg signal. Yanagawa *et al.* (1995) have pro- posed that in Wg-responsive cell lines, Wg signaling generates a hyper- phosphorylated form of Dsh, which is membrane associated. Dsh possibly inacti- vates the Zw3 (Shaggy) serine/threonine protein kinase, which itself acts as a negative regulator of Arm (β-catenin) (see review by Perrimon, 1994b). In re- sponse to the Wg signal, the level of intracellular Arm is stabilized and is correlated with a decrease in the level of Arm phosphorylation. Arm can be detected in the nucleus, where it may control gene expression in cooperation with other factor(s), perhaps the Lef/Tcf transcription factor, which has been recently isolated as a putative partner for β-catenin (Molenaar *et al.*, 1996). In conclusion, as observed for other signaling pathways, it appears that Wg signaling regulates the nuclear translocation of a cytoplasmic molecule (Orsulic and Peifer, 1996; Yost *et al.*, 1996).

Studies on *Drosophila* Wg have pioneered studies on signaling from Wnt proteins in vertebrates. The first member of the *Wnt* family to be identified was *Wnt-1* (previously known as *int-1*), by its induction of mammary carcinomas when ectopically expressed in mice (Nusse and Varmus, 1982). Subsequently, it was found that *Wnt-1* belongs to a large family of conserved molecules, and multiple members of this family have been identified in other vertebrates and invertebrates (see reviews by Nusse and Varmus, 1992, and McMahon, 1992). A number of studies indicate that the Wnt signaling pathway has been conserved between species, as other molecules involved in *Drosophila* Wg signaling have also been identified in vertebrates. The mammalian glycogen synthase kinase-3 (GSK-3) is the mammalian homologue of Zw3 (Siegfried *et al.*, 1992; Ruel *et al.*, 1993). Arm encodes the *Drosophila* homologue of β-catenin and plakoglobin (Peifer and Wieschaus, 1990). Similarly, Dsh has been cloned from *Xenopus* (*Xdsh*, Sokol *et al.*, 1995) and mouse (Sussman *et al.*, 1994), and human homologues of Porc are present in the database (Kadowaki *et al.*, 1996).

B. The Hedgehog Signaling Pathway

A genetic approach has led to the characterization of components of the Hh signaling pathway (Fig. 8D). The observation that *wg* expression disappears in *hh* mutant embryos pointed to the Hh signal as a positive regulator of *wg* transcription. Hh is a rather unusual protein, in that it cleaves itself (see review by Perrimon, 1995). It is synthesized as a precursor that undergoes an autocatalytic internal cleavage to generate two products, HhN (19kDa), the signaling activity, and HhC (25kDa), the processing activity. The current model is that HhN binds to the transmembrane protein Patched (Ptc) in Wg-expressing cells. It has been proposed that binding of Hh to Ptc activates the transmembrane protein Smoothened (Smo) (Marigo *et al.*, 1996; Stone *et al.*, 1996), which signals to the serine/threonine kinase Fused (Fu) (Figure 8D). Fu becomes phosphorylated after activation of the pathway by Hh (Thérond *et al.*, 1996). The signal activates the zinc-finger-containing DNA-binding protein Cubitus interruptus (Ci), which regulates *wg* and *ptc* transcription. The segment-polarity gene, *Costal²* (*Cos²*), has been shown to be required downstream of Fu and upstream of Ci. Genetic analysis has also shown that protein kinase A (PKA) is a negative regulator of the Hh signaling pathway as well. Possibly, in response to the extracellular Hh signal, Smo may regulate the activity of a heterotrimeric G protein complex. According to this model, in the absence of Hh signal, adenylate cyclase activity results in the production of cyclic AMP and activation of PKA. Repression of adenylate cyclase activity in the presence of Hh may in turn down-regulate the level of PKA (see review by Perrimon, 1996).

As is the case for molecules involved in Wnt signaling, vertebrate homologues of all known components involved in Hh signaling exist, and furthermore their mechanisms of action seem to be conserved (Perrimon, 1995, 1996). Interestingly, the human homologue of Ptc was found to be mutated or inactivated in the basal cell naevus syndrome and basal cell carcinoma (Hahn *et al.*, 1996; Johnson *et al.*, 1996). Further, the metastatic capabilities of many human carcinoma cells increase with the level of PKA activity (Young *et al.*, 1995). These observations stress the importance of understanding how this pathway controls cell proliferation.

VIII. Regulation of the Pair-Rule Gene *even skipped* in the Early Embryo: A Paradigm to Dissect the JAK/STAT Signaling Pathway

Recent studies on the regulation of pair-rule gene expression have provided a paradigm to genetically dissect the Janus kinase–signal transducer and activator of transcription (JAK-STAT) pathway in *Drosophila* (see review by Hou and Perrimon, 1997; Fig. 9). This pathway was originally identified in mammalian cells through studies of the transcriptional activation response to a variety of cytokines and growth factors (review by Schindler and Darnell, 1995). These studies have led to the model whereby cytoplasmic JAK proteins, which are bound constitutively to the membrane-proximal domain of cytokine receptors, become activated when the receptor homodimerizes in response to cytokine binding. Dimerization of the receptor brings the receptor-associated JAK proteins into apposition, enabling them to transphosphorylate and thereby to activate each other. The activated JAK proteins phosphorylate a distal tyrosine residue on the receptor, which is subsequently recognized by the SH2 domain present in the STAT proteins. STAT proteins recruited to the receptor complex then become activated by JAKs through phosphorylation on a tyrosine residue. Activated STAT proteins are competent for homo- or heterodimerization and nuclear translocation, where they subsequently activate gene transcription.

A *Drosophila* JAK kinase, encoded by the gene *hopscotch* (*hop*), and a STAT protein, encoded by the gene *marelle* (also known as *Dstat* or *stat92E*), have been characterized (Binari and Perrimon, 1994; Hou *et al.*, 1996; Yan *et al.*, 1996). Mutations in both *hop* and *stat92E* are associated with zygotic lethality, causing death of mutant animals during larval stages. In addition, they exhibit strikingly similar maternal-effect lethal phenotypes that can be observed by examining embryos derived from females carrying homozygous mutant germline clones. In the complete absence of both maternal and zygotic *hop* or *stat92E* activity, embryos exhibit severe segmentation defects. Characterization of the expression pattern of segmentation genes in both *hop* and *stat92E* mutant embryos has revealed that the defective embryonic phenotypes are due to the

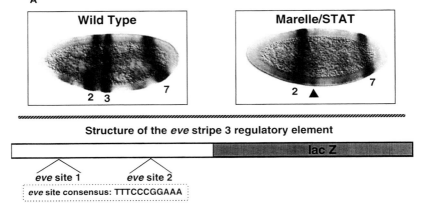

A

Wild Type

Marelle/STAT

2 3   7

2 ▲   7

Structure of the *eve* stripe 3 regulatory element

lac Z

eve site 1   *eve* site 2
eve site consensus: TTTCCCGGAAA

Fig. 9 Regulation of *even-skipped stripe 3* expression: A paradigm to study the JAK-STAT pathway in *Drosophila*. **A.** LacZ expression, driven by the reporter gene construct *eve* 5.2-lacZ, is shown in a wild-type embryo (top left) and a *hop* or *marelle/stat92E* embryo (top right). In wild type, *eve* 5.2-lacZ drives lacZ expression in the second, third, and seventh *eve* stripes. In *hop* and *stat92E* embryos, lacZ expression corresponding to *eve* stripe 3 is almost completely missing. Embryos are oriented with anterior to the left and dorsal at the top. A schematic representation of the 500 bp enhancer with the two STAT92E activator sites is depicted. **B.** A model for Hop-Stat activation in the precellular embryo (see also color insert). See text for details as well as Hou and Perrimon (1997).

defective expression of pair-rule genes such as *runt (run)* and *even-skipped (eve)*. For example, in *hop* or *stat92E* mutant embryos derived from germline clones, the expression of *eve* stripe 3 is greatly diminished. A reporter gene containing a 500-bp fragment of the *eve* promoter has been shown to contain the elements that control the expression of *eve* stripe 3. When the lacZ expression driven by this reporter gene is examined in *hop* or *stat92E* embryos derived from germline clones, there is diminished expression comparable to that observed when the expression of *eve* transcript was assayed (Fig. 9A). In this fragment, two consensus sequences that closely match the mammalian STAT-binding site are present, and activated-STAT proteins bind to these sites. This suggests that STAT activation via phosphorylation by JAKs is likely conserved between mammals and *Drosophila*. If this is true, then one would predict that Hop should be activated by its interaction with a membrane-bound receptor lacking a kinase domain (Fig. 9B—see color insert).

IX. Establishment of Embryonic Dorsoventral Cell Fates: A Paradigm to Analyze the Transforming Growth Factor-β Signaling Pathway

Two of the zygotic genes controlled by the activity of the gradient of Dorsal protein are *decapentaplegic (dpp)*, which encodes a member of the transforming

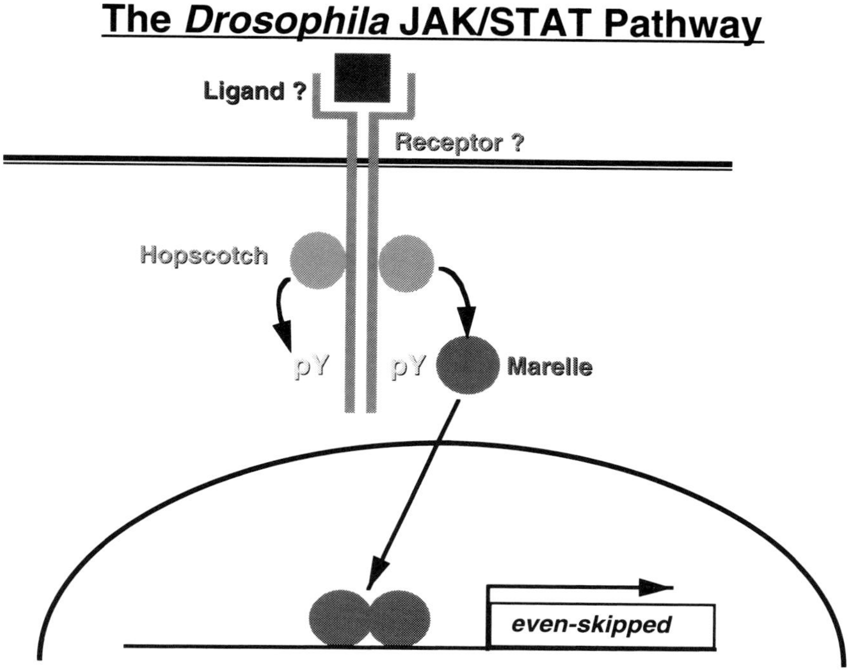

The *Drosophila* JAK/STAT Pathway

Fig. 9 Continued

growth factor-β (TGFβ) family (Padgett *et al.*, 1987; St. Johnston and Gelbart, 1987), and *short gastrulation* (*sog*), which encodes a large secreted protein with cysteine-rich domains (Francois *et al.*, 1994; Holley *et al.*, 1995). While *dpp* is expressed in the dorsal 40% of the embryonic nuclei, *sog* is expressed in the complementary ventral-lateral domain (Fig. 10A; see review by Ferguson, 1996). Dpp acts as a concentration-dependent morphogen to control the differentiation of the embryonic ectoderm. High levels of Dpp promote amnioserosa development, low levels trigger dorsal ectoderm differentiation, and Dpp must be absent for the ventral neurogenic region to develop properly. Secreted Sog appears to antagonize the activity of the Dpp morphogen ventrally, as revealed by the expansion of dorsal structures in *sog* mutants and formation of neurogenic tissues on the dorsal side of the embryo following injection of *sog* mRNAs dorsally (Holley *et al.*, 1995).

Dpp, like other members of the TGFβ family, signals through a heteromeric receptor complex composed of a type I and a type II receptor. Types I and II receptors encode serine/threonine kinase in their cytoplasmic domains. The current model is that binding of Dpp to its receptor induces heterodimerization of the receptors and causes the Type II receptor (Punt) to phosphorylate the Type I

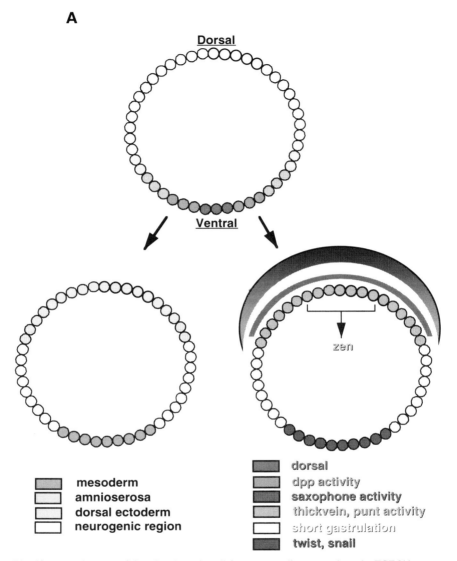

Fig. 10 Establishment of dorsal embryonic cell fates: A paradigm to analyze the TGFβ/decapentaplegic signaling pathway. **A.** Establishment of dorsal/ventral cell fates. The domains of expression of genes involved in D/V axis formation are indicated. In addition, the various cell types that are specified along this axis are shown. Adapted from Rush and Levine (1996). **B.** The Dpp signaling pathway. (See also color insert) See text for details as well as Ferguson (1996) and Derinck and Zhang (1996).

B

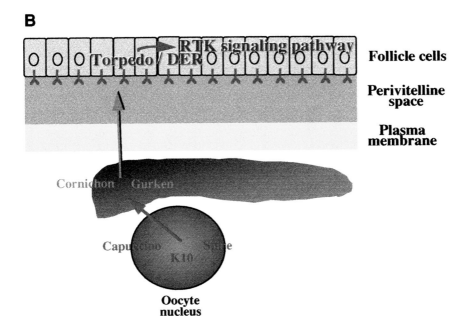

Figure 7B and C   Dorso-ventral patterning during embryogenesis: A paradigm to study RTK and NFkb/Ikb signaling pathways in *Drosophila*. (B) Genes involved in *grk* mRNA localization are indicated. (C, overleaf) Establishment of dorso/ventral embryonic polarity. See text for details and review by Stein (1995).

c

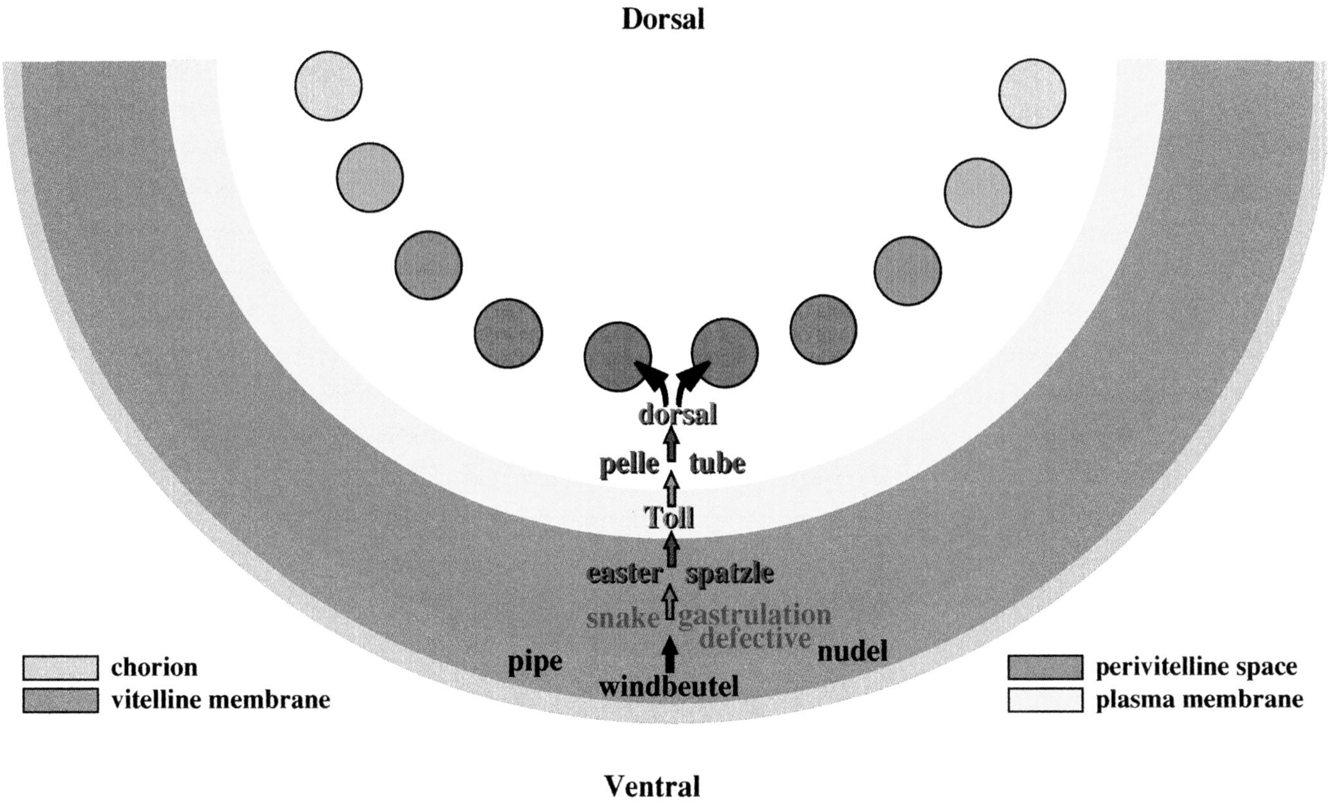

Dorsal

Ventral

pelle   dorsal   tube
         Toll
easter   spatzle   nudel
snake   gastrulation
         defective
pipe   windbeutel

chorion
vitelline membrane

perivitelline space
plasma membrane

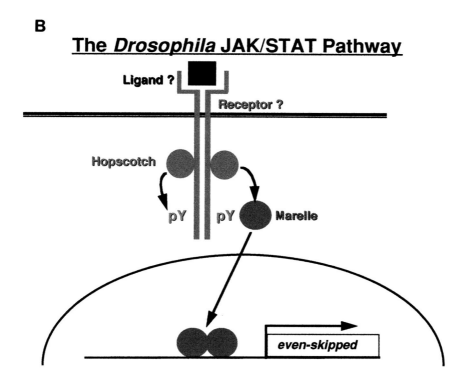

Figure 9B Regulation of *even-skipped stripe* 3 expression: A paradigm to study the JAK/STAT pathway in *Drosophila*. (B) A model for Hop-Stat activation in the precellular embryo. See text for details as well as Hou and Perrimon (1997).

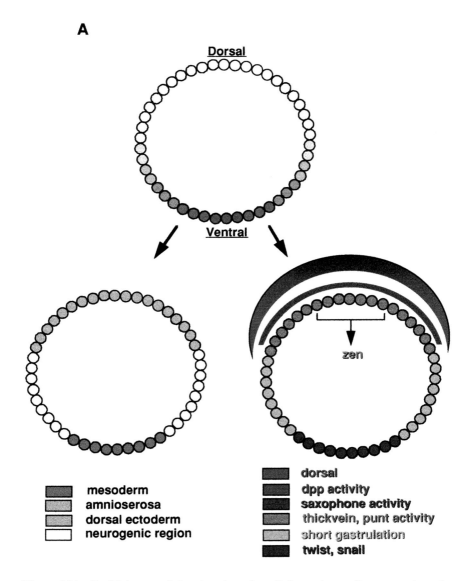

A

Figure 10A Establishment of dorsal embryonic cell fates: A paradigm to analyze the TGFβ/Decapentaplegic signaling pathway. (A) Establishment of dorsal/ventral cell fates. The domains of expression of genes involved in D/V axis formation is indicated. In addition, the various cell types that are specified along this axis are shown. Adapted from Rush and Levine (1996).

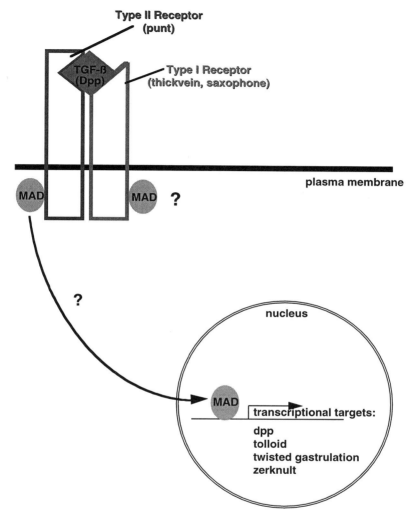

Fig. 10 Continued

receptors (encoded by either *thick veins* or *saxophone*; reviews by Massague *et al.*, 1994; Ferguson, 1996). In this system, the Dpp signal is zygotically expressed, while the receptors are expressed both maternally and zygotically. Genetic analysis of the Dpp signal transduction pathway has identified two new components encoded by *medea* and *mothers against Dpp* (*mad*) (Raftery *et al.*, 1995). Although little is known about the function of *medea*, studies on *mad* have been quite revealing. *mad* encodes a novel evolutionarily conserved protein that and subsequently translocated to the nucleus (Fig. 10B; Hoodless *et al.*, 1996; review by Derinck and Zhang, 1996). It is not clear how many other components are involved in transducing TGFβ signals from the membrane to the nucleus. Indeed, it has been proposed that perhaps this system may be as simple as the JAK/STAT pathway described earlier in this review (Derinck and Zhang, 1996).

X. Neurogenesis in the *Drosophila* Embryo: A Paradigm to Study Notch Signaling

Notch (N) and its cognates, Lin12/Glp1/Xotch, play a central role in cell fate specification of a wide variety of tissues (review by Artavanis-Tsakonas *et al.*, 1994). These large transmembrane receptors have been implicated in cell–cell signaling processes involving groups of equivalent cells or between cells of different types. An example of signaling between cells belonging to a so-called "equivalence group" can be found during the development of the *Drosophila* embryonic central nervous system (CNS). During embryogenesis, an ectodermal monolayer of approximately 1,800 equipotent cells segregate into two distinct cell populations, neuroblasts that form the CNS, and epidermoblasts that give rise to the ventral epidermis. In wild-type embryos, neuroblasts, once they have singled out from the ectodermal layer, inhibit their neighbors from entering the neural pathway. In *N* mutant embryos, all epidermal precursor cells enter the neural fate because of a defective lateral specification process (Fig. 11A), and N was found to act as a receptor in this process (Heitzler and Simpson, 1991).

Formation of the embryonic CNS has provided a model system to identify and analyze the function of components of the evolutionarily conserved N signaling pathway. The current model (Fig. 11B) is that, upon ligand-mediated activation, the receptors multimerize. Receptor activation regulates the activity of the intracellular domain of N that contains a series of tandem cdc10/SW16 repeats (also termed "ankyrin" repeats) (Struhl *et al.*, 1993; Rebay *et al.*, 1993; Lieber *et al.*, 1993). Delta (Dl) and Serrate (Ser) have been identified as ligands for *N*, whereas Suppressor of Hairless (Su(H)) interacts with the cytoplasmic cdc10/SW16 domain. Activated N receptors control the translocation of the Su(H) transcription factors to the nucleus and its interaction with Hairless (see review by Artavanis-Tsakonas *et al.*, 1994), as well as transcription of the bHLH *Enhancer of Split* (*E(spl)*) gene, which has been identified as a putative direct target of Su(H). Subsequently, products of the *E(spl)* gene itself are involved in regulating the expression of genes of the *Ac-S* complex.

XI. Conclusion and Perspectives

In this review we have illustrated how the isolation of mutations affecting either oogenesis or early embryogenesis has led to the identification of simple para-

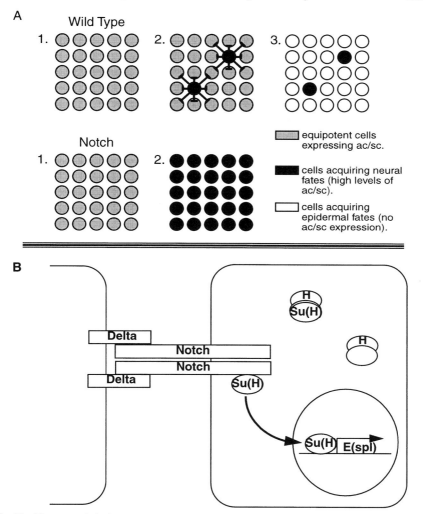

Fig. 11 Neurogenesis in the central nervous system: A paradigm to characterize the Notch pathway. **A.** Role of N in lateral specification. During lateral specification, among a field of initially equivalent cells, a few cells express an inhibitory signal. These cells, which enter the neural fate, as visualized by their expression of genes of the achaete scute (Ac-S) complex, inhibit their neighbors from becoming neuroblasts. In *N* mutant animals, cells lack the proposed receptor for the inhibitory signal, and the entire field of cells enters the neural fate and expresses Ac-S. Adapted from Artavanis-Tsakonas et al. (1994). **B.** The Notch signaling pathway. See text for details as well as Artavanis-Tsakonas et al. (1994).

digms to dissect signal transduction pathways. Signaling pathways employed in *Drosophila* are strikingly similar to those of vertebrates, so these paradigms are clearly valuable model systems to analyze evolutionarily conserved signaling pathways. These model systems can be used for "gene discovery" and subsequent identification of mammalian homologues of novel components identified from genetic studies. Furthermore, the *Drosophila* paradigms can be used as *in vivo* assays to test specific hypotheses drawn from biochemical studies. Studies in mammalian cells have illustrated numerous biochemical interactions between specific molecules. In some instances the meaning of such interactions remains to be validated. Using the insights gained from the use of *Drosophila*, specific experiments can be carried out to address whether such interactions occur *in vivo*.

The examples described provide simple paradigms to identify components that operate in specific signaling pathways and to dissect the details of how signals are transduced. However, in other developmental contexts, multiple pathways may cooperate to establish the proper cell fates. For example, in the *Drosophila* egg chamber, the establishment of the D/V axis of the egg and embryo is linked to the establishment of the A/P axis. Recent studies on the function of the DER RTK and its ligand Grk, as well as on the function of Notch (N) signaling during oogenesis, have demonstrated an intricate relationship between these two pathways in determining the A/P polarity of the egg chamber (Ruohola-Baker *et al.*, 1991; Gonzalez-Reyes *et al.*, 1995; Roth *et al.*, 1995; see review by Lehmann, 1995). In this complicated process, understanding the cooperativity between these two pathways was greatly facilitated by our prior understanding of the DER and N signaling pathways.

Acknowledgments

We are indebted to Dr. A. P. Mahowald for permission to use the scanning electron migrographs shown in Fig. 2. We thank Liz Perkins for comments on the manuscript. Parts of this review were adapted from previously published reviews, in particular, Duffy and Perrimon (1996), Perrimon (1994a,b, 1996), and Hou and Perrimon (1997). E. N. and N. P. are supported by the Howard Hughes Medical Institute.

References

Artavanis-Tsakonas, S., Matsuno, K., and Fortini, M. E. (1994). Notch signaling. *Science* **268,** 225–232.

Bate, M., and Martinez-Arias, S. (1993). "The Development of *Drosophila melanogaster*." Vol II. Cold Spring Harbor Laboratory Press, New York.

Bejsovec, A., and Martinez-Arias, A. (1991). Roles of *wingless* in patterning the larval epidermis of *Drosophila. Development* **113,** 471–485.

Bhanot, P., Brink, M., Harryman Samos, C., Hsieh, J.-C., Wang, Y., Macke, J. P., Andrew, D., Nathans, J., and Nusse, R. (1996). A new member of the *frizzled* family from *Drosophila* functions as a Wingless receptor. *Nature* **382,** 225–230.

Binari, R., and Perrimon, N. (1994). Stripe-specific regulation of pair-rule genes by *hopscotch*, a putative Jak family tyrosine kinase in *Drosophila. Genes Dev.* **8,** 300–312.

Brand, A., and Perrimon, N. (1994). Raf acts downstream of the EGF receptor to determine dorsoventral polarity during *Drosophila* oogenesis. *Genes Dev.* **8,** 629–639.

Bronner, G., Chu-LaGraff, Q., Doe, C. Q., Cohen, B., Weigel, D., Taubert, H., and Jackle, H. (1994). Sp1/egr-like zinc-finger protein required for endoderm specification and germ-layer formation in *Drosophila*. *Nature* **369**, 664–668.

Campos-Ortega, J. A., and Hartenstein, V. (1985). "The Embryonic Development of *Drosophila melanogaster*." Berlin: Springer-Verlag.

Casanova, J., and Struhl, G. (1989). Localized surface activity of *torso*, a receptor tyrosine kinase, specifies terminal body pattern in *Drosophila*. *Genes Dev.* **3**, 2025–2038.

Casanova, J., and Struhl, G. (1993). The *torso* receptor localizes as well as transduces the spatial signal specifying terminal body pattern in *Drosophila*. *Nature* **362**, 152–155.

Chou T.-B., and Perrimon, N. (1992). Use of a yeast site-specific recombinase to produce female germline chimeras in *Drosophila*. *Genetics* **131**, 643–653.

Chou, T.-B., and Perrimon, N. (1996). The autosomal FLP-DFS technique for generating germline mosaics in *Drosophila melanogaster*. *Genetics* **144**, 1673–1679.

Chou, T.-B., Noll, E., and Perrimon, N. (1993). Autosomal *P[ovo^{D1}]* dominant female sterile insertions in *Drosophila* and their use in generating germline chimeras. *Development* **119**, 1359–1369.

Christerson, L. B., and McKearin, D. M. (1994). *orb* is required for anteroposterior and dorsoventral patterning during *Drosophila* oogenesis. *Genes Dev.* **8**, 614–628.

Derinck, R., and Zhang, Y. (1996). Intracellular signalling: The Mad way to do it. *Curr. Biol.* **6**, 1226–1229.

Duffy, J. B., and Perrimon, N. (1994). The Torso pathway in *Drosophila*: Lessons on receptor tyrosine kinase signaling and pattern formation. *Dev. Biol.* **166**, 380–395.

Duffy, J. B., and Perrimon, N. (1996). Recent advances in understanding signal transduction pathways in worms and flies. *Curr. Opin. Cell Biol.* **8**, 231–238.

Emmons, S., Phan, H., Calley, J., Chen, W., James, B., and Manseau, L. (1995). *cappuccino*, a *Drosophila* maternal effect gene required for polarity of the egg and embryo, is related to the vertebrate *limb deformity* locus. *Genes Dev.* **9**, 2482–2494.

Erdelyi, M., and Szabad, J. (1989). Isolation and characterization of dominant female sterile mutations of *Drosophila melanogaster*. I. Mutations on the third chromosome. *Genetics* **122**, 111–127.

Ferguson, E. L. (1996). Conservation of dorsal–ventral patterning in arthropods and chordates. *Curr. Opin. Genet. Dev.* **6**, 424–431.

Francois, V., Solloway, M., O'Neill, J. W., Emery, J., and Bier, E. (1994). Dorsal–ventral patterning of the *Drosophila* embryo depends on a putative negative growth factor encoded by the *short gastrulation* gene. *Genes Dev.* **8**, 2602–2616.

Gans, M., Audit, C., and Masson, M. (1975). Isolation and characterization of sex-linked female-sterile mutants in *Drosophila melanogaster*. *Genetics* **81**, 683–704.

Gillespie, D. E., and Berg, C. A. (1995). *homeless* is required for RNA localization in *Drosophila* oogenesis and encodes a new member of the DE-H family of RNA-dependent ATPases. *Genes Dev.* **9**, 2495–2508.

Golic, K. G. (1991). Site-specific recombination between homologous chromosomes in *Drosophila*. *Science*, **252**, 958–961.

Gonzales, F., Swales, L., Bejsovec, A., Skaer, H., and Martinez-Arias, A. (1991). Secretion and movement of the *wingless* protein in the epidermis of the *Drosophila* embryo. *Mech. Dev.* **35**, 43–54.

Gonzales-Reyes, A., Elliot, H., and St. Johnston, D. (1995). Polarization of both major body axes in *Drosophila* by *gurken-torpedo* signalling. *Nature* **375**, 654–658.

Hahn, S. A., Schutte, M., Hoque, A. T. M. S., Moskaluk, C. A., Da Costa, L. T., Rozenblum, E., Weinstein, C. L., Fischer, A., Yeo, C. J., Hruban, R. H., and Kern, S. E. (1996). DPC4, a candidate tumor suppressor gene at human chromosome 18q21.1. *Science* **271**, 350–353.

Heitzler, P., and Simpson, P. (1991). The choice of cell fate in the epidermis of Drosophila. *Cell* **64**, 1083–1092.

Holley, S. A., Jackson, P. D., Sasai, Y., Lu, B., DeRobertis, E. M., Hoffmann, F. M., and Ferguson, E. L. (1995). A conserved system for dorsal–ventral patterning in insects and vertebrates involving *sog* and *chordin. Nature*, **376**, 249–253.

Hoodless, P. A., Haerry, T., Abdollah, S., Stapleton, M., O'Connor, M. B., Attisano, L., and Wrana, J. L. (1996). MADR1, a MAD-related protein that functions in BMP2 signaling pathways. *Cell* **85**, 489–500.

Hooper, J. E. (1994). Distinct pathways for autocrine and paracrine Wingless signaling in *Drosophila* embryos. *Nature* **372**, 461–464.

Hou, X. S., and Perrimon, S. (1997). The JAK/STAT pathway in *Drosophila. Trends Genet.* **13**, 105–110.

Hou, X. S., Chou, T.-B., Melnick, M. B., and Perrimon, N. (1995). The Torso receptor tyrosine kinase can activate Raf in a Ras-independent pathway. *Cell* **81**, 63–71.

Hou, X. S., Melnick, M. B., and Perrimon, N. (1996). *marelle* acts downstream of the Drosophila HOP/JAK kinase and encodes a protein similar to the mammalian STATs. *Cell* **84**, 411–419.

Johnson, R. L., Rothman, A. L., Xie, J., Goodrich, L. V., Bare, J. W., Bonifas, J. M., Quinn, A. G., Myers, R. M., Cox, D. R., Epstein, E. H.,Jr., and Scott, M. P. (1996). Human homolog of *patched*, a candidate gene for the basal cell nevus syndrome. *Science* **272**, 1668–1671.

Jurgens, G., Wieschaus, E., Nusslein-Volhard, C., and Kluding, H. (1984). Mutations affecting the pattern of the larval cuticle in *Drosophila melanogaster*. II. Zygotic loci on the third chromosome. *Wilhelm Roux's Arch. Dev. Biol.* **193**, 283–295.

Kadowski, T., Wilder, E., Klingensmith, J., Zachary, K., and Perrimon, N. (1996). The segment polarity gene *porcupine* encodes a putative multitransmembrane protein involved in Wingless signaling. *Genes Dev.* **10**, 3116–3128.

King, R. C. (1970). "Ovarian development in *Drosophila melanogaster*." Academic Press, New York.

Kingsley, D. (1994). The TGF-β superfamily: New members, new receptors, and new genetic tests of function in different organisms. *Genes Dev.* **8**, 133–146.

Klingensmith, J., and Nusse, R. (1994). Signaling by *wingless* in *Drosophila. Dev. Biol.* **166**, 396–414.

Klingler, M., Erdelyi, M., Szabad, J., and Nusslein-Volhard, C. (1988). Function of *torso* in determining the terminal anlagen of the *Drosophila* embryo. *Nature* **335**, 275–277.

Komitopoulou, K., Gans, M., Margaritis, L. M., Kafatos, F. C., and Masson, M. (1983). Isolation and characterization of sex-linked female-sterile mutants in *Drosophila melanogaster* with special attention to eggshell mutants. *Genetics* **105**, 897–920.

Lantz, V., Chang, J. S., Horabin, J. I., Bopp, D., and Schedl, P. (1994). The *Drosophila orb* RNA-binding protein is required for the formation of the egg chamber and establishment of polarity. *Genes Dev.* **8**, 598–613.

Lehmann, R. (1995). Cell–cell signaling, microtubules, and the loss of symmetry in the *Drosophila* oocyte. *Cell* **83**, 353–356.

Lewis, E. (1978). A gene complex controlling segmentation in *Drosophila. Nature* **276**, 565–570.

Liaw, G.-J., Rudolph, K. M., Huang, J.-D., Dubnicoff, T., Courey, A. J., and Lengyel, J. A. (1995). The *torso* response element binds GAGA and NTF-1/Elf-1, and regulates *tailless* by relief of repression. *Genes Dev.* **9**, 3163–3176.

Lieber, T., Kidd, S., Alcamo, E., Corbin, V., and Young, M. W. (1993). Antineurogenic phenotypes induced by truncated Notch proteins indicate a role in signal transduction and may point to a novel function for Notch in nuclei. *Genes Dev.* **7**, 1949–1965.

Lindsley, D. L., and Zimm, G. G. (1992). "The genome of *Drosophila melanogaster*." Academic Press, New York.

Mahowald, A. P., and Kambysellis, M. P. (1980). Oogenesis. *In* The genetics and biology of Drosophila, Vol. 2 (M. Ashburner and T. R. F. Wright, Eds.), pp. 141–224. Academic Press, London.

Manoukian, A. S., Yoffe, K., Wilder, E. L., and Perrimon, N. (1995). The *porcupine* gene is required for *wingless* autoregulation in *Drosophila. Development* **121**, 4037–4044.

Manseau, L. J., and Schupbach, T. (1989). *cappuccino* and *spire*: Two unique maternal-effect loci required for both the anteroposterior and dorsoventral patterns in the *Drosophila* embryo. *Genes Dev.* **3**, 1437–1452.

Marigo, V., Davey, R. A., Zuo, Y., Cunningham, J. M., and Tabin, C. J. (1996). Biochemical evidence that Patched is the Hedgehog receptor. *Nature* **384**, 176–179.

Massague, J., Attisano, L., and Wrana, J. L. (1994). The TGF-β family and its composite receptors. *Trends Cell Biol.* **4**, 172–178.

McMahon, A. P. (1992). The Wnt family of developmental regulators. *Trends Genet.* **8**, 236–242.

Mohler, J. D. (1977). Developmental genetics of the *Drosophila* egg. I. Identification of 50 sex-linked cistrons with maternal effects on embryonic development. *Genetics* **85**, 259–272.

Molenaar, M., van de Wetering, M., Oosterwegel, M., Peterson-Maduro, J., Godsave, S., Korinek, V., Roose, J., Destree, O., and Clevers, H. (1996). XTcf-3 transcription factor mediates β-catenin-induced axis formation in *Xenopus* embryos. *Cell* **86**, 391–399.

Noordermeer, J., Klingensmith, J., Perrimon, N., and Nusse, R. (1994). *dishevelled* and *armadillo* act in Wingless signaling pathway in *Drosophila*. *Nature* **367**, 80–83.

Nusse, R., and Varmus, H. E. (1982). Many tumors induced by the mouse mammary tumor virus contain a provirus integrated in the same region of the host genome. *Cell* **31**, 99–109.

Nusse, R., and Varmus, H. E. (1992). *Wnt* genes. *Cell* **69**, 1073–1087.

Nusslein-Volhard, C., and Wieschaus, E. (1980). Mutations affecting segment number and polarity in *Drosophila*. *Nature* **287**, 795–801.

Nusslein-Volhard, C., Wieschaus, E., and Kluding, H. (1984). Mutations affecting the pattern of the larval cuticle in *Drosophila melanogaster*. I. Zygotic loci on the second chromosome. *Roux's Arch. Dev. Biol.* **193**, 267–282.

Nusslein-Volhard, C., Frohnhofer, H. G., and Lehmann, R. (1987). Determination of anteroposterior polarity in *Drosophila*. *Science* **238**, 1675–1681.

Orsulic, S., and Peifer, M. (1996). An *in vivo* structure–function study of armadillo, the beta-catenin homologue, reveals both separate and overlapping regions of the protein required for cell adhesion and for wingless signaling. *J. Cell Biol.* **134**, 1283–1300.

Padgett, R. W., St. Johnston, R. D., and Gelbart, W. M. (1987). A transcript from a *Drosophila* pattern gene predicts a protein homologous to the transforming growth factor-β family. *Nature* **325**, 81–84.

Peifer, M., and Wieschaus, E. (1990). The segment polarity gene *armadillo* encodes a functionally modular protein that is the *Drosophila* homolog of human plakoglobin. *Cell* **653**, 1163–1178.

Peifer, M., Sweeton, D., Casey, M., and Wieschaus, E. (1994). *wingless* signal and zeste-white 3 kinase trigger opposing changes in the intracellular distribution of Armadillo. *Development* **120**, 369–380.

Perrimon, N. (1994a). Signalling pathways initiated by receptor protein tyrosine kinases in *Drosophila*. *Curr. Opin. Cell Biol.* **6**, 260–266.

Perrimon, N. (1994b). The genetic basis of patterned baldness in Drosophila. *Cell* **76**, 781–784.

Perrimon, N. (1995). Hedgehog and beyond. *Cell* **80**, 517–520.

Perrimon, N. (1996). Serpentine proteins slither into the Wingless and Hedgehog fields. *Cell* **86**, 513–516.

Perrimon, N., and Gans, M. (1983). Clonal analysis of the tissue specificity of recessive female-sterile mutations of *Drosophila melanogaster* using a dominant female sterile mutation *Fs(1)K1237*. *Dev. Biol.* **100**, 365–373.

Perrimon, N., Engstrom, L., and Mahowald, A. P. (1984a). The effects of zygotic lethal mutations on female germline functions in *Drosophila*. *Dev. Biol.* **105**, 404–414.

Perrimon, N., Engstrom, L., and Mahowald, A. P. (1984b). Developmental genetics of the 2E–F region of the *Drosophila* X-chromosome: A region rich in "developmentally important" genes. *Genetics* **108**, 559–572.

Perrimon, N., Mohler, J. D., Engstrom, L., and Mahowald, A. P. (1986). X-linked female sterile loci in *Drosophila melanogaster*. *Genetics* **113**, 695–712.

Perrimon, N., Engstrom, L., and Mahowald, A. P. (1989). Zygotic lethals with specific maternal effect phenotypes in *Drosophila melanogaster*. I. Loci on the X-chromosome. *Genetics* **121**, 333–352.

Perrimon, N., Lanjuin, A., Arnold, C., and Noll, E. (1996). Zygotic lethal mutations with maternal effect phenotypes in *Drosophila melanogaster*. II. Loci on the second and third chromosomes identified by P-element-induced mutations. *Genetics* **144**, 1681–1692.

Pignoni, F., Baldarelli, R. M., Steingrimsson, E., Dias, R. J., Patapoutian, A., Merriam, J. R., and Lengyel, J. A. (1990). The *Drosophila* gene *tailless* is expressed at the embryonic termini and is a member of the steroid receptor superfamily. *Cell* **62**, 151–163.

Raftery, L. A., Twombly, V., Wharton, K., and Gelbart, W. M. (1995). Genetic screens to identify elements of the *decapentaplegic* signaling pathway in Drosophila. *Genetics* **139**, 241–254.

Ray, R. P., and Schupbach, T. (1996). Intercellular signaling and the polarization of body axes during *Drosophila* oogenesis. *Genes Dev.* **10**, 1711–1723.

Rebay, I., Fehon, R. G., and Artavanis-Tsakonas, S. (1993). Specific truncations of *Drosophila* Notch define dominant activated and dominant negative forms of the receptor. *Cell* **74**, 319–329.

Roth, S., and Schupbach, T. (1994). The relationship between ovarian and embryonic dorsoventral patterning in *Drosophila*. *Development* **120**, 2245–2257.

Roth, S,. Shira Neuman-Silberberg, F. S., Barcelo, G., and Schupbach, T. (1995). *cornichon* and the EGF receptor signaling process are necessary for both anterior-posterior and dorsal–ventral pattern formation in *Drosophila*. *Cell* **81**, 967–978.

Ruel, L., Pantesco, V., Lutz, Y., Simpson, P., and Bourouis, M. (1993). Functional significance of a family of protein kinases encoded at the shaggy locus in Drosophila. *EMBO J.* **12**, 1657–1669.

Ruohola-Baker, H., Bremer, K. A., Baker, D., Swedlow, J. R., Jan, L. Y., and Jan, Y. N. (1991). Role of neurogenic genes in establishment of follicle cell fate and oocyte polarity during oogenesis in *Drosophila*. *Cell* **66**, 433–449.

Rusch, J., and Levine, M. (1996). Threshold responses to the dorsal regulatory gradient and the subdivision of primary tissue territories in the *Drosophila* embryo. *Curr. Opin. Genet. Dev.* **6**, 416–423.

St. Johnston, R. D., and Gelbart, W. M. (1987). Decapentaplegic transcripts are localized along the dorsal–ventral axis of the *Drosophila* embryo. *EMBO J.* **6**, 2785–2791.

St. Johnston, D., and Nusslein-Volhard, C. (1992). The origin of pattern and polarity in the *Drosophila* embryo. *Cell* **68**, 201–219.

Schindler, C., and Darnell, J. (1995). Transcriptional responses to polypeptide ligands: The JAK/STAT pathway. *Ann. Rev. Biochem.* **64**, 621–651.

Schupbach, T., and Wieschaus, E. (1986). Maternal-effect mutations altering the anterior–posterior pattern of the *Drosophila* embryos. *Roux's Arch. Dev. Biol.* **195**, 302–317.

Schupbach, T., and Wieschaus, E. (1989). Female sterile mutations on the second chromosome of *Drosophila melanogaster*. I. Maternal effect mutations. *Genetics* **121**, 101–117.

Siegfried, E., and Perrimon, N. (1994). *Drosophila* Wingless: A paradigm for the function and mechanism of Wnt signaling. *Bioessays* **16**, 395–404.

Siegfried, E., Chou, T-B., and Perrimon, N. (1992). *wingless* signaling acts through *zeste-white 3*, the *Drosophila* homolog of *glycogen synthase kinase-3*, to regulate *engrailed* and establish cell fate. *Cell* **71**, 1167–1179.

Siegfried, E., Wilder, E., and Perrimon, N. (1994). Components of *wingless* signaling in *Drosophila*. *Nature* **367**, 76–80.

Sokol, S. Y., Klingensmith, J., Perrimon, N., and Itoh, K. (1995). Dorsalizing and neutralizing properties of Xdsh, a maternally expressed *Xenopus* homolog of *dishevelled*. *Development* **121**, 1637–1647.

Spradling, A. C. (1993). Developmental genetics of oogenesis. *In* "The development of *Drosophi-*

la melanogaster" (M. Bate and A. Martinez-Arias, Eds.), pp. 1–70. Cold Spring Harbor Laboratory Press, Plainview, N.Y.

Sprenger, F., and Nusslein-Volhard, C. (1993). The terminal system of axis determination in the *Drosophila* embryo. *In* "The development of Drosophila melanogaster," Vol. 1 (M. Bate and A. Martinez-Arias, Eds.), pp. 365–386. Cold Spring Harbor Laboratory Press, Plainview, NY.

Stein, D. (1995). The link between ovary and embryo. *Curr. Biol.* **5**, 1360–1363.

Stone, D. M., Hynes, M., Armanini, M., Swanson, T. A., Gu, Q., Johnson, R. L., Scott, M. P., Pennica, D., Goddard, A., Phillips, H., Noll, M., Hooper, J. E., de Sauvage, F., and Rosenthal, A. (1996). The tumor-suppressor gene *patched* encodes a candidate receptor for Sonic hedgehog. *Nature* **384**, 129–133.

Struhl, G. (1989). Morphogen gradients and the control of body pattern in insect embryos. *In* "Cellular Basis of Morphogenesis," pp. 65–91. Wiley, Chichester (*Ciba Found. Symp.* 144).

Struhl, G., Fitzgerald, K., and Greenwald, I. (1993). Intrinsic activity of the Lin-12 and Notch intracellular domains in vivo. *Cell* **74**, 331-345.

Sussman, D. J., Klingensmith, J., Salinas, P., Adams, P. S., Nusse, R., and Perrimon, N. (1994). Isolation and characterization of a mouse homolog of the *Drosophila* segment polarity gene *dishevelled. Dev. Biol.* **166**, 73–86.

Szabad, J., Erdelyi, M., Goffmann, G., Szidonya, J., and Wright, T. R. F. (1989). Isolation and characterization of dominant female sterile mutations of *Drosophila melanogaster*. II. Mutations on the second chromosome. *Genetics* **122**, 823–835.

Thérond, P. P., Knight, J. D., Kornberg, T. B., and Bishop, J. M. (1996). Phosphorylation of the fused protein kinase in response to signaling from hedgehog. *Proc. Natl. Acad. Sci. USA* **93**, 4224–4228.

Therrien, M., Chang, H. C., Solomon, N. M., Karim, F. D., Wassarman, D. A., and Rubin, G. M. (1995). KSR, a novel protein kinase required for RAS signal transduction. *Cell* **83**, 879–888.

van den Heuvel, M., Nusse, R., Johnston, P., and Lawrence, P. A. (1989). Distribution of the *wingless* gene product in *Drosophila* embryos: A protein involved in cell–cell communication. *Cell* **59**, 739–749.

Wang, Y., Macke, J. P., Abella, B. S., Andreasson, K., Worley, P., Gilbert, D. J., Copeland, N. G., Jenkins, N. A., and Nathans, J. (1996). A large family of putative transmembrane receptors homologous to the product of the *Drosophila* tissue polarity gene *frizzled. J. Biol. Chem.* **271**, 4468–4476.

Weigel, D., Jurgens, G., Klingler, M. and Jackle, H. (1990). Two gap genes mediate maternal terminal pattern information in *Drosophila. Science* **248**, 495–498.

Wieschaus, E., Nusslein-Volhard, C., and Jurgens, G. (1984). Mutations affecting the pattern of the larval cuticle in *Drosophila melanogaster*. III. Zygotic loci on the X-chromosome and 4th chromosome. *Wilhelm Roux's Arch. Dev. Biol.* **193**, 296–307.

Yan, R., Small, S., Desplan, C., Dearolf, C. R., and Darnell, J. E., Jr., (1996). Identification of a *Stat* gene that functions in *Drosophila* development. *Cell* **84**, 421–430.

Yanagawa, S., van Leeuwen, F., Wodarz, A., Klingensmith, J., and Nusse, R. (1995). The Dishevelled protein is modified by Wingless signaling in *Drosophila. Genes Dev.* **9**, 1087–1097.

Yoffe, K., Manoukian, A., Wilder, E., Brand A., and Perrimon, N. (1995). Evidence for *engrailed*-independent *wingless* autoregulation in *Drosophila. Dev. Biol.* **170**, 636–650.

Yost, C., Torres, M., Miller, J. R., Huang, E., Kimelmah, D., and Moon, R. T. (1996). The axis-inducing activity, stability, and subcellular distribution of β-catenin is regulated in *Xenopus* embryos by glycogen synthase kinase 3. *Genes Dev.* **10**, 1443–1454.

Young, M. R., Montpetit, M., Lozano, Y., Djordjevic, A., Devata, S., Matthews, J. P., Yedavalli, S., and Chejfec, G. (1995). Regulation of Lewis lung carcinoma invasion and metastasis by protein kinase A. *Intl. J. Cancer* **61**, 104–109.

Index

Contents to Previous Volumes

Section II
Cytoskeletal Mechanisms in Chordate Development

Contents of Volume 32

Contents of Volume 33

Contents of Volume 34